Droplet-Based Microfluidics: Design, Fabrication and Applications

Droplet-Based Microfluidics: Design, Fabrication and Applications

Editor

Pingan Zhu

MDPI • Basel • Beijing • Wuhan • Barcelona • Belgrade • Manchester • Tokyo • Cluj • Tianjin

Editor
Pingan Zhu
Department of Mechanical
Engineering
City University of Hong Kong
Hong Kong
China

Editorial Office
MDPI
St. Alban-Anlage 66
4052 Basel, Switzerland

This is a reprint of articles from the Special Issue published online in the open access journal *Micromachines* (ISSN 2072-666X) (available at: www.mdpi.com/journal/micromachines/special_issues/Droplet_Based_Microfluidics_Design_Fabrication_and_Applications).

For citation purposes, cite each article independently as indicated on the article page online and as indicated below:

LastName, A.A.; LastName, B.B.; LastName, C.C. Article Title. *Journal Name* **Year**, *Volume Number*, Page Range.

ISBN 978-3-0365-8467-6 (Hbk)
ISBN 978-3-0365-8466-9 (PDF)

Contents

About the Editor

Pingan Zhu

Dr. Pingan Zhu is currently an assistant professor in the Department of Mechanical Engineering at the City University of Hong Kong. He received his Ph.D. degree in Mechanical Engineering from the University of Hong Kong (HKU) in 2017 and a Bachelor's degree from the University of Science and Technology of China in 2013. His research focuses on the fundamentals and applications of micro-/nano-scale fluid flows, including microfluidics, fluid dynamics, surface wettability, micro/nanorobots, and biomimetics. Dr. Zhu has developed a versatile platform referred to as microfluidics-enabled soft manufacture (MESM) for engineering fluid templates into solid materials with tailored properties and interdisciplinary applications, which has been widely featured by local, national, and international media, such as *The Times, Science News, News Scientist,* and Phys.org. He has published more than forty papers in prestigious journals, including *Science, Chemical Reviews, Nature Communications, Advanced Materials,* and *The Innovation*, one book invited by Springer Nature, and one filed PCT patent. His work has been recognized by a number of awards, such as Micromachines Young Investigator Award (2022), IAAM Young Scientist Medal (2022), Outstanding reviewer for Lab on a Chip (2022), Outstanding Achievement Award for China Rising Star in Science and Technology (2021), TechConnect Global Innovation Award (2018), honorable mention of Hong Kong Young Scientist Award (2017), and HKU Mechanical Engineering Outstanding Thesis Award (2016–2017).

micromachines

Editorial

Editorial for the Special Issue on Droplet-Based Microfluidics: Design, Fabrication, and Applications

Pingan Zhu

Department of Mechanical Engineering, City University of Hong Kong, Hong Kong 999077, China; pingazhu@cityu.edu.hk

Citation: Zhu, P. Editorial for the Special Issue on Droplet-Based Microfluidics: Design, Fabrication, and Applications. *Micromachines* **2023**, *14*, 693. https://doi.org/10.3390/mi14030693

Received: 20 March 2023
Accepted: 20 March 2023
Published: 21 March 2023

Microfluidics is a rapidly growing field of research that involves the manipulation and analysis of fluids in small-scale channels, usually with dimensions ranging from sub-micrometer to sub-millimeter. This technology has widespread applications in fields such as chemistry, biology, medicine, physics, engineering, and the environment. One particularly appealing subcategory of microfluidics is droplet-based microfluidics, which permits the generation, manipulation, and analysis of tiny droplets and bubbles with exceptional precision and accuracy. Advancing this field necessitates innovative techniques for droplet/bubble generation and manipulation, as well as the development of novel materials and technologies for microfluidic devices. The aim of this Special Issue is to showcase the latest developments in droplet-based microfluidics, with a focus on the fundamentals of fluid mechanics, fabrication of microfluidic devices, and the generation, manipulation, and applications of droplets. The fundamentals of fluid mechanics continue to be important in the design and optimization of microfluidic devices, while advances in fabrication techniques have enabled the creation of increasingly complex and precise devices. Droplet generation and manipulation are also critical for many microfluidic applications, ranging from materials synthesis to chemical and biological analysis.

1. Fundamentals of Fluid Mechanics

Fluid mechanics is the cornerstone of droplet-based microfluidics as it governs the behavior of fluids at the microscale. A comprehensive comprehension of the principles of fluid mechanics is vital to devise and create effective droplet-based microfluidic systems.

Frolov et al. [1] explore the interaction of shock waves with bubbly water for generating a propulsive force. The study aims to investigate two potential directions for improving underwater propulsion: (1) replacing chemically inert gas bubbles with chemically reactive ones and (2) increasing the pulsed detonation frequency from tens of hertz to kilohertz. This study could offer valuable insights into the behavior of bubbles in microfluidic devices that utilize shock waves, thereby paving the way for designing microfluidic devices that leverage shock waves for various applications, including droplet generation and manipulation.

In addition to bubble dynamics, Frolov et al. [2] investigate the self-ignition of triethylaluminum (TEA) and triethylborane (TEB) microdroplets in air. The authors propose a novel mechanism of the heterogeneous interaction of gaseous oxygen with liquid TEA/TEB microdroplets to calculate the self-ignition of a spatially homogeneous mixture of fuel microdroplets in ambient air under normal pressure and temperature conditions. The findings provide insights into the combustion behavior of fuel droplets, which can aid in designing microfluidic devices that involve reactions with highly reactive droplets for micro-combustion.

Concerning solid–fluid interactions, Sheidaei et al. [3] present a numerical study that predicts the dispersion rate of nanoparticles in a gas–liquid dual-microchannel separated by a porous membrane. The dispersion rate of airborne nanoparticles can be regulated by adjusting the fluid flow velocity, membrane porosity, and particle diameter. This study offers fundamental insights into the mechanisms of nanoparticle dispersion, which can be utilized to optimize the design of lung-on-a-chip microfluidic devices for nanoparticle separation and filtration with implications for health monitoring.

2. Fabrication of Microfluidic Devices

The fabrication of microfluidic devices requires an understanding of material properties and manufacturing techniques that can be used to create structures with desired characteristics. Atmakuri et al. [4] explored the effect of filler materials on the wettability and mechanical properties of basalt/E-glass woven fabric-reinforced composites, which could be used to create microfluidic devices with high mechanical strength and tunable wettability. They found that the incorporation of graphite improves both the hydrophobicity and mechanical properties of the composites.

In the field of 4D printing, Long et al. [5] developed variable stiffness conductive composites that can change their stiffness and conductivity properties in response to external stimuli by alternately printing liquid metals and silicone elastomers. This technology has potential applications in microfluidics, where materials with programmable properties can be used to fabricate droplet-based microfluidic devices with on-demand manipulation of droplets and fluid flows.

Thermal imprinting is a common technique for creating microfluidic structures with high aspect ratios, but the accuracy of microstructures can be compromised due to the thermal behavior of the materials during imprinting. Ciganas et al. [6] developed a finite element model for accurately predicting the thermal imprint of high-aspect-ratio microfluidic structures fabricated from different polymers, providing insights into the behavior of these materials during imprinting. The models can provide guidelines for optimizing imprinting conditions and parameters, which is conducive to the fabrication of microfluidic devices with improved accuracy of microstructures.

Finally, Cai et al. [7] develop a method for the fabrication of transparent and flexible digital microfluidic devices with improved optical properties and mechanical flexibility by laser ablation. The device can perform different droplet manipulation functions and provides an alternative to conventional digital microfluidic devices that are based on glass or silicon wafers.

3. Droplet Generation

The fabrication of microfluidic devices for droplet generation has been a major focus of recent research in the field. Anyaduba et al. [8] demonstrate a novel approach to generating picoliter-sized droplets with high precision and efficiency using microfluidic devices fabricated via 3D printed molds. By utilizing complex geometries made possible by 3D printing, the authors provide a promising avenue for high-throughput biological or chemical assays.

The control of droplet generation in microfluidic devices heavily relies on the surface wettability of microfluidic channels. In this regard, Warr et al. [9] investigate the surface modification of 3D printed microfluidic devices to achieve controlled wetting in two-phase flow. By utilizing hydrophobic monomers, the authors have developed a technique to render the surfaces of microfluidic devices more hydrophobic, providing a means to regulate droplet generation in microfluidic devices. This advancement in surface modification technology is expected to have significant implications in the field of chemical and biological assays.

Dai et al. [10] delved into the dynamic behavior of double emulsion formation in a tri-axial capillary device. The study provides insights into the mechanisms of double emulsion generation in microfluidic devices by developing a semi-analytical model for predicting the droplet size distribution with the wall effect and various flow conditions. By utilizing a one-step process and controlling the size of the droplets, the authors provided a promising avenue for monodisperse double emulsion generation in microfluidic devices.

Trossbach et al. [11] presented a portable, negative-pressure actuated, dynamically tunable microfluidic droplet generator. The authors demonstrate the capabilities of the device by using it to produce monodisperse droplets with varied volumes, dynamically tune the droplet composition, and create droplet-templated cell spheroids from primary cells. This device's portability and easy-to-use nature make it a valuable tool for a range of applications, especially for non-specialists.

4. Droplet Manipulation

Microfluidic droplet manipulation is a rapidly growing field that has potential applications in various scientific and technological domains. In this regard, researchers have focused on investigating the fundamental mechanisms behind droplet manipulation and exploring new techniques for precise control over droplet generation, sliding, and transfer.

Chen et al. [12] present a study on the asymmetric jetting phenomenon that occurs during the impact of liquid drops on superhydrophobic concave surfaces. The study reveals that the deformation of the liquid-liquid interface during the droplet impact causes asymmetric jetting, with the droplet's impact position and surface curvature playing crucial roles. These findings provide insights into the underlying mechanisms of droplet manipulation and have significant implications for designing open-space microfluidic devices with controlled droplet manipulation and generation.

Yonemoto et al. [13] investigate the sliding behavior of droplets on an inclined solid substrate. The study highlights the dependence of the onset of droplet sliding on factors such as droplet size, substrate inclination angle, and contact angle. The results contribute to the fundamental understanding of droplet manipulation mechanisms and provide a basis for designing open-space microfluidic devices in applications such as droplet-based assays and droplet-based microreactors.

In addition to droplet impact and sliding, precise control over the amount and location of liquid deposition is another critical aspect of droplet manipulation. Liu et al. [14] present a novel technique that utilizes a micropipette to transfer ultra-small volumes of liquid adhesive onto a substrate in the femtoliter to picoliter range. The method enables precise control over the amount and location of adhesive, opening up new avenues for designing and developing microfluidic devices with controlled material distribution.

5. Applications of Droplet-Based Microfluidics

Droplet-based microfluidics has found numerous applications in various fields, including medical diagnostics, materials science, and wearable electronics. In this regard, Tiemeijer et al. [15] present a single-cell droplet microfluidics platform for analyzing the functional heterogeneity of cytotoxic T-cells (CTLs). The platform utilizes soluble stimuli and artificial antigen-presenting cells (APCs) to activate CTLs and identify functional heterogeneity based on various parameters. The tool proposed provides a means for high-throughput and single-cell analysis of CTLs, paving the way for the selection of potent CTLs for cell-based therapeutic strategies.

In continuous crystallization processes, the suitability of material systems is critical to achieving high-quality crystals. Kufner et al. [16] propose a strategy for fast decision making on the suitability of material systems for continuous crystallization using a microfluidic slug flow crystallizer. The approach involves pre-selection of the solvent/solvent mixture, verifying slug flow stability, and modeling temperature-dependent solubility in the material system. The strategy represents a general approach for optimizing the design and operation of continuous crystallization processes.

Choe et al. [17] present a novel wearable strain sensor that utilizes droplet-based technology. The sensor is made of ultrasoft and ultrastretchable silicone elastomers filled with conductive liquid-metal droplets, exhibiting anisotropic conductivity, and maintaining metallic conductivity when strained. The sensors can be integrated into clothing and conform to the body, making them suitable for use in healthcare and sports applications, including the development of wearable electronics and soft robotics.

In summary, the articles presented in this Special Issue showcase the extensive range of research being conducted in the field of droplet-based microfluidics and emphasize the significance of droplet-based microfluidics as a versatile and robust tool for scientific research and technological advancement. From fundamental studies of fluid mechanics to the design of innovative devices and their applications in various fields, droplet-based microfluidics provides a plethora of exciting research opportunities. Further exploration of this field will undoubtedly uncover new discoveries and applications in the near future.

Lastly, I extend my sincere gratitude to all the authors for their valuable contributions to this Special Issue, and to the reviewers for their dedicated efforts and time spent enhancing the quality of the papers.

Funding: This research received no external funding.

Conflicts of Interest: The authors declare no conflict of interest.

References

1. Frolov, S.M.; Avdeev, K.A.; Aksenov, V.S.; Sadykov, I.A.; Shamshin, I.O.; Frolov, F.S. Interaction of Shock Waves with Water Saturated by Nonreacting or Reacting Gas Bubbles. *Micromachines* **2022**, *13*, 1553. [CrossRef] [PubMed]
2. Frolov, S.M.; Basevich, V.Y.; Belyaev, A.A.; Shamshin, I.O.; Aksenov, V.S.; Frolov, F.S.; Storozhenko, P.A.; Guseinov, S.L. Kinetic Model and Experiment for Self-Ignition of Triethylaluminum and Triethylborane Droplets in Air. *Micromachines* **2022**, *13*, 2033. [CrossRef] [PubMed]
3. Sheidaei, Z.; Akbarzadeh, P.; Guiducci, C.; Kashaninejad, N. Prediction of Dispersion Rate of Airborne Nanoparticles in a Gas-Liquid Dual-Microchannel Separated by a Porous Membrane: A Numerical Study. *Micromachines* **2022**, *13*, 2220. [CrossRef] [PubMed]
4. Atmakuri, A.; Kolli, L.; Palevicius, A.; Urbaite, S.; Janusas, G. Influence of Filler Materials on Wettability and Mechanical Properties of Basalt/E-Glass Woven Fabric-Reinforced Composites for Microfluidics. *Micromachines* **2022**, *13*, 1875. [CrossRef] [PubMed]
5. Long, F.; Xu, G.; Wang, J.; Ren, Y.; Cheng, Y. Variable Stiffness Conductive Composites by 4D Printing Dual Materials Alternately. *Micromachines* **2022**, *13*, 1343. [CrossRef] [PubMed]
6. Ciganas, J.; Griskevicius, P.; Palevicius, A.; Urbaite, S.; Janusas, G. Development of Finite Element Models of PP, PETG, PVC and SAN Polymers for Thermal Imprint Prediction of High-Aspect-Ratio Microfluidics. *Micromachines* **2022**, *13*, 1655. [CrossRef] [PubMed]
7. Cai, J.; Jiang, J.; Jiang, J.; Tao, Y.; Gao, X.; Ding, M.; Fan, Y. Fabrication of Transparent and Flexible Digital Microfluidics Devices. *Micromachines* **2022**, *13*, 498. [CrossRef] [PubMed]
8. Anyaduba, T.D.; Otoo, J.A.; Schlappi, T.S. Picoliter Droplet Generation and Dense Bead-in-Droplet Encapsulation via Microfluidic Devices Fabricated via 3D Printed Molds. *Micromachines* **2022**, *13*, 1946. [CrossRef] [PubMed]
9. Warr, C.A.; Crawford, N.G.; Nordin, G.P.; Pitt, W.G. Surface Modification of 3D Printed Microfluidic Devices for Controlled Wetting in Two-Phase Flow. *Micromachines* **2023**, *14*, 6. [CrossRef] [PubMed]
10. Dai, Y.; Cha, H.; Nguyen, N.-K.; Ouyang, L.; Galogahi, F.; Yadav, A.S.; An, H.; Zhang, J.; Ooi, C.H.; Nguyen, N.-T. Dynamic Behaviours of Monodisperse Double Emulsion Formation in a Tri-Axial Capillary Device. *Micromachines* **2022**, *13*, 1877. [CrossRef] [PubMed]
11. Trossbach, M.; de Lucas Sanz, M.; Seashore-Ludlow, B.; Joensson, H.N. A Portable, Negative-Pressure Actuated, Dynamically Tunable Microfluidic Droplet Generator. *Micromachines* **2022**, *13*, 1823. [CrossRef] [PubMed]
12. Chen, C.; Zhong, H.; Liu, Z.; Wang, J.; Wang, J.; Liu, G.; Li, Y.; Zhu, P. Asymmetric jetting during the impact of liquid drops on superhydrophobic concave surfaces. *Micromachines* **2022**, *13*, 1521. [CrossRef] [PubMed]
13. Yonemoto, Y.; Fujii, Y.; Sugino, Y.; Kunugi, T. Relationship between Onset of Sliding Behavior and Size of Droplet on Inclined Solid Substrate. *Micromachines* **2022**, *13*, 1849. [CrossRef] [PubMed]
14. Liu, H.; Chen, X.; Wang, S.; Jiang, S.; Chen, Y.; Li, F. An Ultra-Micro-Volume Adhesive Transfer Method and Its Application in fL–pL-Level Adhesive Distribution. *Micromachines* **2022**, *13*, 664. [CrossRef] [PubMed]
15. Tiemeijer, B.M.; Descamps, L.; Hulleman, J.; Sleeboom, J.J.F.; Tel, J. A Microfluidic Approach for Probing Heterogeneity in Cytotoxic T-Cells by Cell Pairing in Hydrogel Droplets. *Micromachines* **2022**, *13*, 1910. [CrossRef] [PubMed]
16. Kufner, A.C.; Krummnow, A.; Danzer, A.; Wohlgemuth, K. Strategy for Fast Decision on Material System Suitability for Continuous Crystallization Inside a Slug Flow Crystallizer. *Micromachines* **2022**, *13*, 1795. [CrossRef] [PubMed]
17. Choe, M.; Sin, D.; Bhuyan, P.; Lee, S.; Jeon, H.; Park, S. Ultrasoft and Ultrastretchable Wearable Strain Sensors with Anisotropic Conductivity Enabled by Liquid Metal Fillers. *Micromachines* **2023**, *14*, 17. [CrossRef] [PubMed]

micromachines

Article

Interaction of Shock Waves with Water Saturated by Nonreacting or Reacting Gas Bubbles

Sergey M. Frolov [1,2,3,*], Konstantin A. Avdeev [1], Viktor S. Aksenov [1,2], Illias A. Sadykov [1], Igor O. Shamshin [1,3] and Fedor S. Frolov [1,3]

1 Department of Combustion and Explosion, Semenov Federal Research Center for Chemical Physics of the Russian Academy of Sciences, 119991 Moscow, Russia
2 Institute of Laser and Plasma Technologies, National Research Nuclear University "Moscow Engineering Physics Institute", 115409 Moscow, Russia
3 Department of Computational Mathematics, Federal State Institution "Scientific Research Institute for System Analysis of the Russian Academy of Sciences", 117218 Moscow, Russia
* Correspondence: smfrol@chph.ras.ru

Abstract: A compressible medium represented by pure water saturated by small nonreactive or reactive gas bubbles can be used for generating a propulsive force in large-, medium-, and small-scale thrusters referred to as a pulsed detonation hydroramjet (PDH), which is a novel device for underwater propulsion. The PDH thrust is produced due to the acceleration of bubbly water (BW) in a water guide by periodic shock waves (SWs) and product gas jets generated by pulsed detonations of a fuel–oxidizer mixture. Theoretically, the PDH thrust is proportional to the operation frequency, which depends on both the SW velocity in BW and pulsed detonation frequency. The studies reported in this manuscript were aimed at exploring two possible directions of the improvement of thruster performances, namely, (1) the replacement of chemically nonreacting gas bubbles by chemically reactive ones, and (2) the increase in the pulsed detonation frequency from tens of hertz to some kilohertz. To better understand the SW-to-BW momentum transfer, the interaction of a single SW and a high-frequency (≈7 kHz) sequence of three SWs with chemically inert or active BW containing bubbles of air or stoichiometric acetylene–oxygen mixture was studied experimentally. Single SWs and SW packages were generated by burning or detonating a gaseous stoichiometric acetylene–oxygen or propane–oxygen mixture and transmitting the arising SWs to BW. The initial volume fraction of gas in BW was varied from 2% to 16% with gas bubbles 1.5–4 mm in diameter. The propagation velocity of SWs in BW ranged from 40 to 580 m/s. In experiments with single SWs in chemically active BW, a detonation-like mode of reaction front propagation ("bubbly quasidetonation") was realized. This mode consisted of a SW followed by the front of bubble explosions and was characterized by a considerably higher propagation velocity as compared to the chemically inert BW. The latter could allow increasing the PDH operation frequency and thrust. Experiments with high-frequency SW packages showed that on the one hand, the individual SWs quickly merged, feeding each other and increasing the BW velocity, but on the other hand, the initial gas content for each successive SW decreased and, accordingly, the SW-to-BW momentum transfer worsened. Estimates showed that for a small-scale water guide 0.5 m long, the optimal pulsed detonation frequency was about 50–60 Hz.

Keywords: bubbly water; bubbly detonation; shock wave-to-bubbly water momentum transfer; bubble explosion; underwater propulsion

Citation: Frolov, S.M.; Avdeev, K.A.; Aksenov, V.S.; Sadykov, I.A.; Shamshin, I.O.; Frolov, F.S. Interaction of Shock Waves with Water Saturated by Nonreacting or Reacting Gas Bubbles. *Micromachines* **2022**, *13*, 1553. https://doi.org/10.3390/mi13091553

Academic Editor: Pingan Zhu

Received: 31 August 2022
Accepted: 16 September 2022
Published: 19 September 2022

1. Introduction

A novel type of underwater propulsion device for producing hydrojet thrust, referred to as the pulsed detonation hydroramjet (PDH), was proposed in [1]. This propulsion device is a flow-through pulsed detonation tube inserted in a flow-through water guide. The pulsed detonation tube periodically (with a frequency of tens of hertz) generates shock waves (SWs) and produces gas jets due to ignition and deflagration-to-detonation transition in the combustible mixture filling the tube, and it transmits these SWs and product gas jets to water in the water guide, thus creating the compressible flow of bubbly water (BW) in it. The water guide takes pure outboard water through an intake and ejects BW through a nozzle. Hydrojet thrust is produced due to the acceleration of BW in the water guide by periodic SWs and product gas jets. Theoretically, the PDH thrust is proportional to the operation frequency depending on both the SW velocity in BW and on the frequency of pulsed detonations. The PDH can contain no moving parts and can be used for generating a propulsive force in large-, medium-, and small-scale underwater thrusters. Similar principles can potentially be used in microrobotics to provide the driving force [2] and in medicine for needle-free injections [3].

There are many publications on the processes inherent in PDH operation, including SW propagation in BW, submerged gas jet penetration and bubble formation, SW coalescence, etc. The basic phenomenon in the PDH is SW propagation in compressible BW. This phenomenon was earlier studied experimentally in a vertical hydroshock tube [4,5] containing a high-pressure chamber (HPC) and low-pressure chamber (LPC) separated by a bursting diaphragm, and a measuring section filled with BW with a certain volumetric gas content α. In experiments, after the rupture of the bursting diaphragm, an SW of known intensity (very weak, weak, or strong depending on the gas pressure in the HPC) was formed in the LPC and transmitted to BW. The evolution of SW velocity and other parameters in BW was monitored with pressure sensors mounted along the measuring section and with a high-speed video camera through optical windows. Experiments on very weak (quasiacoustic) SW propagation in BW with different α [6] showed that the speed of sound passed through a deep minimum as α increased from zero to 100% and was much less than the speed of sound in pure water (1500 m/s) and in pure air (340 m/s). This result corresponded well to the known theoretical relationships [7,8]. Experiments on weak SW propagation in water with air bubbles at α ranging from 1 to 20% showed [9] that the SW velocity in BW was supersonic and varied from 150 to 100 m/s when α increased from 2 to 5%, and it dropped to 50 m/s when α increased to 20%. The measured velocity of an SW reflected from a rigid wall appeared to be higher than that of the incident wave because of bubble shrinking in the BW compressed by the incident SW and therefore the lowering of α. Video recording of single bubble motion behind a weak SW in BW with $\alpha \approx 1$–3% showed [10] that the shock-induced bubble velocity attained a value of 3–4 m/s. Experiments with strong SWs in BW with α ranging from 0.5 to 6% showed [11] that the measured SW velocities were considerably greater than those registered in [9], other conditions being equal, and it attained 400 m/s at $\alpha = 2\%$ and 250 m/s at $\alpha = 5\%$. Depending on BW parameters such as bubble size, gas thermal conductivity, liquid viscosity, etc., the SWs propagating in BW were found to exhibit various pressure profiles [12,13], namely, with a smooth or oscillatory pressure time history [14]. Experiments with SWs propagating in BW with $\alpha = 8\%$ and air bubbles 0.1, 0.48, and 0.69 mm in diameter indicated [15] that the pressure oscillation frequency behind SWs decreased with bubble diameter. The authors of [16] used high-speed photography and image post-processing to register time-resolved structural changes in a submerged gaseous jet emanating from a Laval nozzle. In [17], the results of experimental study on gas jetting by an underwater detonation tube were reported, and the mechanism of shock wave propagation and bubble deformation was discussed. The effect of the nozzle attached to a detonation tube on the underwater SW and gas detonation bubble was investigated in [18]. Three types of nozzles (converging, straight, and diverging) were examined. The converging nozzle was shown to enhance water–gas mixing and increase the peak pressure of the SW compared with the straight nozzle as well as to essentially inhibit the bubble pulsation process. The diverging nozzle was shown to suppress water–gas mixing, increase the gas jet velocity, and enhance the bubble pulsation process. High-speed photography, digital particle image velocimetry, underwater pressure field measurements, and CFD calculations were used in [19,20] to study the two-phase flow nearby the open end of the detonation tube submerged in water. Stoichiometric explosive mixtures of methane, hydrogen, and acetylene with oxygen were detonated in the tube under the same fill conditions. The oscillation frequencies and directional growth of the detonation gas bubble were investigated. The dynamic behavior of the bubble in the first oscillation was found to be very similar to that of a conventional underwater explosion.

Theoretical and computational studies of SW propagation in BW were commonly based on one-dimensional conservation equations of mass, momentum, and energy for two mutually penetrating continua, water and gas. The simplest isothermal models of SWs in BW [15,21] assumed a small gas volume fraction in incompressible water and noninteracting spherical gas bubbles moving in water without velocity slip. The range of validity of the governing equations used in [15,21] was studied in [22]. Shock wave propagation in BW with both compressible phases was considered in [23]. A more detailed model of SW propagation in bubbly liquid accounting for the thermal conductivity of phases was proposed in [24]. The authors of [25] generalized the nonisothermal model formulation to an arbitrary number of fluids with interphase mass and energy transfer. The surface tension force at a curved interface of phases in BW was introduced in [26,27] to account for different pressures in phases. Shock-induced chemical energy release in SWs propagating through BW saturated with reactive gas bubbles was investigated numerically in [28,29]. The effect of bubble polydispersity on shock wave propagation in a bubbly liquid was investigated numerically in [30]. The averaged shock structure in one-dimensional calculations was shown to become less oscillatory and tending to monotonic when the bubble size distribution broadened. The authors of [31] computationally studied the interaction of shock waves generated by two underwater detonation tubes. The dynamics of detonation gas bubbles and spectral characteristics of pressure field were analyzed, and the formation of a high-pressure zone in the region between the tubes was revealed. In the up-to-date models, the governing equations are supplemented with the various semiempirical relationships for interphase mass, momentum, and energy exchange caused by shock-induced phase velocity slip and temperature differences.

Momentum transfer from a single strong SW to BW with chemically inert air bubbles was studied computationally [32,33] and experimentally [34,35]. It turned out that the highest efficiency of momentum transfer was achieved at a volumetric gas content α of about 20–25%. With such a gas content, the increment of the absolute velocity of BW behind a strong traveling SW attained a value as high as 30 m/s. The first low-frequency valveless and valved models of PDH were designed, manufactured, and tested in [36,37]. The performance of PDH models operating at a pulse generation frequency of up to 20 Hz was investigated in [38,39]. The time-averaged specific impulse of PDH models attained 550 s [40], which was higher than that of the most advanced liquid propellant rocket engines.

The possible directions of the improvement of PDH thrust performances are to use BW with bubbles of chemically active rather than inert gas (i.e., to increase the SW velocity in BW) and to increase the frequency of pulsed detonations from tens of hertz to some kilohertz (i.e., to replace pulsed detonations by continuously rotating detonations [41]). The former direction is substantiated by the following findings. When an SW penetrates a liquid containing bubbles of a chemically reactive gas uniformly distributed over the volume, "bubbly detonation", that is a self-sustaining detonation-like solitary pressure wave propagating quasi-steadily at a supersonic velocity, may occur. Bubbly detonation was apparently observed for the first time in experiments [42] conducted in a hydroshock tube of a square 50 $\times$ 50 mm cross-section 1985 mm long and aimed at studying the interaction of SWs with a chain of reactive gas bubbles (a mixture of 70% Ar + 30% ($2H_2$ + O_2)) placed in glycerin. The bubble chain length was $\approx$670 mm, and the mean bubble diameter was $\approx$10 mm. Later, systematic experimental studies of bubbly detonation in water with bubbles of a reactive gas (stoichiometric acetylene–oxygen mixture) were conducted in a vertical hydroshock tube with an inner diameter of 35 mm and a total height of 5635 mm (the height of the BW column was 4195 mm) [43]. In experiments [44,45], bubbly detonation was initiated by transmitting gaseous detonation into the BW column with α of up to 10%.

Bubbly detonation occurred only at α less than 6% (upper limit) at a considerable distance from the gas–BW interface attaining 2.5 to 3.5 m. At larger gas contents, bubbly detonation failed to occur. At very low gas contents (less than 0.5%), bubbly detonation was also not observed (lower limit). The influence of liquid viscosity on the limits of the existence of bubbly detonation in terms of α was studied experimentally in [46]. Experiments were conducted under conditions similar to those in [44,45]. An increase in liquid viscosity (by adding up to 50%vol glycerol to water) at α ranging from 1 to 6% made it possible to initiate bubbly detonation at a lower SW intensity (less than 1.7 MPa instead of 4–5 MPa). The main specific features of bubbly detonation were outlined in these studies. Firstly, its propagation velocity was always higher than the SW propagation velocity in a liquid with bubbles of a chemically inert gas under similar conditions and higher than the speed of sound in bubbly liquid. For example, in [44] at $\alpha \approx 2\%$, the propagation velocities of bubbly detonation and SW, as well as the speed of sound in BW were $\approx$560 m/s, $\approx$425 m/s, and $\approx$85 m/s, respectively. Secondly, the process of propagation of bubbly detonation was self-sustaining [44], whereas SWs in a liquid with bubbles of a chemically inert gas gradually decayed. For using bubbly detonations in the PDH (e.g., for boosting thrust), it is necessary to know whether it is possible to obtain bubbly detonation in water without thickening additives at short distances (less than about 1 m) in a wide range of α and whether bubbly detonation gives any noticeable effect as compared to a SW. As for the increase in the frequency of pulsed detonations from tens of hertz to some kilohertz, one has to take into account that the volumetric gas content and mean bubble diameter in BW filling the PDH water guide may depend on the operation frequency, so that the initial parameters of BW in each subsequent operation cycle will be determined by the parameters of previous cycles. Therefore, there is a need in detailed studies to measure the effect of the SW generation frequency on the flow pattern in the PDH water guide and on the efficiency of SW-to-BW momentum transfer. No such studies are available in the literature.

In view of these two possible directions of the improvement of PDH thrust performances, the objective of the present work was twofold. On the one hand, the phenomenology of single SW propagation in pure water saturated with nonreacting or reacting bubbles, specific features of SW-to-BW momentum transfer, and the phenomenology of propagating bubbly detonations was studied experimentally. On the other hand, the specific features of the interaction of high-frequency ($\approx$7 kHz) SW pulses with water saturated with air bubbles was studied experimentally. These objectives, as well as the obtained results, are the novel and distinctive features of the present work.

2. Materials and Methods

2.1. Test Rig for Studies of Single Shock Wave Propagation in Bubbly Water

Figure 1a shows a schematic of the test rig used in experimental studies of single SW propagation in BW. A vertical hydroshock tube with a cross-section of 50 × 100 mm consisted of the HPC and LPC separated by a bursting diaphragm, two optical sections with Plexiglas windows (a slit 10 × 200 mm in the upper section of LPC and six rectangular windows 55 × 55 mm in the lower section of LPC), and a bubble generator. The overpressure in HPC is measured by the low-frequency pressure sensor P0 (Metronic KURANT-DI200). The error of determining the overpressure in HPC was estimated at 1%. Along the LPC, 6 piezoceramic pressure sensors P1 to P6 were installed (four KISTLER 211B2 and two PCB 113B24) flush with the inner surface of the tube. The natural frequency of pressure sensors P1 to P6 was 500 kHz. The error of determining the pressure amplitude in experimental conditions was estimated at 10%. To maintain atmospheric pressure, $P_{0,LPC}$ = 0.1 MPa, above the BW column surface, the LPC had a hole 3 mm in diameter at a distance of 1 cm below the diaphragm. To obtain contrast video frames, video recording was carried out in backlight.

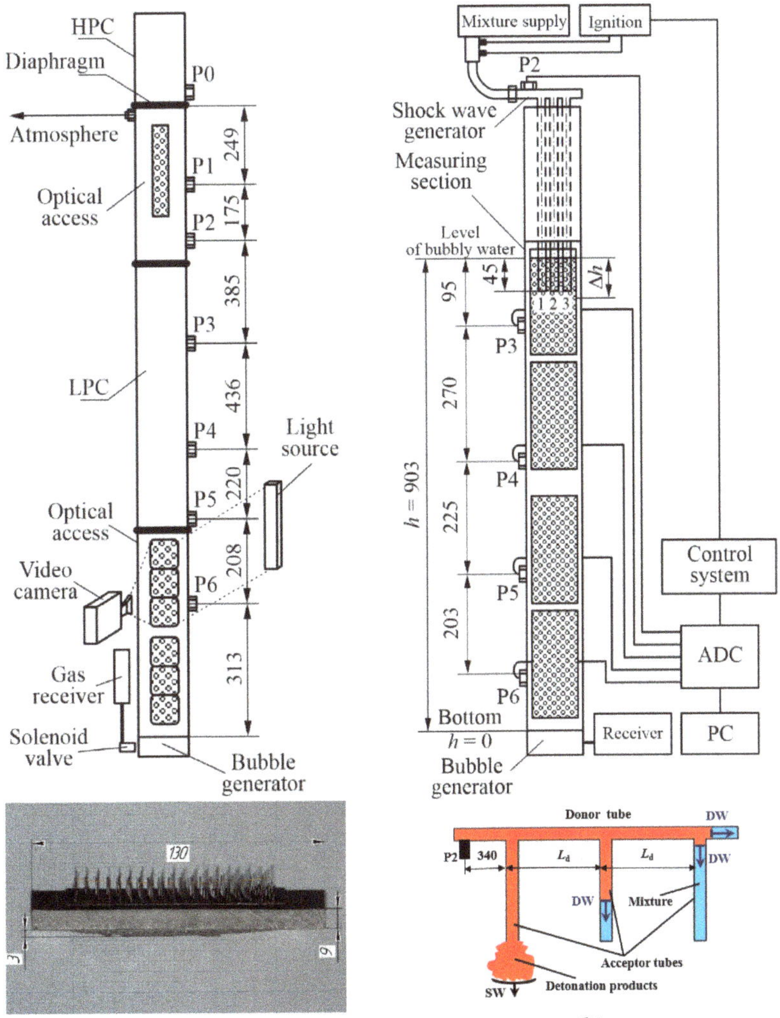

Figure 1. Schematics of test rigs for studies of (**a**) single shock wave and (**b**) shock wave package propagation in bubbly water. Inserts show the photo of bubble generator and schematic of shock wave generator. Dimensions are given in millimeters.

The bubble generator was made in the form of a steel plate 9 mm thick, in which 50 capillaries with an inner diameter of 0.26 mm were mounted in the nodes of a square grid with a step of 10 mm. Gas (air or a premixed C_2H_2 + 2.5 O_2) was supplied to the bubble generator from the receiver with a volume of 6 L at a given pressure through the solenoid valve. The volumetric content of gas, α, in BW depended on the initial pressure in the receiver, the value of which was determined during preliminary experiments. For the sake of convenience, in what follows, water with air bubbles will be referred to as inert BW, and water with bubbles of a reactive gas mixture will be referred to as active BW.

The experimental procedure was as follows. Based on the required α, the height of the water column, $h - \Delta h = h(1 - \alpha)$, was calculated. Thereafter, the LPC was filled with water up to the required level. Next, a bursting diaphragm consisting of several sheets of a 50-micron polyethylene terephthalate film was installed between the HPC and LPC. The number of film sheets in experiments varied from 2 to 10 depending on the target pressure in the HPC, P0, HPC. After evacuation, the HPC was filled with a combustible mixture (C_2H_2 + 2.5 O_2) up to pressure P0,HPC. Then, the control and data acquisition systems were put into a standby mode, and a signal was sent from the remote control to the control system, which first opened the solenoid valve for gas supply from the receiver to the bubble generator for a time sufficient to form uniform BW over the entire water column height (≈8 s in the present experiments), and then, it simultaneously issued a synchronization signal to ignition, video camera (Phantom Miro LC310), and analog-to-digital converter (ADC, R-Technology QMBox QMS20).

2.2. Test Rig for Studies of Shock Wave Package Propagation in Bubbly Water

Figure 1b shows a schematic of the test rig used in experimental studies of SW package propagation in inert BW. In general, this test rig was somewhat similar to that shown in Figure 1a but differed from it in that the HPC was replaced by the SW generator and the LPC was equipped with more windows for optical access. The SW generator was installed in the upper part of the measuring section and was used for generating a series of three successive SWs in inert BW. For the sake of clarity, Figure 2 shows the photograph with the explosive view of the SW generator. The SW generator consisted of a curved donor detonation tube 20 mm in inner diameter and 1250 mm long and three acceptor detonation tubes attached to it, which were each 0.9 m long. The donor detonation tube included a predetonator installed at the tube inlet, a KISTLER 211B2 piezoceramic overpressure sensor installed at a distance of 200 mm from the tube inlet, and a bursting diaphragm installed at the tube outlet for preventing water suction into the tube after the expansion of detonation products to BW.

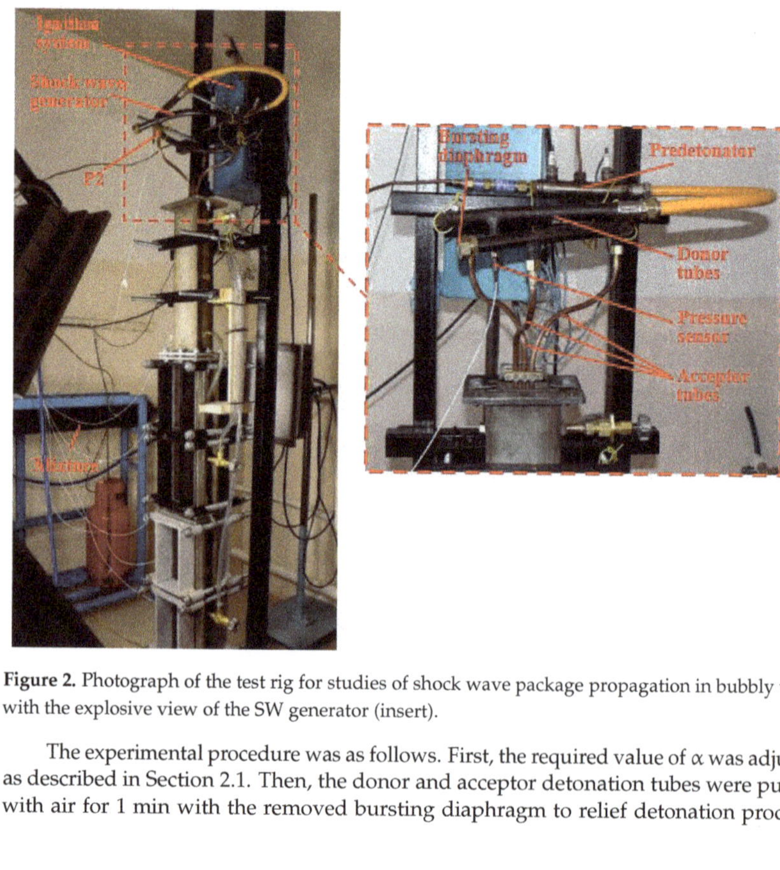

Figure 2. Photograph of the test rig for studies of shock wave package propagation in bubbly water with the explosive view of the SW generator (insert).

The experimental procedure was as follows. First, the required value of α was adjusted as described in Section 2.1. Then, the donor and acceptor detonation tubes were purged with air for 1 min with the removed bursting diaphragm to relief detonation products remaining from the previous experiment and filled with the stoichiometric propane–oxygen mixture. Next, the bursting diaphragm was installed, and the tube was blown with a combustible mixture until a control signal was sent by the control system to simultaneously trigger the ignition system, the data acquisition system (based on the QMBox QMS20 ADC), and the high-speed video camera. After ignition, a detonation wave was initiated in the predetonator, which then propagated along the donor detonation tube and, passing through the attachment points of acceptor detonation tubes 1–3 (see insert in Figure 1b), branched into a detonation wave running along the donor detonation tube and a detonation wave running along the corresponding acceptor detonation tube. Taking into account that the tube branching did not actually affect the detonation velocity D in its different parts, detonation in the second acceptor detonation tube occurred with a delay $\tau = L_d/D$ with respect to the first acceptor detonation tube, whereas in the third acceptor detonation tube, it occurred with the same delay with respect to the second acceptor detonation tube (here, L_d is the distance between the corresponding cross-sections of the donor detonation tube branching), providing the same delay in the release of detonation from the acceptor tubes into BW. The experimental value of detonation velocity in the acceptor detonation tubes was obtained by processing the pressure records of sensor P2 and video records (see below).

3. Results and Discussion

3.1. Single Shock Wave Propagation in Bubbly Water

To obtain the gas pressure in the HPC at a level feasible for hydrojet propulsion (up to 6–8 MPa [1]), the HPC was filled by the stoichiometric acetylene–oxygen mixture, which was ignited and burned to rupture the bursting diaphragm. The initial pressure of the mixture in the HPC was $P_{0,HPC}$ = 0.4–0.6 MPa. In all experiments, prior to diaphragm rupture, the LPC was filled with an inert or active BW to a height of ≈196 cm with bubbles of initial diameter d_0 = 1.5–4 mm. The remaining part of the LPC with a height of 2–5 cm above the BW column surface was filled with gas at atmospheric pressure. The initial volume fraction of gas in BW, α, varied from 2.0 ± 0.1% to 10.0 ± 0.5%. The air and water were at room temperature T_0 = 298 K.

Figures 3 and 4 show the records of pressure sensors P1 to P6 in two experiments conducted under the same initial conditions with inert and active BW, respectively. The time was counted from the launch of synchronization signal. At the initial stage of SW propagation (see records of sensor P1 in the bottom of Figures 3 and 4), the amplitude and shape of the curves in both cases were very similar: the average SW amplitude was ≈7 MPa, which was superimposed by pressure fluctuations with a frequency of 23–27 kHz. In the SW propagating through inert BW, the amplitude of pressure fluctuations reached 15 MPa at sensor P2 and decreased with time attaining ≈5 MPa at sensor P6, while the characteristic frequency of pressure fluctuations was 25–50 kHz. This frequency was close to the frequency of transverse acoustic oscillations of the BW column. The SW front in inert BW on all sensors was gentle with a duration of 0.2–0.3 ms. Pressure fluctuations in the records of Figure 3 were mainly observed within ≈200 μs after the passage of the SW front. When the SW propagated through active BW (see records of sensors P2 to P6 in Figure 4), the peak pressure fluctuations in the wave were significantly higher (up to 22 MPa) than in Figure 3 and did not decrease with time. The characteristic frequency of pressure fluctuations was very high (100–500 kHz) and approached the natural frequency of pressure sensors. As in Figure 3, pressure fluctuations in the records of Figure 4 were mainly observed within ≈200 μs after the passage of the SW front; however, the duration of the peak intensity of pressure pulsations was only 20–40 μs. The pressure wave front in this case had a much shorter duration (0.02–0.1 ms vs. 0.2–0.3 ms) than in Figure 3.

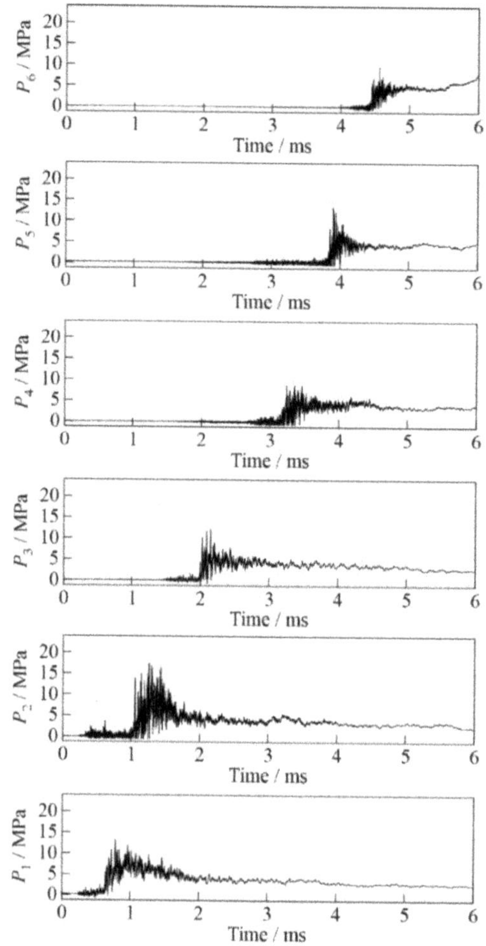

Figure 3. Records of pressure sensors P1 (bottom) to P6 (top) in the experiment with inert bubbly water (α = 2%, $P_{0,HPC}$ = 0.6 MPa).

Figure 5 shows a sequence of video frames illustrating SW propagation through active BW at α = 2%. Video recording was made through the top three windows of the lower optical section of the LPC (see Figure 1a). Shock wave propagation in active BW is accompanied with shock-induced explosions of individual bubbles, which appear as multiple luminous spots behind the propagating SW. The apparent propagation velocity of the luminocity front determined by its slope to the horizontal line is seen to be nearly constant. When such a SW–luminosity front complex reflects from the bottom of the hydroshock tube, the arising pressure is so high that the 9 mm thick steel plate of bubble generator is bent down by 3 mm (see insert in Figure 1a).

Figure 6 compares the measured dependences of the average SW velocity on the distance traveled in inert and active BW at two values of α: 2% (Figure 6a) and 10% (Figure 6b). Although the SW structure in active BW did not contain a solitary wave inherent in bubbly detonation [42–46] (see Figure 4), the SW velocity in active BW was approximately 100 m/s (at α = 2%) and 50 m/s (at α = 10%) higher than in inert BW with other conditions being equal. In other words, the replacement of inert bubbles with active bubbles increased the SW velocity significantly: by 20–30%. The increase in the SW propagation velocity in active BW was obviously caused by chemical energy deposition due to shock-induced explosions of individual bubbles. At a segment between 600 and 1200 mm, the SW velocity in active BW was about constant and equal to ≈500 m/s at α = 2% and ≈270 m/s at α = 10% (horizontal dashed lines in Figure 6a,b, respectively). A further decrease in the SW velocity with distance could be attributed to the nonuniform structure of BW closer to the bubble generator near the tube bottom, in particular at α = 2%. As compared to the bubbly detonation velocity measured in [44] for active BW with bubbles of acetylene–oxygen mixture at α = 2% (≈650 m/s), a constant propagation velocity of 500 m/s in the present experiments was 23% less, but it was attained at a considerably shorter distance: 600 mm vs. 2300 mm in [44]. The latter is important for practical applications in the small-scale PDH, because the elevated SW velocity could allow increasing the PDH operation frequency and thrust. As the constant-velocity mode of SW propagation in active BW is associated with shock-induced explosions of individual bubbles, this mode will be further referred to as bubbly quasidetonation.

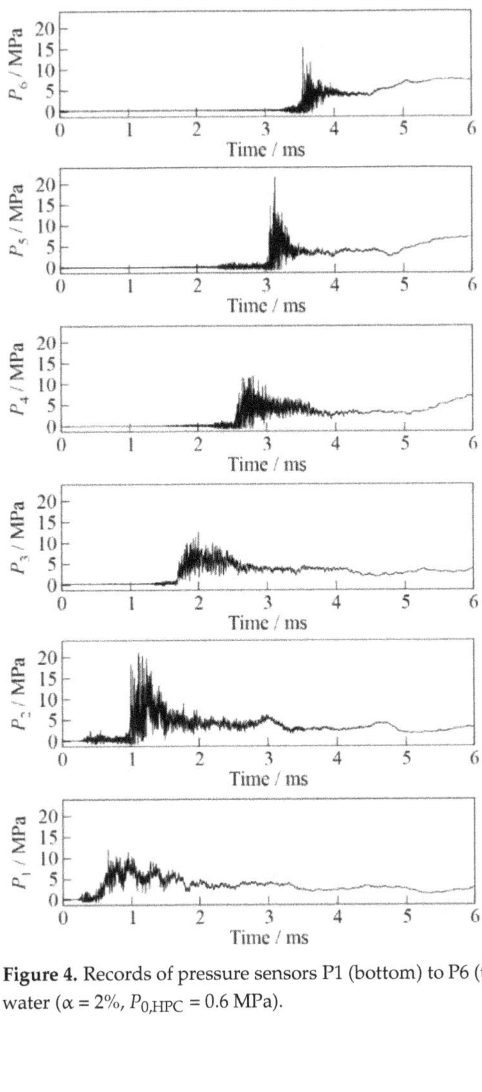

Figure 4. Records of pressure sensors P1 (bottom) to P6 (top) in the experiment with active bubbly water (α = 2%, $P_{0,HPC}$ = 0.6 MPa).

Figure 5. Video frames of propagation of bubbly quasidetonation in active bubbly water at $\alpha = 2\%$ and $P_{0,HPC} = 0.6$ MPa. Frames 1 to 30 are numbered sequentially from the moment of ignition. Frame size 160 × 448 pixels, video recording rate 40,000 fps.

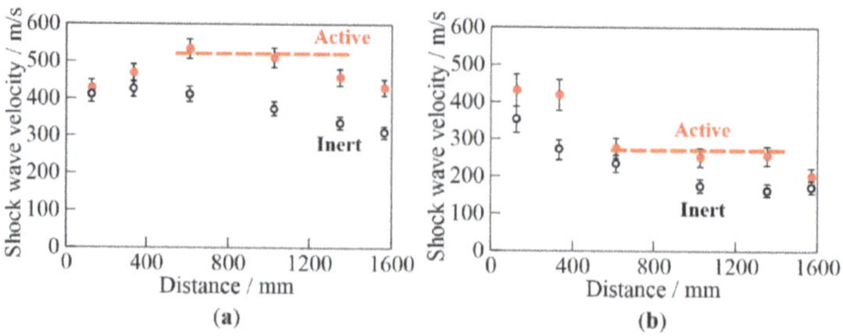

Figure 6. The SW velocity vs. distance ($P_{0,HPC}$ = 0.6 MPa) in inert and active bubbly water at (**a**) $\alpha = 2\%$ and (**b**) 10%.

Figure 7 is plotted for better understanding the difference between SW propagation velocities in inert and active BW. It compares video frames of experiments with inert and active BW at $\alpha = 2\%$. As in Figure 5, video recording was made through the top three windows of the lower optical section of the LPC. Note that pressure sensor P5 was installed 65 mm above the upper edge of the upper window, and sensor P6 was in the field of view in the lower window 10 mm below the horizontal bridge separating the lower window from the middle one. The red dashed lines mark the SW positions determined from the moments of the onset of bubble deformation between two successive video frames (at time intervals of 25 μs). During SW propagation in inert BW (upper row with frames (a) to (d)), the deformation and collapse of bubbles behind its front, as expected, did not lead to any glow. When SW propagated in active BW, bright flashes of light from explosions of individual bubbles were clearly visible. The average SW velocity in the tube section under consideration, obtained from the processing of video records of experiments with inert and reactive gas bubbles, was 320 and 400 m/s, respectively. These values were very close to the SW velocities determined from the shock front arrival time at sensors P5 and P6: 310 and 430 m/s, respectively (see Figure 6a). In Figure 7, during the exposure time of one frame Δt = 25 μs, the SW front traveled a distance $\Delta x \approx$10 mm. Individual frames show all the bubbles reacted over a time interval of Δt and were located in a band with a width of Δx. It is interesting that the luminous exploded bubbles were located at a distance of 10–20 mm behind the SW front; i.e., their self-ignition delay was 25–50 μs. Thus, the pressure fluctuations in the records of Figures 3 and 4 were associated with shock-induced fluctuations of gas bubbles, and the peak pressure fluctuations in Figure 4 were associated with bubble explosions.

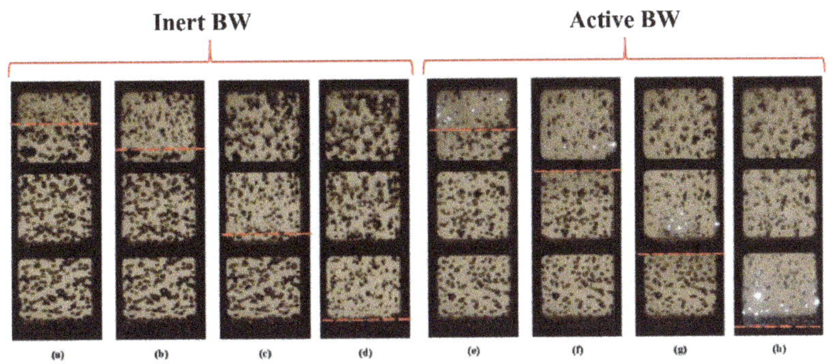

Figure 7. Video frames of SW propagation in inert (**a–d**) and active (**e–h**) bubbly water at α = 2% and $P_{0,HPC}$ = 0.6 MPa. Time countdown is from the moment of ignition. Frame size 160 × 448 pixels, video recording rate 40,000 fps. The red dashed lines show the positions of the SW front at (**a**) t = 4.091 ms; (**b**) 4.166 ms; (**c**) 4.366 ms; (**d**) 4.591 ms; (**e**) 3.250 ms; (**f**) 3.325 ms; (**g**) 3.475 ms; and (**h**) 3.650 ms.

Figure 8 shows a sequence of video frames made at a speed of 500,000 fps in one of the experiments with active BW. Based on these frames, one can trace the dynamics of compression of an individual gas bubble behind the SW and determine the moment of its explosion and subsequent thermal expansion. Red circles mark two bubbles used to trace the entire process of bubble–SW interaction. Bubbles with an initial diameter $d_0 \approx$ 2.5–3.5 mm (frames 1 and 15 in Figure 8) decrease in size to $d \approx$ 0.5–1.0 mm before explosion (frames 13 and 28), i.e., the bubble size decreases by a factor of $d_0/d \approx$ 3.5–5 before explosion. After explosion, bubbles thermally expand to sizes exceeding the initial size: $d \approx (1.5\text{–}2)d_0$. The bubbles which explode in frames 14 and 15 are compressed in a propagating SW during 18–26 μs, which is in good agreement with the estimate made earlier in this paper based on the positions of the SW front and the glow. The time of intense chemical transformation (luminescence time) does not exceed the exposure time of a single frame (2 μs). The average rate of bubble collapse can be estimated from the change in the bubble size over the time interval from the moment of SW arrival at a bubble to bubble explosion, and it reaches 40–60 m/s. After explosion, the rate of bubble expansion reaches 100–125 m/s. It is worth noting that the error in calculating the sizes and rates is estimated at 30%–50% due to the limited spatial resolution. If one assumes that bubble compression is adiabatic, the gas temperature, T, inside bubbles can be estimated as $T = T_0(d_0/d)^{3(\gamma-1)}$. At $d_0/d \approx$ 3.5–5, the estimated temperature is higher than 1100 K. At such a temperature and a pressure of $\approx$7 MPa in the SW (see Figures 3 and 4), the self-ignition delay of undiluted stoichiometric acetylene–oxygen mixture is several microseconds [47]. It should be noted, however, that judging by the registered glow, not all bubbles explode, which indicates the presence of heat losses and nonadiabatic bubble compression. One of the possible reasons for this effect is the instability of the bubble surface due to its uneven shock loading, which leads to the intensification of interfacial heat transfer. In order to register the development of instability, it was necessary to increase the spatial resolution of video filming and conduct experiments with weaker SWs.

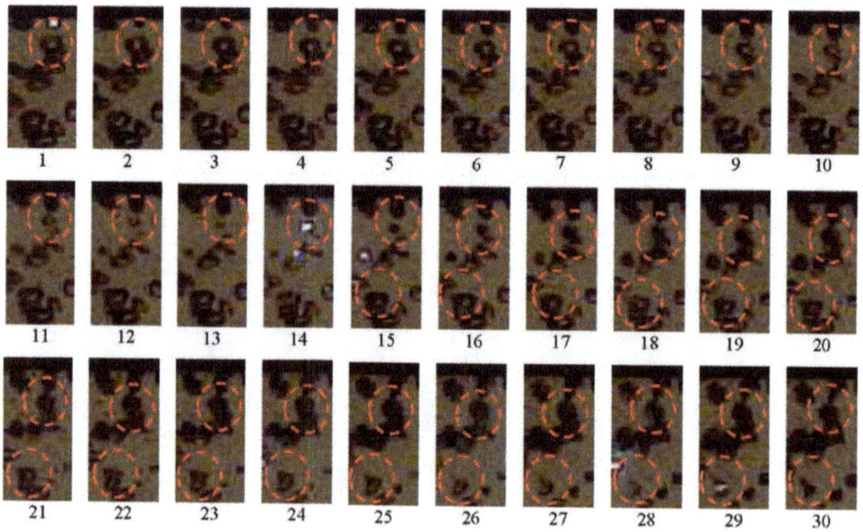

Figure 8. Video frames of SW propagation in active bubbly water at $\alpha = 3\%$ and $P_{0,HPC} = 0.55$ MPa. The first frame corresponds to $t_1 = 4.9$ ms, the following frames (numbered from 1 to 30 are made with an interval of $\Delta t = 2$ µs. The frame size is 24×48 pixels. The top edge of the images is 6.5 cm below pressure sensor P5.

As an example, Figure 9 shows video frames of the behavior of a single air bubble in inert BW with $\alpha = 2\%$ when interacting with a relatively weak SW (amplitude $\Delta P = 0.13$ MPa, propagation velocity $D = 100$ m/s, low-frequency speed of sound $C = 85$ m/s [48], Mach number M = 1.2). The passage of the SW through the bubble is seen to induce the formation of a cumulative jet that penetrates the bubble (frames 14 to 20). This phenomenon was discussed earlier in the literature, e.g., in [49]. Obviously, under such conditions, the intensity of interfacial heat transfer increases and the self-acceleration of chemical reactions leading to bubble explosion becomes more difficult.

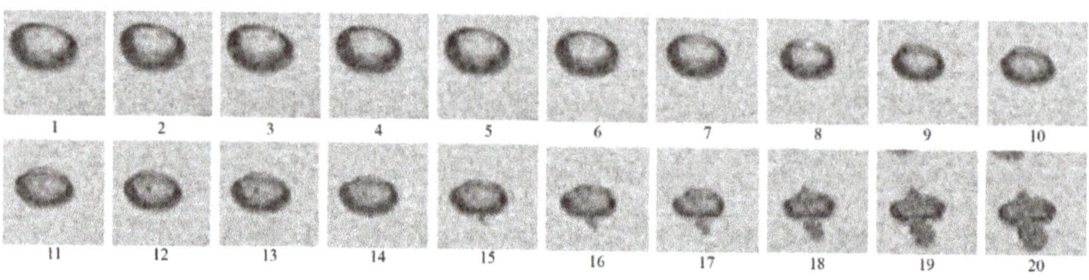

Figure 9. Video frames of SW propagation in inert bubbly water at $\alpha = 2\%$, $\Delta P = 0.13$ MPa and $D = 100$ m/s. Bubble size $d_0 = 2.8$ mm. Frames are numbered sequentially from 1 to 20 with an interval of $\Delta t = 50$ µs.

Another possible reason for the observed effect is the nonuniform distribution of bubbles in water. Closely spaced bubbles influence each other even at a low volumetric gas content. As an example, Figure 10 shows video frames of deformation of two closely spaced bubbles in the same experiment as in Figure 8. The size of the bubble in the foreground is about 2.5 mm (the average size of the ellipsoid along two semiaxes), and there is a larger bubble with a diameter of about 4.5 mm in the background. In this case, the effect of bubble "piercing" by cumulative jets is much less pronounced, and after a single compression–expansion phase, the bubbles acquire the shape of a "sphere" with a strongly perturbed surface.

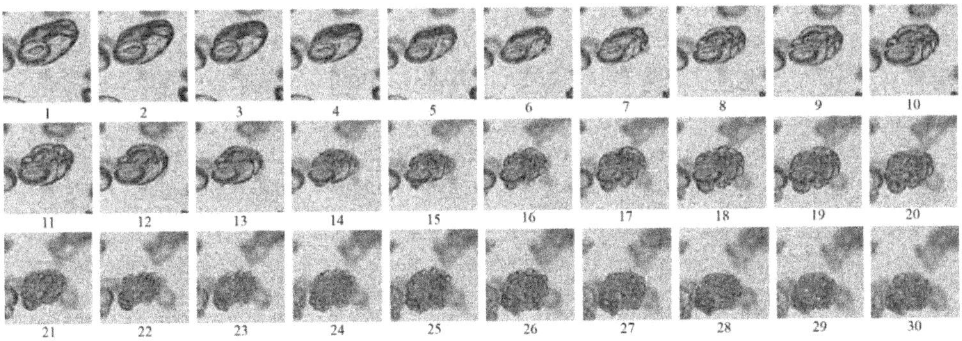

Figure 10. Video frames of SW propagation in inert bubbly water at α = 2%, ΔP = 0.13 MPa, and D = 100 m/s. The sizes of the bubbles in the foreground and background are d_0 = 2.5 mm and 4.5 mm, respectively. Frames are numbered sequentially from 1 to 30 with an interval of Δt = 50 μs.

Figure 11 shows fragments of video frames for the same experiment as shown in Figure 7e–h relevant to active BW. As the SW propagated from top to bottom, several closely spaced bubbles were first compressed (frames 1–4). Then, three bubbles with a diameter of $d \approx 2$ mm exploded almost simultaneously (frame 5). Thereafter, a larger bubble with a diameter of $d \approx 3.5$ mm exploded with a time delay (frame 6), despite it being subjected to shock compression earlier, being somewhat upstream with respect to the two bubbles that exploded earlier. Furthermore, this bubble expanded (frame 7), shrank again (frame 8) and expanded again (frames 9–11), taking the shape of a parachute (frames 12–14), and then broke up into small fragments (frames 15–20).

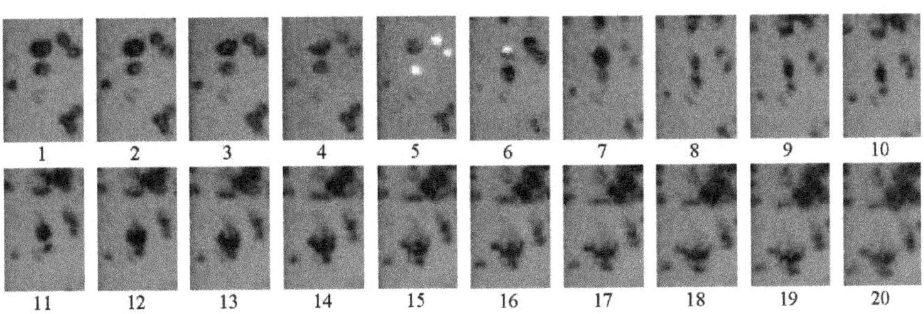

Figure 11. Fragments of video frames of SW propagation in active bubbly water at α = 2% and $P_{0,HPC}$ = 0.6 MPa. The first frame corresponds to t_1 = 3.175 ms. Frames are numbered sequentially from 1 to 20 with an interval of Δt = 25 μs. Time countdown starts from the moment of ignition in the HPC.

The processing of video records made it possible to gain information on the shock-induced motion of individual bubbles behind the traveling SW. Thus, Figure 12a,b show the measured time histories of the bubble velocity in inert and active BW, respectively. In both cases, the bubble initial position was chosen near pressure sensor P6. The corresponding records of pressure sensor P6 are also shown in Figure 12a,b. The pressure records were averaged over time intervals equal to the exposure time of video frames (25 μs). The pressure profiles behind the SWs in both cases were almost identical. The initial ascending parts of curves U_b(t) and their maxima (100–200 μs after SW arrival) corresponded to the velocity of the initial bubble, whereas the plateaus following the decline of the curves corresponded to the velocity of bubble microfragments formed after shock-induced bubble fragmentation. The plateau at the $U_b(t)$ curves represented the equilibrium velocity of phases (gas and water) behind the traveling SW. The bubble velocity fluctuations on the plots were the results of errors in determining bubble coordinates during data digitization. Quantitatively, in both cases, the velocities of initial bubbles in the SWs reached the value of ≈30 m/s, while the velocities of bubble microfragments in Figure 12a,b decreased to ≈17 m/s and ≈13 m/s, respectively, after about 1 ms after the passage of the SW. With additional averaging over time (in the interval of 0.25 ms after SW arrival) and over the ensemble of bubbles, the average bubble velocity in the tests of Figure 12a,b was 24 m/s and 25 m/s, respectively. Thus, in relatively weak SWs (at $P_{0,HPC}$ = 0.6 MPa), the characteristic velocities of chemically inert and active bubbles in BW with α = 2% were approximately identical; i.e., bubble explosions did not contribute much to the SW-to-BW momentum transfer.

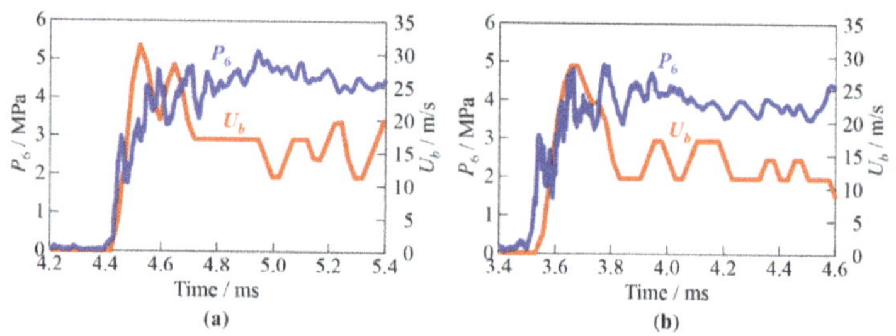

Figure 12. Measured time histories of pressure P_6 registered by pressure sensor P6 and bubble velocity U_b near this sensor at α = 2% and $P_{0,HPC}$ = 0.6 MPa: (**a**) inert; (**b**) active bubbly water.

3.2. *Shock Wave Package Propagation in Bubbly Water*

In all experiments with SW package propagation in inert BW, a column ≈903 mm high with α ranging from 2% to 16% and air bubbles of mean diameter d_0 = 3–4 mm were formed in the measuring section. The air and water were at room temperature. Figure 13 shows typical records of several signals: a signal triggering the video camera (a), the signal of electronic shutter (b), the record of pressure sensor P2 (c), and the record of pressure sensor P3 (d) installed in BW at a depth of 95 mm from the free surface of the BW column and at a distance of 50 mm from the exit sections of acceptor detonation tubes. The blue vertical line corresponds to the moment of time t_0 when the countdown of video frames is launched. The error of determining this point in time depends on the sampling frequency (±1 μs at 1 MHz) and the trigger response time (±5 μs). The average detonation velocity in the acceptor detonation tubes was 2100 ± 150 m/s and the pressure behind the front of the detonation wave reached 4–6 MPa (estimated from the maximum pressure recorded by sensor P2 after applying the low-pass filter with a cutoff frequency of 100 kHz; see Figure 13).

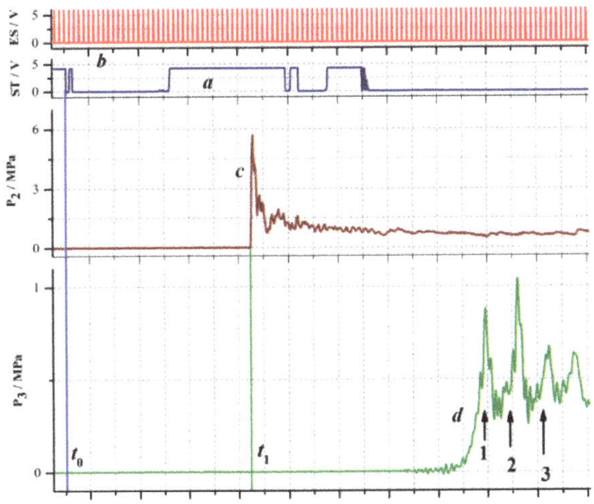

Figure 13. Typical recorded signals in an experiment with SW package propagation in BW at $\alpha = 2\%$: (a) a signal triggering the video camera, ST; (b) electronic shutter, ES; (c) pressure sensor P2, and (d) pressure sensor P3; numbers 1, 2, and 3 indicate the successive SWs.

The procedure of determining the detonation velocity in the acceptor detonation tubes was as follows. The pressure record (see signal (c) in Figure 13) was used to determine the moment of time when a detonation wave reached pressure sensor P2 (see Figure 1b). Then, using the signal of the electronic shutter (signal (b) in Figure 13), the number of the video frames corresponding to this moment in time was determined. Based on the known number of video frames (37 pcs) from the initial moment of time and the time interval between frames (20 μs at a shooting rate of 50,000 fps), the time t_1 when the detonation wave arrived at sensor P2 (green vertical line in Figure 13) was determined as $t_1 = 37 \times 20 = 740$ μs. The time moment when the detonation wave exited from the tube (t_2) was determined from the video record (see below).

For the experiment presented in Figure 13, $t_2 = 68 \times 20 = 1360$ μs. Knowing the distance between pressure sensor P2 and the tube cut, $L = 1265$ mm, it is possible to determine the average detonation propagation velocity $D = L/(t_2 - t_1)$, which in this case was equal to 2040 ± 60 m/s. The time interval between SW pulses can be estimated from the distance ($\approx$300 mm) between adjacent acceptor detonation tubes and from the average detonation velocity ($\approx$2100 m/s). Calculation gives a value approximately equal to 140 μs (frequency $\approx$ 7 kHz). In the experiment of Figure 13, the same time interval separates the pressure maxima at sensor P3 (signal (d) in Figure 13). Note that the fourth pressure maximum in the record of sensor P3 arises due to SW propagation not only in the longitudinal but also in the transverse direction in the BW column.

Table 1 shows the measured values of the velocity of the leading front of the SW package in BW with different α at three measuring segments: between pressure sensors P3 and P4, P4 and P5, and P5 and P6. The front velocity Dij was calculated as the known distance L_{ij} between the pressure sensors (see Figure 1b) divided by the time interval required for the front to travel between the sensors, Δt_{ij}: $D_{ij} = L_{ij}/\Delta t_{ij}$. The maximum error in measuring the front velocity is estimated at 3%. It follows from Table 1 that the velocity of the leading front of the SW package exceeds the corresponding low-frequency speed of sound in BW by a factor of 1.5–2; i.e., the front propagation velocity is supersonic. In addition, it is seen that the velocity of the leading front of the wave package at the position of sensor P3 decreases with α.

Table 1. Measured values of the SW package front velocity, D_{ij}, at measuring segments ij in bubbly water with different α.

α, %	C, m/s	D_{34}, m/s	D_{45}, m/s	D_{56}, m/s
2	85	180	140	126
4	61	115	92	82
8	44	75	60	55
16	32	50	40	40

Figure 14 shows examples of records of pressure sensor P3 for different values of α. The following specific features of SW package propagation in BW can be highlighted. Firstly, the shape of the leading front of the wave package in BW differs significantly from the shape of an SW front in gas: the time of pressure rise in the front of the SW package in BW varies from 0.2 (Figure 14a–c) to 0.4 ms (Figure 14d). Secondly, as α increases, the time intervals between the individual pressure maxima in the successive SWs decrease, i.e., the SWs catch up with each other and merge, and the profile of the SW package is smoothed and approaches a triangular shape. These effects are associated with a stepwise decrease in the volumetric gas content after each successive SW and, accordingly, with an increase in the low-frequency speed of sound in BW ahead of each successive SW (see Table 1). Thirdly, with an increase in α, the SW amplitudes in the SW package decrease, which is consistent with the decrease in the propagation velocity of its leading front noted above (see Table 1).

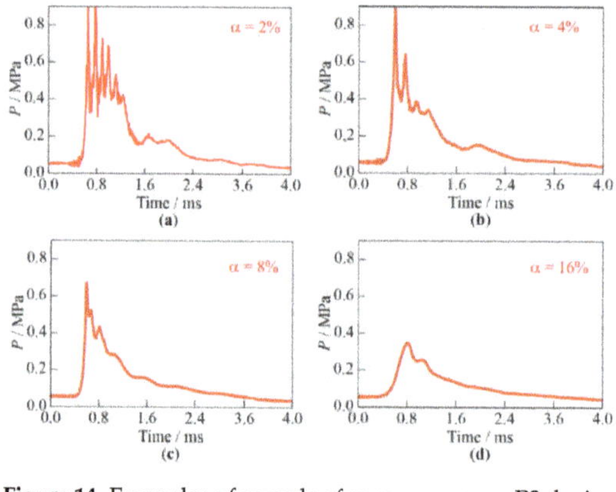

Figure 14. Examples of records of pressure sensor P3 during the propagation of an SW package of three SWs in bubbly water with different α: **(a)** 2%; **(b)** 4%; **(c)** 8%; and **(d)** 16%.

High-speed video filming makes it possible to measure the penetration depth of gaseous detonation products, L_p, and the velocity of BW behind the SW package, u_{cs}, by tracking the motion of the contact surface "detonation products–BW" [35]. Table 2 shows the measured depths of product gas jets and the contact surface velocities at different distances (20 mm ($u_{sc,20\text{-}mm}$), 40 mm ($u_{sc,40\text{-}mm}$), and 60 mm ($u_{sc,60\text{-}mm}$)) from the cutoff of acceptor detonation tubes, which was determined from video frames. It can be seen that with an increase in α from 2% to 16%, the longitudinal velocity of the contact surface near the cut of the acceptor detonation tubes (20 mm) decreases. However, at greater depths (40 and 60 mm), the longitudinal velocity of the contact surface increases by a factor of 2 with such an increase in α. Interestingly, as α increases, the depth of penetration of product gas jets changes insignificantly.

Table 2. Measured depths of product gas jet penetration, L_p, and velocities of the contact surface "detonation products–bubbly water" for SW packages in bubbly water with different α.

α, %	L_p, mm	$u_{sc,20\text{-mm}}$, m/s	$u_{sc,40\text{-mm}}$, m/s	$u_{sc,60\text{-mm}}$, m/s
2	66	35	13	3
4	56	39	13	4
8	56	35	20	4
16	57	26	26	6

For illustration, Figure 15 shows video frames of the penetration of detonation products from three successive detonation waves into BW at α = 2%. The first detonation wave enters BW from the central acceptor tube (frame #70); then, with a delay of ≈140 μs (frame #77), the second detonation wave leaves the right acceptor tube, and after another ≈140 μs (frame #84), the third detonation wave leaves the left acceptor tube. With the growth of the product gas bubble, the height of the BW column noticeably increases (in frame #200, it comes close to the upper border of the image). It is interesting that the shape of the product gas bubble does not show any signs of asymmetry: over time, the product gas bubble is elongated mainly in the longitudinal direction, while its transverse dimensions change insignificantly.

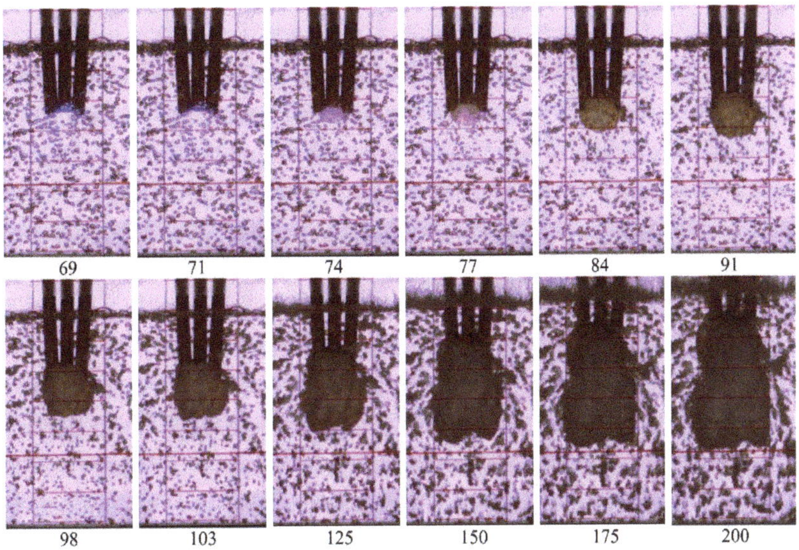

Figure 15. Video frames of the outflow of pulsed jets from three acceptor detonation tubes to bubbly water with α = 2%; frame size 168 × 320 pixels, frame rate 50,000 fps; shutter speed 18 μs.

The results of experiments show that the use of high-frequency SW pulses in the PDH is pointless because the interference of pulses impairs the SW-to-BW momentum transfer. For efficient operation of the PDH, it is necessary to maintain the required SW intensity and the required level of volumetric gas content (≈20–25% [35]) in its water guide. The SW intensity is mainly determined by the fuel mixture used. Thus, the detonation velocity in propane–oxygen and propane–air mixtures is ≈2360 and ≈1800 m/s, respectively, and the overpressure at the Chapman–Jouguet point is ≈3.52 and ≈1.73 MPa, respectively. As for the volumetric gas content, its required level can be provided either by the forced aeration of seawater or by using gaseous detonation products of the previous cycle. In the former case, for implementing the operation process, the frequency of SW pulses must be such as to exclude cycle interference; i.e., by the beginning of the next cycle, both the SW and the product gas bubble from the previous cycle must leave the water guide, whereas the water guide must be filled with a new charge of BW. With a small-scale water guide length of 0.5 m and a BW velocity in the water guide of 25 m/s [39] (at an approaching water velocity of 10 m/s), the required operation frequency should not exceed ≈50 Hz. In the latter case, the required PDH operation frequency must exceed ≈50 Hz, but a new cycle must begin after the water guide is filled (more or less uniformly) with the detonation products of the previous cycle. If one takes into account that the volume occupied by detonation products in the water guide should be at a level of ≈20–25%, then the operation frequency should not exceed ≈60 Hz.

4. Conclusions

Compressible bubbly water can be used for generating a propulsive force in large, medium, and small-scale underwater thrusters referred to as pulsed detonation hydroramjets. Thrust in such thrusters is produced due to the acceleration of bubbly water in a flow-through water guide by periodic shock waves and product gas jets generated by pulsed detonations of a fuel–oxidizer mixture in a detonation tube inserted into the water guide. The experimental studies reported herein were aimed at exploring two possible directions of the improvement of thruster performances, namely, (1) the replacement of chemically nonreacting gas bubbles by chemically reactive bubbles, and (2) the increase in the pulsed detonation frequency from tens of hertz to some kilohertz.

Experiments on single shock wave propagation in bubbly water with chemically reactive gas bubbles revealed the possibility of obtaining a bubbly quasidetonation wave propagating at the velocity, exceeding by up to 50% the velocity of shock wave propagation in bubbly water with chemically nonreacting gas due to shock-induced energy release caused by bubble explosions. Such quasidetonation waves were obtained in bubbly water with bubbles of acetylene–oxygen mixture 1.5–4 mm in diameter without thickening additives, at volumetric gas content up to 10% and at distances less than 1 m. The registered increase in the shock wave velocity can be used for increasing the thruster operation frequency and thrust.

Experiments on high-frequency (≈7 kHz) shock wave package propagation in bubbly water with air bubbles of different diameters (3–4 mm) at volumetric gas content up to 16% showed that the use of high-frequency shock wave pulses is pointless for the thrusters under consideration: the resulting interference of pulses worsens the shock wave-to-bubbly water momentum transfer. On the one hand, the shock waves penetrating bubbly water quickly merge, feeding each other and increasing the velocity of bubbly water, while on the other hand, the initial gas content of bubbly water for each subsequent shock wave decreases and, accordingly, the efficiency of shock wave-to-bubbly water momentum transfer decreases. Estimates show that for a small-scale water guide of 0.5 m length, the optimal frequency of detonation pulses is about 50–60 Hz.

Author Contributions: Conceptualization, S.M.F.; methodology, S.M.F. and V.S.A.; investigation, K.A.A., V.S.A., F.S.F., I.A.S. and I.O.S.; data curation, K.A.A., I.A.S. and I.O.S.; writing—original draft preparation, S.M.F.; writing—review and editing, S.M.F.; visualization, I.A.S. and I.O.S.; supervision, S.M.F.; project administration, S.M.F.; funding acquisition, S.M.F. All authors have read and agreed to the published version of the manuscript.

Funding: This research was supported by a subsidy given to the Semenov Federal Research Center for Chemical Physics of the Russian Academy of Sciences to implement the state assignment on the topic No. 0082-2019-0006 (Registration No. AAAA-A21-121011990037-8) and a subsidy given to the Federal State Institution "Scientific Research Institute for System Analysis of the Russian Academy of Sciences" to implement the state assignment on the topic No. FNEF-2022-0005 (Registration No. 1021060708369-1-1.2.1).

Conflicts of Interest: The authors declare no conflict of interest.

Abbreviations

ADC	Analog-to-digital converter
BW	Bubbly water
ES	Electronic shutter
HPC	High-pressure chamber
LPC	Low-pressure chamber
PC	Personal computer
PDH	Pulsed detonation hydroramjet
SW	Shock wave
ST	Signal trigger
C	Low-frequency speed of sound in bubbly water
D	Detonation velocity
D_{ij}	Detonation velocity at measuring segment ij
d	Bubble diameter
d_0	Initial bubble diameter
h	Height of bubbly water column
L	Distance
L_d	Distance between the cross sections of the donor detonation tube branching
L_{ij}	Distance between measuring ports i and j
L_p	Penetration depth of gaseous detonation products into bubbly water
M	Mach number
$P_{0,HPC}$	Initial pressure in the HPC
$P_{0,LPC}$	Initial pressure in the LPC
T	Gas temperature
T_0	Room temperature
t_1	Time taken to detonation wave to arrive at sensor P2
t_2	Time taken to detonation wave to exit a tube
u_{cs}	Contact surface velocity
α	Volumetric gas content
γ	Specific heat ratio
Δh	Cumulative height of gas volume
ΔP	Pressure amplitude
Δt	Time interval
Δt_{ij}	Time taken to the wave front to travel between measuring ports i and j
Δx	Distance interval
τ	Delay time

References

1. Frolov, S.M.; Frolov, F.S.; Aksenov, V.S.; Avdeev, K.A. Hydrojet Pulsed Detonation Engine (Variants) and the Method of Hydrothrust Development. Patent Application PCT/RU2013/001148 of 23.12.2013. Available online: https://patentscope.wipo.int/search/en/detail.jsf?docId=WO2015099552 (accessed on 30 August 2022).
2. Zhou, Y.; Dai, L.; Jiao, N. Review of Bubble Applications in Microrobotics: Propulsion, Manipulation, and Assembly. *Micromachines* **2022**, *13*, 1068. [CrossRef]
3. Golub, V.V.; Bazhenova, T.V.; Baklanov, D.I.; Ivanov, K.V.; Krivokorytov, M.S. Using of hydrogen-air mixture detonation in needle-free injection devices. *High Temp.* **2013**, *51*, 138–140. [CrossRef]
4. Gelfand, B.E.; Gubin, S.A.; Kogarko, B.S.; Kogarko, S.M. Investigations of compression waves in a mixture of liquid with gas bubbles. *Sov. Phys. Dokl.* **1973**, *18*, 787–791.
5. Borisov, A.A.; Gelfand, B.E.; Timofeev, E.I. Shock waves in liquids containing gas bubbles. *Int. J. Multiph. Flow* **1983**, *9*, 531–543. [CrossRef]
6. Nakoryakov, V.E.; Pokusaev, B.G.; Shraiber, I.R.; Kuznetsov, V.V.; Malykh, N.V. *Wave Processes in Two-Phase Systems*; Kutateladze, S.S., Ed.; Inst. Therm. Phys. Sib. Br. USSR Academy of Sciences: Moscow, Russia, 1975; pp. 54–64.
7. Silberman, E. Sound velocity and attenuation in bubbly mixtures measured in standing wave tubes. *J. Acoust. Soc. Am.* **1957**, *29*, 925. [CrossRef]
8. Van Wijngaarden, L. One-dimensional flow of liquids containing small bubbles. *Annu. Rev. Fluid Mech.* **1972**, *4*, 369–396. [CrossRef]
9. Mori, J.; Hijikata, K.; Komine, A. Propagation of pressure waves in two-phase flow. *Int. J. Multiph. Flow* **1975**, *2*, 139–152. [CrossRef]
10. Kalra, S.P.; Zvirin, Y. Shock wave-induced bubble motion. *Int. J. Multiph. Flow* **1981**, *7*, 115–119. [CrossRef]

11. Sytchev, A.I. Strong waves in bubbly liquids. *Techn. Phys.* **2010**, *55*, 783–788.
12. Noordzij, L.V. Shock waves in bubble-liquid mixture. *Phys. Commun. Amst.* **1971**, *3*, 51–56.
13. Burdukov, A.P.; Kuznetsov, V.V.; Kutateladze, S.S.; Nakoryakov, V.E.; Pokusaev, B.G.; Shreiber, I.R. Shock wave in a gas-liquid medium. *J. Appl. Math. Tech. Phys.* **1973**, *3*, 65–69. [CrossRef]
14. Campbell, I.J.; Pitcher, A.S. Shock waves in a liquid containing gas bubbles. *Proc. Royal Soc. A* **1958**, *243*, 534–545.
15. Kutateladze, S.S.; Nakoryakov, V.E. *Heat and Mass Transfer and Waves in Gas-Liquid Systems*; Nauka Publisher: Novosibirsk, Russia, 1984.
16. Zhang, X.; Li, S.; Yu, D.; Yang, B.; Wang, N. The evolution of interfaces for underwater supersonic gas jets. *Water* **2020**, *12*, 488. [CrossRef]
17. Wang, C.; Li, N.; Huang, X.; Liu, W.; Weng, C. Shock wave and bubble pulsation characteristics in a field generated by single underwater detonation. *Phys. Fluids* **2022**, *34*, 066108. [CrossRef]
18. Wang, C.; Li, N.; Huang, X.; Weng, C. Investigation of the effect of nozzle on underwater detonation shock wave and bubble pulsation. *Energies* **2022**, *15*, 3194. [CrossRef]
19. Liu, W.; Li, N.; Huang, X.; Kang, Y.; Li, C.; Qiang, W.; Weng, C. Experimental study of underwater pulse detonation gas jets: Bubble velocity field and time–frequency characteristics of pressure field. *Phys. Fluids* **2021**, *33*, 083324. [CrossRef]
20. Liu, W.; Li, N.; Weng, C.; Huang, X.; Kang, Y. Bubble dynamics and pressure field characteristics of underwater detonation gas jet generated by a detonation tube. *Phys. Fluids* **2021**, *33*, 023302. [CrossRef]
21. Van Wijngaarden, L. On equations of motion for mixtures of liquid and gas bubbles. *J. Fluid Mech.* **1968**, *33*, 465–474. [CrossRef]
22. Caflisch, R.E.; Miksis, M.J.; Papanicolaou, G.C.; Ting, L. Effective equations for wave propagation in bubbly liquids. *J. Fluid Mech.* **1985**, *153*, 259–273. [CrossRef]
23. Stuhmiller, J.H. The influence of interfacial pressure forces on the character of two-phase flow model equations. *Int. J. Multiph. Flow* **1977**, *3*, 551–560. [CrossRef]
24. Nigmatulin, R.I. *Dynamics of Multiphase Media*; Hemisphere, N.Y., Ed.; CRC Press: Boca Raton, FL, USA, 1990.
25. Saurel, R.; Lemetayer, O. A multiphase model for compressible flows with interfaces, shocks, detonation waves and cavitation. *J. Fluid Mech.* **2001**, *431*, 239–271. [CrossRef]
26. Radvogin, Y.B.; Posvyanskii, V.S.; Frolov, S.M. Stability of 2D two-phase reactive flows. *J. Phys. IV Fr.* **2002**, *12*, 437–444. [CrossRef]
27. Lidskii, B.V.; Posvyanskii, V.S.; Semenov, I.V.; Tukhvatullina, R.R.; Frolov, S.M. Well-posedness of the mixed evolutionary-boundary value problem and its discrete analog for multiphase flows. *Combus. Explos.* **2013**, *6*, 137–144.
28. Tukhvatullina, R.R.; Frolov, S.M. Shock waves in liquid containing inert and reactive gas bubbles. *Combust. Explos.* **2017**, *10*, 52–61.
29. Tukhvatullina, R.R.; Frolov, S.M. Numerical simulation of shock and detonation waves in bubbly liquids. *Shock. Waves* **2020**, *30*, 263–271. [CrossRef]
30. Ando, K.; Colonius, T.; Brennen, C.E. Numerical simulation of shock propagation in a polydisperse bubbly liquid. *Int. J. Multiph. Flow* **2011**, *37*, 596–608. [CrossRef]
31. Hou, Z.; Li, N.; Huang, X.; Li, C.; Kang, Y.; Weng, C. Three-dimensional numerical simulation on near-field pressure evolution of dual-tube underwater detonation. *Phys. Fluids* **2022**, *34*, 033304. [CrossRef]
32. Avdeev, K.A.; Aksenov, V.S.; Borisov, A.A.; Tukhvatullina, R.R.; Frolov, S.M.; Frolov, F.S. Numerical simulation of momentum transfer from the shock waves to the bubble medium. *Combust. Explos.* **2015**, *8*, 57–67.
33. Avdeev, K.A.; Aksenov, V.S.; Borisov, A.A.; Tukhvatullina, R.R.; Frolov, S.M.; Frolov, F.S. Numerical simulation of momentum transfer from a shock wave to bubbly medium. *Russ. J. Phys. Chem. B* **2015**, *9*, 363–374. [CrossRef]
34. Avdeev, K.A.; Aksenov, V.S.; Borisov, A.A.; Frolov, S.M.; Frolov, F.S.; Shamshin, I.O. Momentum transfer from a shock wave to a bubbly liquid. *Russ. J. Phys. Chem. B* **2015**, *9*, 895–900. [CrossRef]
35. Frolov, S.M.; Avdeev, K.A.; Aksenov, V.S.; Borisov, A.A.; Frolov, F.S.; Shamshin, I.O.; Tukhvatullina, R.R.; Basara, B.; Edelbauer, W.; Pachler, K. Experimental and computational studies of shock wave-to-bubbly water momentum transfer. *Int. J. Multiph. Flow* **2017**, *92*, 20–38. [CrossRef]
36. Frolov, S.; Avdeev, K.; Aksenov, V.; Frolov, F.; Sadykov, I.; Shamshin, I.; Tukhvatullina, R. Pulse-detonation hydrojet. In *Scientific-Practical Conference "Research and Development-2016"*; Springer: Berlin/Heidelberg, Germany, 2016; p. 709.
37. Frolov, S.M.; Aksenov, V.S.; Sadykov, I.A.; Avdeev, K.A.; Shamshin, I.O. Testing of experimental models of a hydrojet with pulsed-detonation combustion of liquid fuel. *Combust. Explos.* **2017**, *10*, 73–82.
38. Frolov, S.M.; Aksenov, V.S.; Sadykov, I.A.; Avdeev, K.A.; Shamshin, I.O. Hydrojet engine with pulse detonation combustion of liquid fuel. *Dokl. Phys. Chem.* **2017**, *475*, 129–133. [CrossRef]
39. Frolov, S.M.; Avdeev, K.A.; Aksenov, V.S.; Frolov, F.S.; Sadykov, I.A.; Shamshin, I.O.; Tukhvatullina, R.R. Pulsed detonation hydroramjet: Simulations and experiments. *Shock. Waves* **2020**, *30*, 221–234. [CrossRef]
40. Frolov, S.M.; Avdeev, K.A.; Aksenov, V.S.; Frolov, F.S.; Sadykov, I.A.; Shamshin, I.O. Pulsed detonation hydroramjet: Design optimization. *J. Mar. Sci. Eng.* **2022**, *10*, 1171. [CrossRef]
41. Bykovski, F.A.; Zhdan, S.A. *Continuous Spin Detonation*; Siberian Branch of RAS: Novosibirsk, Russia, 2013.
42. Hasegawa, T.; Fujiwara, T. Detonation oxyhydrogen bubbled liquids. *Proc. Combust. Inst.* **1982**, *19*, 675–683. [CrossRef]
43. Sychev, A.I. Shock-wave ignition of liquid–gas bubble system. *Combust. Explo. Shock. Waves* **1985**, *21*, 250–254. [CrossRef]
44. Sychev, A.I. Detonation waves in a liquid–gas bubble system. *Combust. Explo. Shock. Waves* **1985**, *21*, 365–372. [CrossRef]

45. Sychev, A.I.; Pinaev, A.V. Self-sustaining detonation in liquids with bubbles of explosive gas. *J. Appl. Mech. Tech. Phys.* **1986**, *27*, 119–123. [CrossRef]
46. Pinaev, A.V.; Sychev, A.I. Effects of gas and liquid properties on detonation-wave parameters in liquid-bubble systems. *Combust. Explo. Shock. Waves* **1987**, *23*, 735–742. [CrossRef]
47. Tereza, A.M.; Slutskii, V.G.; Severin, E.S. Ignition of acetylene-oxygen mixtures behind shock waves. *Russ. J. Phys. Chem. B* **2009**, *3*, 99–108. [CrossRef]
48. Kedrinskiy, V.K. *Hydrodynamics of Explosion: Models and Experiment*; Siberian Branch of RAS: Novosibirsk, Russia, 2000.
49. Mitropetros, K.; Fomin, P.A. Shock-induced ignition of fuel-lean gas bubbles in organic liquids. In *Transient Combustion and Detonation Phenomena: Fundamentals and Applications*; Roy, G.D., Frolov, S.M., Eds.; Torus Press: Moscow, Russia, 2014; pp. 101–112.

micromachines

MDPI

Article

Kinetic Model and Experiment for Self-Ignition of Triethylaluminum and Triethylborane Droplets in Air

Sergey M. Frolov [1,2,*], Valentin Y. Basevich [1], Andrey A. Belyaev [1], Igor O. Shamshin [1], Viktor S. Aksenov [2], Fedor S. Frolov [1], Pavel A. Storozhenko [3] and Shirin L. Guseinov [3]

1 Department of Combustion and Explosion, Semenov Federal Research Center for Chemical Physics of the Russian Academy of Sciences, 119991 Moscow, Russia
2 Institute of Laser and Plasma Technologies, National Research Nuclear University "Moscow Engineering Physics Institute", 115409 Moscow, Russia
3 State Research Center "State Scientific Research Institute of Chemistry and Technology of Organo-Element Compounds", 105118 Moscow, Russia
* Correspondence: smfrol@chph.ras.ru

Abstract: Triethylaluminum $Al(C_2H_5)_3$, TEA, and triethylborane, $B(C_2H_5)_3$, TEB, are transparent, colorless, pyrophoric liquids with boiling points of approximately 190 °C and 95 °C, respectively. Upon contact with ambient air, TEA, TEB, as well as their mixtures and solutions, in hydrocarbon solvents, ignite. They can also violently react with water. TEA and TEB can be used as hypergolic rocket propellants and incendiary compositions. In this manuscript, a novel scheme of the heterogeneous interaction of gaseous oxygen with liquid TEA/TEB microdroplets accompanied by the release of light hydrocarbon radicals into the gas phase is used for calculating the self-ignition of a spatially homogeneous mixture of fuel microdroplets in ambient air at normal pressure and temperature (NPT) conditions. In the primary initiation step, TEA and TEB react with oxygen, producing an ethyl radical, which can initiate an autoxidation chain. The ignition delay is shown to decrease with the decrease in the droplet size. Preliminary experiments on the self-ignition of pulsed and continuous TEA–TEB sprays in ambient air at NPT conditions are used for estimating the Arrhenius parameters of the rate-limiting reaction. Experiments confirm that the self-ignition delay of TEA–TEB sprays decreases with the injection pressure and provide the data for estimating the activation energy of the rate-limiting reaction, which appears to be close to 2 kcal/mol.

Keywords: triethylaluminum; triethylborane; oxygen intrusion reaction; rate constant; activation energy; droplet; self-ignition delay; formation of radicals; detailed kinetics; computational code

Citation: Frolov, S.M.; Basevich, V.Y.; Belyaev, A.A.; Shamshin, I.O.; Aksenov, V.S.; Frolov, F.S.; Storozhenko, P.A.; Guseinov, S.L. Kinetic Model and Experiment for Self-Ignition of Triethylaluminum and Triethylborane Droplets in Air. *Micromachines* **2022**, *13*, 2033. https://doi.org/10.3390/mi13112033

Academic Editor: Pingan Zhu

Received: 23 September 2022
Accepted: 18 November 2022
Published: 21 November 2022

1. Introduction

Triethylaluminum $Al(C_2H_5)_3$ (TEA) and triethylborane $B(C_2H_5)_3$ (TEB) are transparent, colorless, pyrophoric liquids with boiling points of approximately 190 °C and 95 °C, freezing points of approximately −46 °C and −93 °C, and densities of 0.832 and 0.677 g/cm^3 (at 25 °C), respectively [1,2]. Their solutions remain stable when stored away from heat sources in a dry, inert atmosphere, but, at elevated temperatures, they slowly decompose to form hydrogen, ethylene, and elemental aluminum and boron. Upon contact with air, TEA and TEB and their solutions in hydrocarbon solvents ignite. They also react violently with heated water [3–5]. TEA, TEB, and their solutions should only be handled under a dry, inert atmosphere such as nitrogen or argon. TEA is used as a component of the Ziegler–Natta catalyst for the polymerization of olefins [6–8]. It is also used in reactions with ethylene for the growth of hydrocarbon radicals at the aluminum atom and, with the subsequent hydrolysis of the resulting higher aluminum alkyls, to obtain fatty a-alcohols. In addition, TEA is used as an alkylating agent in the synthesis of other organoelement and organic compounds. TEB is used in organic chemistry as an initiator in low-temperature radical reactions [9], in the deoxygenation of alcohols [10], and in other processes. Both

TEA and TEB are used as a hypergolic rocket propellant, in napalm and incendiary compositions [1,11,12], as well as in micropropulsion [13] and microrobotics [14]. The SpaceX Falcon 9 rocket is known to use a TEA–TEB mixture as a first- and second-stage hypergolic ignitor [15]. TEA–TEB mixtures were also used for motor ignition in the Atlas and Delta commercial launch vehicles. According to [16], "Triethylaluminum market valued at 225.1 million USD in 2020 is expected to reach 255.2 million USD by the end of 2026, growing at a Compound Annual Growth Rate of 1.8% during 2021–2026".

In experiments with liquid TEA sprayed through a nozzle in air in the form of microdroplets, spontaneous combustion of the mixture is observed. Complete combustion of TEA in air corresponds to the overall reaction [17]

$$2Al(C_2H_5)_3 + 21O_2 \rightarrow Al_2O_3 + 12CO_2 + 15H_2O + Q_A \quad (1)$$

where Q_A is the heat of reaction (1). The value of Q_A is approximately 2444 kcal [2], 2293 kcal [18], and 1955 kcal [19]. The latter value is based on the quantum-mechanical calculation of the structures and energy characteristics of all molecular complexes involved in the TEA self-ignition process. Complete combustion of TEB in air corresponds to the overall reaction

$$2B(C_2H_5)_3 + 21O_2 \rightarrow B_2O_3 + 12CO_2 + 15H_2O + Q_B \quad (2)$$

where Q_B is the heat of reaction (2). The value of Q_B is approximately 2100 kcal [20], i.e., it is close to the value of Q_A. In a complex chemical process of transformation of the initial components into the products of overall reactions (1) and (2), in which the reacting components participate in many heterogeneous and gas-phase elementary reactions, two stages can be distinguished: the stage of self-ignition and the stage of rapid explosive combustion [21]. In applied terms, the kinetic analysis of the former stage seems to be the most important, since it is this stage that determines the time of the entire process. According to the literature, the self-ignition of TEA and TEB in air occurs through radical reactions [17,22,23]. In the primary initiation step, TEA and TEB react with oxygen, producing an active ethyl radical, which can initiate an autoxidation chain in competition with termination or other pathways.

This work deals with the development of a theoretical model and preliminary experimental studies of the first stage, i.e., the self-ignition. Based on the model, the self-ignition delays of TEA–TEB droplets in air at normal pressure and temperature (NPT) conditions are calculated, whereas experiments are used for estimating the Arrhenius parameters of the rate-limiting reaction. For the sake of definiteness, the kinetic analysis is performed for TEA. However, the model proposed herein can be directly applied to the self-ignition of TEB and TEA–TEB mixtures.

2. Materials and Methods

2.1. Kinetic Model of TEA Droplet Self-Ignition in Air at NPT Conditions

For a kinetic analysis of the self-ignition stage, it is necessary to draw up a scheme of elementary reactions. By definition, the primary reaction in the scheme should be a heterogeneous reaction that occurs when oxygen molecules available in air collide with TEA droplets. Such a reaction, leading to the self-ignition of a mixture occupying a limited volume, should be characterized by a sufficiently low activation energy. Presumably, this may be a heterogeneous reaction of the intrusion of an O_2 molecule to TEA with the formation of the $(C_2H_5)_2$Al–O–O–(C_2H_5) molecule directly in the collision of TEA and O_2 molecules

$$Al(C_2H_5)_3 + O_2 = (C_2H_5)_2Al\text{–}O\text{–}O\text{–}(C_2H_5) + Q_I \quad (3)$$

or by forming an intermediate complex $Al(C_2H_5)_3O_2$ according to the scheme [17]

$$Al(C_2H_5)_3 + O_2 = Al(C_2H_5)_3O_2 \rightarrow (C_2H_5)_2Al\text{–}O\text{–}O\text{–}(C_2H_5) + Q_I \quad (4)$$

where Q_I = 113 kcal/mol. Reaction (3) or (4) can be followed by the monomolecular decomposition reaction through two channels:

$$(C_2H_5)_2Al\text{–}O\text{–}O\text{–}(C_2H_5) \rightarrow (C_2H_5)_2Al\text{–}O + O\text{–}(C_2H_5) + Q_{II}; \quad (5)$$

$$(C_2H_5)_2Al\text{–}O\text{–}O\text{–}(C_2H_5) \rightarrow (C_2H_5)_2Al\text{–}O\text{–}O + C_2H_5 + Q_{III} \quad (6)$$

where Q_{II} = −77.9 kcal/mol and Q_{III} = −90.5 kcal/mol [17]. Reactions (3 or 4) + (5) and (3 or 4) + (6) can be considered as exothermic bimolecular reactions. These reactions occur during collisions of gas-phase molecules with the surfaces of TEA droplets (heterogeneous reactions). In this case, volatile active radicals C_2H_5 and C_2H_5O enter the gas phase (air) and interact with oxygen, releasing heat and giving rise to other sequential and parallel reactions, the same as in the gas-phase kinetics of the oxidation, self-ignition, and combustion of light alkanes (methane, ethane, and butane) and their derivatives. Kinetic schemes, corresponding equations of chemical kinetics, algorithms, and codes that describe similar gas-phase processes exist, and they can be readily used as subroutines for the numerical solution of the problem under consideration. It is worth emphasizing that we consider only the initial stage of the self-ignition process, rather than the entire process of TEA droplet combustion. At this stage, the size and chemical composition of droplets, as well as the oxygen concentration in the gas, change only a little, and can be considered constant.

Based on this prerequisite, the following kinetic model of TEA self-ignition in air at NPT conditions is proposed. The rate constant of reactions (3 or 4) + (5) and (3 or 4) + (6) is approximated as

$$K = A\exp\left(-\frac{\varepsilon}{RT}\right) \quad (7)$$

where A is the preexponential factor; T is the temperature; and ε is the activation energy. The consumption rate of TEA molecules per unit volume of the mixture can be expressed by the formula

$$\frac{dn_{TEA}}{dt} = -\frac{n_{O_2}u_{O_2}}{4}SNw,\ w \approx \lambda\exp\left(-\frac{\varepsilon}{RT}\right) \quad (8)$$

where n_{TEA} is the number of TEA molecules per unit volume; n_{O_2} is the number of oxygen molecules per unit volume; u_{O_2} is the thermal velocity of oxygen molecules; $\frac{n_{O_2}u_{O_2}}{4}$ is the number of collisions of oxygen molecules with a unit surface of a droplet per unit time; $S = 4\pi r_d^2$ is the surface area of a TEA droplet; r_d is the TEA droplet radius; N is the number of TEA droplets per unit volume; w is the reaction probability in one collision; and λ is the steric factor. This latter factor is unknown. Its value is probably in the range of 0.1–0.01. According to the kinetic theory of gases, the thermal velocity u_{O_2} at temperature T is

$$u_{O_2} = (8RT/32\pi)^{1/2} \quad (9)$$

By definition, the derivative $\frac{dn_{TEA}}{dt}$ can be also expressed as

$$\frac{dn_{TEA}}{dt} = -Kn_{O_2}n_{TEA,s} = -KSd_1n_{O_2}n_{TEA,d}N \quad (10)$$

where d_1 is the thickness of the outer molecular monolayer in a TEA droplet; Sd_1 is the volume of the outer molecular monolayer in a TEA droplet; $n_{TEA,d}$ is the number of TEA molecules per unit droplet volume; $n_{NEA,s}$ is the number of TEA molecules in the volume Sd_1N. Substituting Equation (7) into Equation (8) and comparing Equation (8) with Equation (10), one obtains

$$A = \frac{\lambda u_{O_2}}{4n_{TEA,d}d_1} \quad (11)$$

Equation (11) has a simple physical meaning. The values of d_1 and $n_{\text{TEA,d}}$ are expressed in terms of the effective radius, r_1, of the TEA molecule: $d_1 = 2r_1$, $\frac{1}{n_{\text{TEA,d}}} = 4\pi r_1^3/3$. From here and from Equation (11), one obtains

$$A = \left(\frac{2}{3}\right)\lambda u_{O_2}\pi r_1^2 = \left(\frac{2}{3}\right)\lambda u_{O_2}\sigma \tag{12}$$

where $\sigma = \pi r_1^2$ is the effective collision cross-section. Formula (12) coincides with the definition of the preexponential factor in the thermal theory of the rate constants of bimolecular reactions in gases [24] up to a factor of 2/3. It follows from Equations (7), (10), and (11) that

$$\frac{dn_{\text{TEA}}}{dt} = -\frac{\lambda}{4}u_{O_2}SNn_{O_2}\exp\left(-\frac{\varepsilon}{RT}\right) \tag{13}$$

It follows from Equation (13) that the rate of reactions (3 or 4) + (5) and (3 or 4) + (6) depends on the TEA droplet size, r_d, the oxygen concentration in the environment, n_{O_2}, and the local instantaneous air temperature, T:

$$\frac{dn_{\text{TEA}}}{dt} \sim \frac{n_{O_2}}{r_d}\exp\left(-\frac{\varepsilon}{RT}\right) \tag{14}$$

Therefore, it could be expected that the self-ignition delay of TEA spray in ambient air could be a function of the spray injection pressure, as the droplet diameter generally depends on the injection pressure: the higher the injection pressure, the smaller the droplet size and the shorter the self-ignition delay. According to Equation (14), the self-ignition delay is shorter if the environment contains more oxygen, and if the local instantaneous air temperature is higher.

In addition to the uncertainty in the value of λ, the rate constant (7) contains an unknown activation energy ε. There exist empirical formulae establishing the relationship between ε and the heat of the exothermic reaction, Q, in a linear approximation:

$$\varepsilon = a - bQ \tag{15}$$

with positive parameters a and b. The values of a and b vary depending on the type (set) of reactions. These formulae include the well-known Polanyi–Semenov rule [24]:

$$\varepsilon = 11.5 - 0.25Q \text{ kcal/mol} \tag{16}$$

This rule, when applied to bimolecular reactions (3 or 4) + (5) and (3 or 4) + (6), gives, respectively,

$$\varepsilon = 2.7 \text{ and } 5.9 \text{ kcal/mol} \tag{17}$$

As noted in [21], the formulae such as (16) must be used with great caution. The same is true for estimates (17). They can only be considered as a rough approximation, which must be verified and refined experimentally. In experiments, the parameters of the reaction rate in Equation (13) are not measured directly. However, the induction period before the self-ignition of TEA droplets and some other kinetic and thermodynamic parameters can be measured, which depend on the rate constant (7). In this case, the activation energy ε can be found by solving the inverse problem. To do this, one must first solve the direct problem, which consists in calculating the self-ignition delay with a variation in activation energy ε.

2.2. Self-Ignition Delays for C_2H_5–Air and C_2H_5O–Air Mixtures

The mathematical statement of the problem is the statement of the standard problem of the self-ignition of a gas mixture [25] with a given detailed kinetics [26], which is supplemented by a heterogeneous mechanism for the formation of C_2H_5 or C_2H_5O radicals. The equations for the conservation of the energy and mass of the components have the form

$$c_p \rho \frac{dT}{dt} = \Phi \tag{18}$$

$$\rho \frac{dY_j}{dt} = w_j + \Psi,\ j = 1, 2, \ldots, M \tag{19}$$

where t is time; M is the number of components in the gas mixture; Y_j is the mass fraction of the jth component; c_p is the heat capacity of the gas mixture at constant pressure; ρ is the density of the mixture; Φ is the heat release in chemical reactions; w_j is the component consumption in chemical reactions; and Ψ is the formation of C_2H_5 or C_2H_5O in reaction (6) or (5), respectively. The system of Equations (18) and (19) is supplemented by the ideal gas equation of state, expressions for Φ and w_j [27], and by the polynomial relationship for the heat capacity, while the polynomial coefficients are taken from [28].

Compared to the standard problem formulation for gas mixture self-ignition, the expression for Ψ and all other considerations associated with this circumstance are new. It is assumed that the mixture is initially represented by pure air, and radicals C_2H_5 or C_2H_5O appear in the gas due to the heterogeneous reaction (6) or (5), respectively. Self-ignition delays depend on the rates of formation of C_2H_5 and C_2H_5O radicals in heterogeneous reactions (6) and (5) and on the rates of their interaction with oxygen in the gas phase. As both reactions, (6) and (5), are possible, for determining the effect of these radicals on the self-ignition delay, the problem must be solved for two options: (*i*) for a C_2H_5–air mixture and (*ii*) for a C_2H_5O–air mixture.

2.2.1. Option *i*: C_2H_5–Air Mixture

The expression for Ψ is

$$\Psi = \begin{cases} W_{C_2H_5} \frac{dn_{C_2H_5}}{dt} \text{ for } C_2H_5 \\ 0 \text{ for other species} \end{cases} \tag{20}$$

Here, $W_{C_2H_5}$ is the molecular mass of C_2H_5; $\frac{dn_{C_2H_5}}{dt}$ is the rate of change in the concentration (mol/cm^3/s) of C_2H_5 radicals in the heterogeneous reaction (6), which, in accordance with reaction (6) and Equation (13), satisfies the equation

$$\frac{dn_{C_2H_5}}{dt} = \frac{\lambda}{4} u_{O_2} S N n_{O_2} \exp\left(-\frac{\varepsilon}{RT}\right) \tag{21}$$

2.2.2. Option *ii*: C_2H_5O–Air Mixture

For this option, the same system of equations is solved as for option *i*, but in Equations (20) and (21) everywhere in the rows and in the indices, C_2H_5 must be replaced by C_2H_5O. To solve the problem for these two options, a special computational code has been developed.

2.3. Experimental Setup

Before designing the experimental setup, it was implied that the self-ignition delay, τ_i, of the TEA/TEB spray in ambient air could be estimated experimentally either by measuring the time delay between the start of spray injection and the appearance of self-ignition luminosity, if TEA/TEB is sprayed in a short pulse mode, or by measuring the width of the dark zone between the injector nozzle face and the luminous combustion plume, if the TEA/TEB is sprayed continuously. It could be assumed that ignition occurs much later than the pulsed injection of TEA/TEB. During this time, TEA/TEB droplets slow down, causing the ignition to occur in a virtually quiescent and spatially homogeneous mixture. With an average droplet path of around 10 cm and a speed of escape from the injector nozzle of the order of 10^4 cm/s, the deceleration time is around 1 ms. This time is much shorter than the expected self-ignition delay time, even at the lowest estimated value of activation energy in Equation (17), $\varepsilon \approx 2$ kcal/mol. With the continuous spraying of TEA/TEB into the ambient air, one could expect the appearance of a quasi-stationary luminous combustion plume. In this case, TEA/TEB droplets ignite due to the air flow around them, and the self-ignition delay, τ_i, of the spray is determined by the time given to a droplet to enter the zone of the luminous combustion plume. The value of τ_i can be estimated experimentally based on the width, L, of the dark zone between the injector nozzle face and luminous combustion plume measured along the spray axis. The value τ_i is related to L and the speed of the TEA/TEB spray at the nozzle exit, U, as $\tau_i \approx L/U$.

Based on these implications, the experimental setup for fuel spraying in air was designed and manufactured. The experimental setup consisted of an electromagnetic fuel injector (BOSCH 0 280 158 017) and a system for ensuring its operation (hydraulic system and microprocessor control unit), an optical system, a high-speed camera (Phantom Miro LC310), and a safety system. The elements of the setup are shown in Figure 1a. The injector nozzle had 4 holes with a diameter of 0.2 mm. The microprocessor control unit monitored the current through the injector and the voltage applied, and issued synchronization and trigger signals for the high-speed video camera. In the preliminary experiments, the standard 13%TEA–87%TEB mixture provided by the production company was used. The density of the TEA–TEB mixture was 0.703 g/cm^3. To prevent clogging of the setup communications by the condensed reaction products of the TEA–TEB mixture with air, the injector was sprayed with n-heptane before and after each experiment. n-heptane (density 0.684 g/cm^3) was also used for estimating the flow rate and characteristic droplet size in the spray at different injection pressures. The nominal flow rate of n-heptane at an overpressure of 3 and 6 atm was 2.55 $\pm$ 0.08 mL/s (1.74 $\pm$ 0.05 g/s) and 5.1 $\pm$ 0.2 mL/s (3.5 $\pm$ 0.1 g/s), respectively. The droplet diameter in n-heptane sprays at the injection overpressure of 3 and 6 atm measured by the slide sampling method [29,30] was ~80–120 and ~30–50 μm, respectively. The operation frequency of the injector in the pulsed mode as well as the injection duration time were varied (see below).

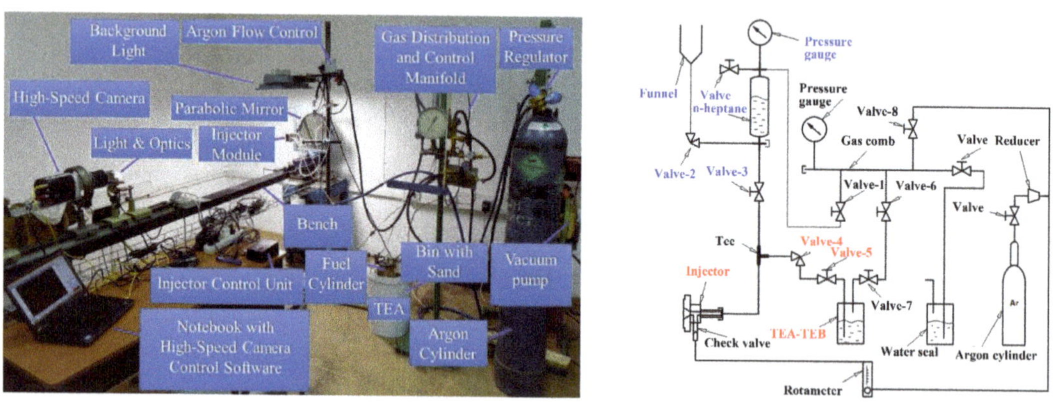

(**a**) (**b**)

Figure 1. (**a**) Experimental setup and (**b**) the hydraulic scheme of fuel supply.

Figure 1b shows the hydraulic scheme of the experimental setup. The experimental procedure was as follows. (i) Before supplying the TEA–TEB mixture to the injector, all communications were thoroughly purged with argon; (ii) liquid n-heptane was poured into the transparent measuring tank through the funnel and pressurized by argon to an overpressure of 3 atm using valve 1; (iii) with valves 2, 4, and 5 closed, valve 3 open, and the injector turned on, the communications were spilled with n-heptane; (iv) valve 3 was closed and the pressure was relieved by briefly turning on the injector; (v) valves 6 and 7 were opened and the TEA–TEB tank was pressurized by argon to the overpressure of 3 to 6 atm; valve 8 was used to control the pressure level; (vi) valves 6 to 8 were closed and valves 1, 4, and 5 opened; (vii) using a rotameter, a small flow rate of argon (around 1 L/min) was established around the injector nozzle to avoid nozzle clogging by the condensed reaction products of the TEA–TEB mixture with air available in the vicinity of the nozzle face; and (viii) the injector was turned on and operated for a preset time either in the pulsed or continuous injection mode.

Figure 2 shows two options used for fastening the injector in the housing, which differed in the means of supplying argon and the geometry of the insulating cavity. The first series of experiments was performed using the recessed injector of Figure 2a and direct video registration of spray self-luminosity during ignition and combustion in ambient air (see Section 3.3). As the dark zone between the flame and the nozzle mouth could be quite short, the second series of experiments was performed using the flat Injector of Figure 2b (see Section 3.3). In the latter case, spray self-ignition and combustion was registered both by direct video registration of self-luminosity and by the schlieren method.

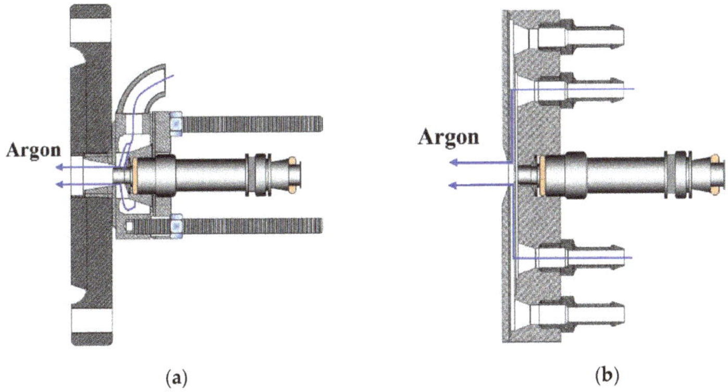

Figure 2. Two options of injector housing: (**a**) recessed injector and (**b**) flat injector.

3. Results and Discussion

3.1. Self-Ignition of Stoichiometric C_2H_5–Air and C_2H_5O–Air Mixtures at NPT Conditions

Before conducting the calculations for options *i* and *ii*, we calculated the self-ignition delays for stoichiometric C_2H_5–air and C_2H_5O–air mixtures in the absence of heterogeneous reactions, i.e., at $\Psi_{C_2H_5} = \Psi_{C_2H_5O} = 0$. The overall reaction of C_2H_5 with oxygen reads

$$4C_2H_5 + 13O_2 \rightarrow 8CO_2 + 10H_2O.$$

The stoichiometric composition of the C_2H_5–air mixture is $X_{C_2H_5} = 0.061$; $X_{O_2} = 0.197$; and $X_{N_2} = 0.742$. Self-ignition delays were calculated for the C_2H_5–air mixture of stoichiometric composition for NPT conditions: $T_0 = 300$ K and $p = 1$ atm.

The self-ignition delays obtained using the developed kinetic code were compared with the results of calculations using the standard kinetic code KINET, developed by M. G. Neigauz at the Institute of Chemical Physics of the Russian Academy of Sciences for homogeneous gas mixtures [31]. In both cases, the same kinetic mechanism was used: the detailed kinetic mechanism of combustion and oxidation of alkanes up to C4 [26]. It should be noted that the KINET code applies somewhat different thermodynamic data than [28]: the polynomial dependence of heat capacity on temperature has fewer terms.

Table 1 shows the values of the self-ignition delays τ_i, the temperature of the reaction products T_e, volume fractions of C_2H_5 radical, $(X_{C_2H_5})_e$, oxygen, $(X_{O_2})_e$, and water, $(X_{H_2O})_e$, in the reaction products, as well as the maximum value of methane volume fraction, $(X_{CH_4})_{max}$, obtained by calculations using the two indicated codes. Figure 3a shows the calculated time histories of temperature obtained by the two indicated codes.

Both the data in Table 1 and the curves in Figure 3a are in satisfactory agreement with each other. The slight differences in the results seem to be caused by the differences in the thermodynamic data used.

Table 1. Estimated values of the main variables.

Code	τ_i, s	T_e, K	$(X_{C_2H_5})_e$	$(X_{O_2})_e$	$(X_{H_2O})_e$	$(X_{CH_4})_{max}$
New code	0.10	2455	0	0.007	0.14	0.0053
KINET	0.09	2336	0	0.010	0.13	0.0057

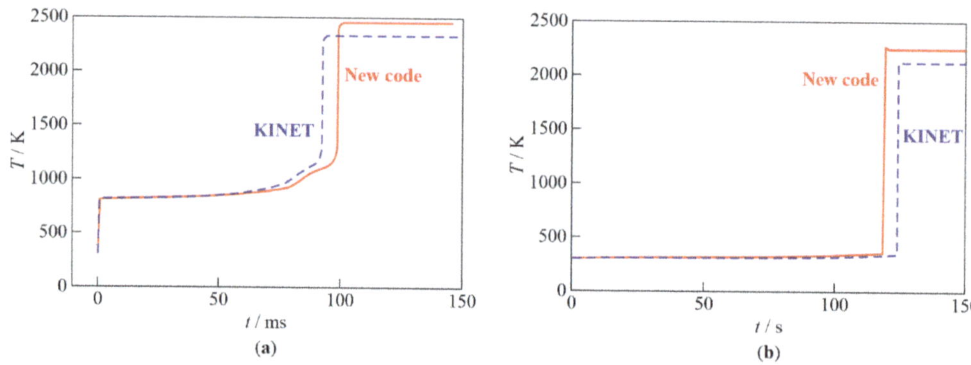

Figure 3. Calculated time histories of temperature during self-ignition of C_2H_5–air (**a**) and C_2H_5O–air (**b**) mixtures of stoichiometric composition at NPT conditions using two codes: new code and KINET [31].

Then, the self-ignition delays of the stoichiometric C_2H_5O–air mixture were calculated under NPT conditions: T_0 = 300 K and p = 1 atm. The overall reaction of C_2H_5O with oxygen reads

$$4C_2H_5O + 11O_2 \rightarrow 8CO_2 + 10H_2O.$$

The stoichiometric composition of the C_2H_5O–air mixture is $X_{C_2H_5O} = 0.063$; $X_{O_2} = 0.172$; and $X_{N_2} = 0.765$. Figure 3b shows the calculated time histories of temperature for the stoichiometric C_2H_5O–air mixture. The self-ignition delays obtained with the new code and KINET are 118 and 124 s, respectively, i.e., the results differ by less than 5%. Thus, calculations show that the self-ignition delay of a stoichiometric C_2H_5O–air mixture is much longer than that of a stoichiometric C_2H_5–air mixture at NPT conditions. This indicates the much greater reactivity of the C_2H_5 radical in air compared to the C_2H_5O radical. In view of this, the option with the C_2H_5O radical can be omitted.

3.2. *Self-Ignition of TEA Droplets in Air at NPT Conditions*

The problem for option *i* (see Section 2.2) was solved numerically for several values of ε, from 2 to 6 kcal/mol (see Equation (17)), to compare the calculated self-ignition delays with the experiment. For the sake of definiteness, the parameters entering Equation (21) were assumed to have the following values: $\lambda = 0.1$; $u_{O_2} = 4.46 \cdot 10^4$ cm/s; $r_d = 50$ µm; $S = 3.14 \cdot 10^{-4}$ cm^2; $N = 210$ cm^{-3}; $c_p = 1.33 \cdot 10^{-3}$ J/(cm^{3}K) (see Appendices A and B). Initial NPT conditions are $t = 0$; $T = T_0 = 300$ K; pressure $p = 1$ atm; volume fractions of oxygen and nitrogen $X_{O_2} = 0.21$ and $X_{N_2} = 0.79$. All other components: $X_j = 0$; $n_{C_2H_5} = 0$; $n_{O_2} = X_{O_2}/22{,}400 = 9.37 \cdot 10^{-6}$ mol/cm^3.

Calculations show that the self-ignition delay τ_i depends very strongly on ε (Table 2). Figure 4 shows the calculated time histories of temperature within $0 \leq t \leq \tau_i$ at $\varepsilon = 5$ and 6 kcal/mol. The upper limit of the specified time interval is chosen equal to τ_i because, at $t > \tau_i$, the reactions leading to the complete burnout of TEA, which are not included in the reaction scheme, become significant. Figure 5 shows the calculated time history of the C_2H_5 radical concentration, $n_{C_2H_5}(t)$, at ε = 6 kcal/mol.

Table 2. Self-ignition delays for option *i*.

ε, kcal/mol	τ_i, s
2	0.045
4	0.62
5	2.6
6	11.1

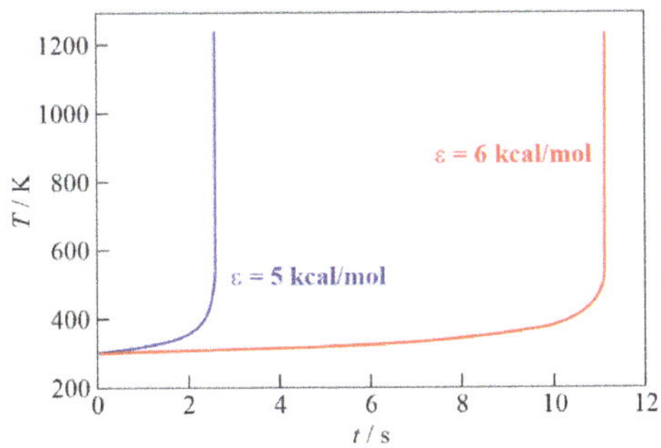

Figure 4. Calculated time histories of temperature during self-ignition of stoichiometric C_2H_5–air mixture with the heterogeneous reaction at NPT conditions and ε = 5 and 6 kcal/mol.

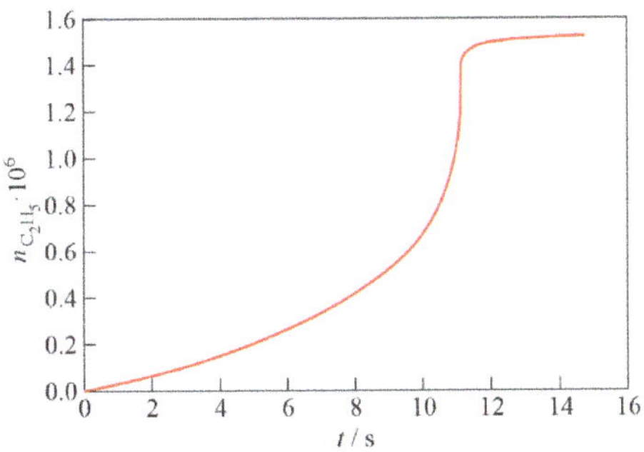

Figure 5. Calculated time history of the C_2H_5 radical concentration in a heterogeneous reaction at NPT conditions; ε = 6 kcal/mol.

3.3. Experimental Results

Figures 6 and 7 show selected video frames of TEA–TEB spray self-ignition during pulsed (Figure 6) and continuous (Figure 7) spraying from the recessed injector of Figure 2a at injection overpressure 3 atm. The duration of the single-shot pulsed spray in the experiment of Figure 6 is 20 ms. After the termination of spray injection in Figure 6, one can see the successive appearance of haze and smoke in the spray core, followed by the formation of a luminous self-ignition spot of a green color, characteristic of TEB combustion (see frame #35 in Figure 6). The first hot spot is located at a distance of around 130 mm from the nozzle face. Subsequently, while this hot spot rapidly grows with time, several other hot spots appear, grow, and overlap with each other, moving in lateral, downstream, and upstream directions and forming a luminous combustion plume. The evolution of the continuous spray in Figure 7 has much in common with that of Figure 6, but the first hot spot is located somewhat closer to the nozzle face (at around 100 mm) and the luminous combustion plume looks more elaborated. The characteristic spray velocity at the nozzle exit estimated based on the mass flow rate and injector nozzle diameter is approximately 20 m/s.

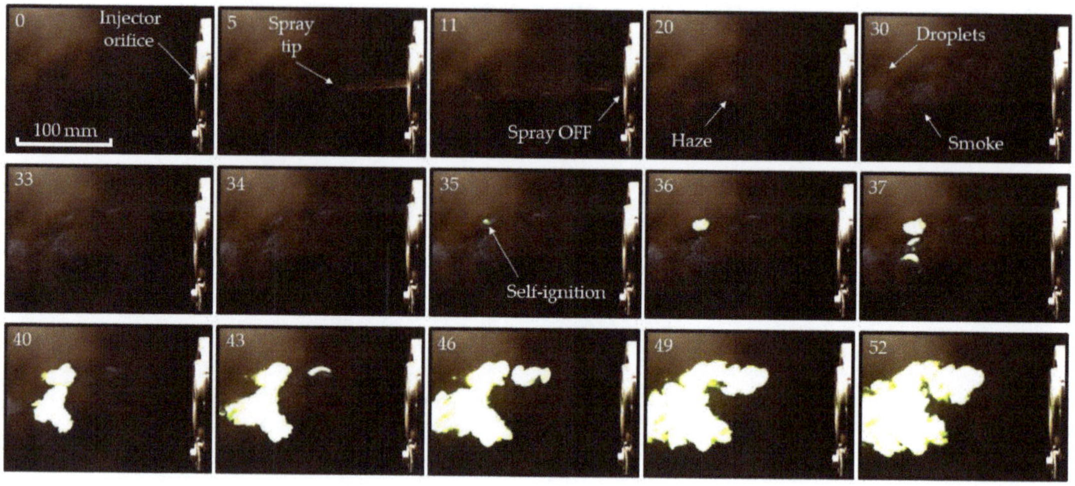

Figure 6. Sequence of video frames of pulsed TEA–TEB spray self-ignition and combustion in ambient air at NPT conditions: recessed injector. Frame numbers correspond to time in milliseconds from the start of injection. Injection overpressure is 3 atm. Frame size is 672 × 456 pixels (213 × 145 mm^2), frame rate 1000 fps, shutter speed 400 μs.

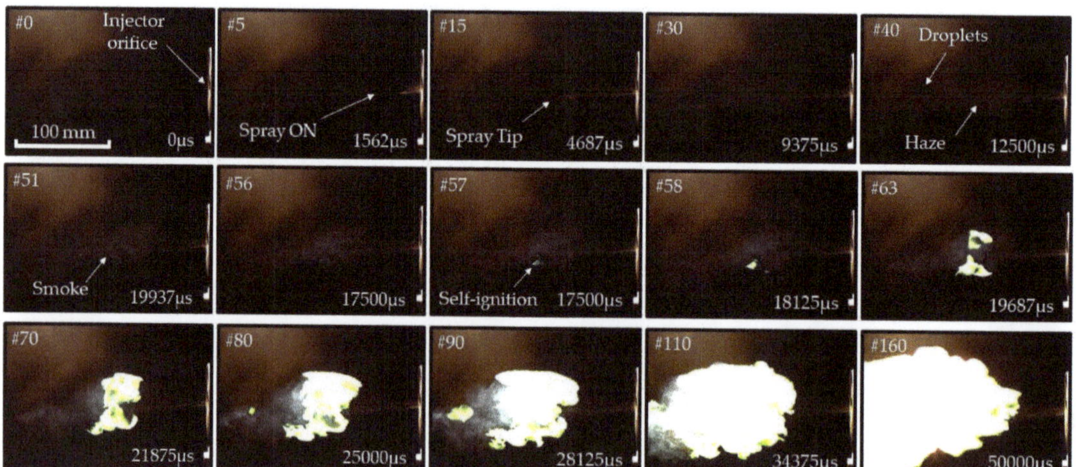

Figure 7. Sequence of video frames of continuous TEA–TEB spray self-ignition and combustion in ambient air at NPT conditions: recessed injector. Injection overpressure is 3 atm. Frame size is 672 × 456 pixels (213 × 145 mm^2), frame rate 3200 fps, shutter speed 300 μs, injection duration 500 ms.

The measured self-ignition delays in Figures 6 and 7 are approximately 30 and 20 ms, respectively. The distance between the nozzle face and luminous plume in a continuous spray appeared to be variable with time (Figure 8) rather than quasi-stationary, as was expected. Thus, while the first ignition event in Figure 8 appeared at a distance of ~100 mm from the nozzle face, it dropped to zero in 100 ms after the start of injection (see Figure 8).

This means that the apparent velocity of self-ignition spreading toward the nozzle face was higher than the spray velocity (~20 m/s).

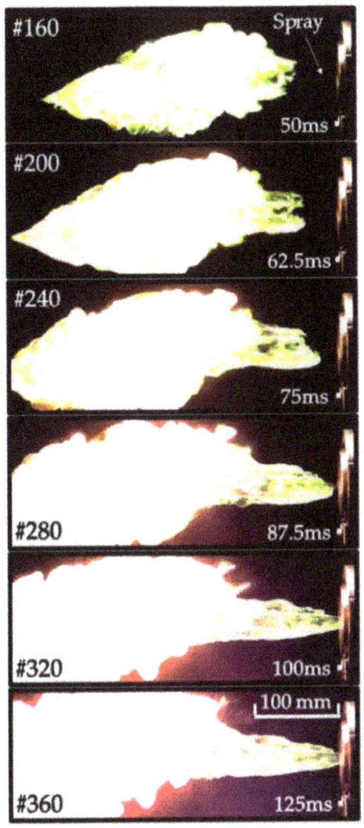

Figure 8. Sequence of video frames of continuous TEA–TEB spray combustion in ambient air at NPT conditions: recessed injector, the same test fire as in Figure 7. Injection overpressure is 3 atm. Frame size is 1280 × 456 pixels (406 × 145 mm^2), frame rate 3200 fps, shutter speed 300 μs, injection duration 500 ms.

Consider Figure 9 for a better understanding of the circumstances of TEA–TEB spray self-ignition. This figure shows a sequence of synchronized schlieren (left column) and self-luminosity (right column) video frames of TEA–TEB spray injection by the flat injector of Figure 2b, followed by spray self-ignition and combustion. The injection overpressure is 6 atm, and the spray velocity is ~40 m/s. The schlieren images in the left column are superimposed by the self-luminosity images reflected from the mirror of the schlieren system. Schlieren images contain thin droplet trajectories. After a certain delay, the trajectories inflate, forming cloud-like swellings, which can be attributed to the liquid mist and gas, self-heated due to the spontaneous reaction in air. The latter is substantiated by the fact that the luminous self-ignition plume is located exactly in these swellings of the droplet trajectories. Furthermore, Figure 10 shows the pulsed injection of the TEA–TEB spray into the ambient air at NPT conditions with a very short pulse duration of 0.7 ms. Similar to Figure 9, the schlieren images here are also superimposed by the self-luminosity images reflected from the mirror. One can see the appearance of the cloud-like trajectory swellings lagging behind the moving droplets and growing with time in the form of conical tongues of hot matter. It is seen from the schlieren images that the luminous self-ignition plume is also located exactly in these swellings.

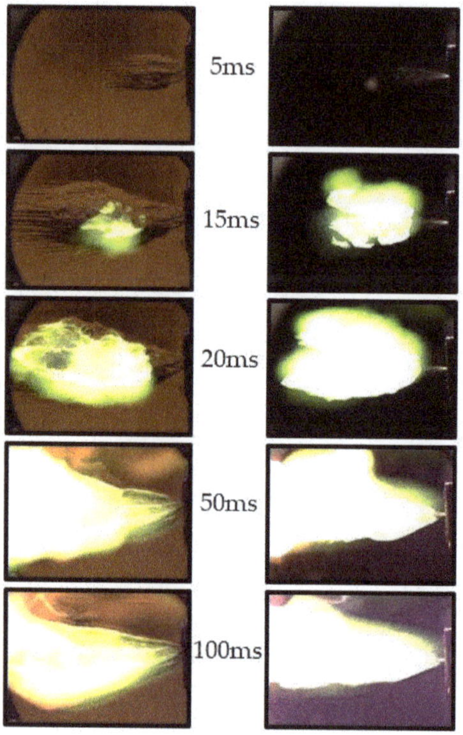

Figure 9. Sequence of schlieren ((**left**) column) and self-luminosity ((**right**) column) video frames of continuous TEA–TEB spray self-ignition and combustion in ambient air at NPT conditions: flat injector. Injection overpressure is 6 atm. Frame size is 590 × 392 pixels (192 × 127 mm^2), frame rate 12,000 fps, shutter speed 81.5 μs, injection duration 100 ms.

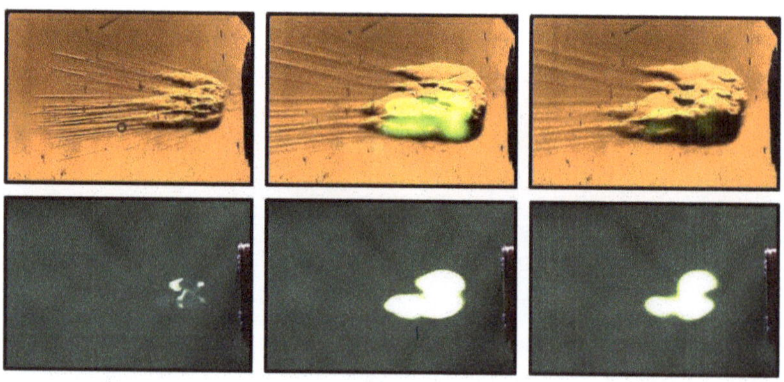

Figure 10. Sequence of schlieren ((**upper**) raw) and self-luminosity ((**lower**) raw) video frames of self-ignition and combustion of the pulsed TEA–TEB spray in ambient air at NPT conditions: flat injector. Injection overpressure is 6 atm. Frame size is 480 × 320 pixels (156 × 104 mm^2), frame rate 5000 fps, shutter speed 190 μs, injection duration 0.7 ms.

Tables 3 and 4 show the statistics of self-ignition events in the case of four successive injection pulses with a pulse duration of 10 ms and time interval between pulses of 1000 ms

(Table 3) and six successive pulses with a pulse duration of 5 ms and time interval between pulses of 400 ms (Table 4). The mean self-ignition delay is seen to be shorter for the conditions of Table 4, thus indicating that the shorter interval between pulses (400 ms vs. 1000 ms) promotes self-ignition, presumably due to the availability of hot residual air on the spray path, which is in line with Equation (14). The results of Figures 6–8, as well as Tables 3 and 4, indicate that the activation energy of TEA–TEB mixture self-ignition in Equation (7) is approximately $\varepsilon \approx 2$ kcal/mol (see Table 2).

Table 3. Measured self-ignition delays in the test fire with 4 successive injection pulses with a pulse duration of 10 ms and time interval between pulses 1000 ms; injection overpressure is 3 atm.

Test No.	1	2	3	4	Mean	RMS
τ_i, ms	35	32	28	28	31	3

Table 4. Measured self-ignition delays in the test fire with 6 successive injection pulses with a pulse duration of 5 ms and time interval between pulses 400 ms; injection overpressure is 3 atm.

Test No.	1	2	3	4	5	6	Mean	RMS
τ_i, ms	24	20	19	19	17	17	19	2

Tables 5 and 6 show the statistics of self-ignition events in the case of 15 successive injection pulses with a pulse duration of 1 and 2 ms, respectively, and a time interval between pulses of 100 ms. Contrary to Tables 3 and 4, the injection overpressure for the test fires of Tables 5 and 6 was 6 atm rather than 3 atm. The mean self-ignition delay is seen to be around 6 ms, which is shorter than that for the lower injection overpressure by a factor of 3 to 5. This result is also in line with Equation (14) as the characteristic droplet size decreases with the injection pressure.

Table 5. Measured self-ignition delays in the test fire with 15 successive injection pulses with a pulse duration of 1 ms and time interval between pulses 100 ms; injection overpressure is 6 atm.

Test No.	1	2	3	4	5	6	7	8	9	10	11	12	13	14	15	Mean	RMS
τ_i, ms	5	5.6	5.8	5.8	5.4	5.6	5.8	5.6	5.4	5.6	5.6	6	6.4	6	6.8	5.8	0.4

Table 6. Measured self-ignition delays in the test fire with 15 successive injection pulses with a pulse duration of 2 ms and time interval between pulses 100 ms; injection overpressure is 6 atm.

Test No.	1	2	3	4	5	6	7	8	9	10	11	12	13	14	15	Mean	RMS
τ_i, ms	5	5	4.8	5.2	5.4	5.2	5.2	5.8	5.8	6	5.4	6.2	6.2	6.2	6.2	5.6	0.5

4. Discussion

The proposed kinetic model of the self-ignition of TEA–TEB droplets at NPT conditions, implying the intrusion of oxygen to the condensed phase with the formation of $(C_2H_5)_2Al–O–O–(C_2H_5)$ and $(C_2H_5)_2B–O–O–(C_2H_5)$ molecules, seems plausible due to the indisputable experimental fact of the self-ignition of liquid TEA, TEB, or TEA–TEB mixture upon contact with air. The model shows that the rate of decomposition of these molecules with the formation of the active ethyl radical depends on the size of the liquid droplets, as well as the local instantaneous oxygen concentration and temperature in the environment, implying that the ignition delay is shorter for the higher injection pressure and hotter air. The results of preliminary experiments on the self-ignition of pulsed and continuous TEA–TEB sprays in air at NPT conditions have confirmed these implications and provided the data for estimating the activation energy of the rate-limiting reaction, which appeared to be close to 2 kcal/mol.

It must be noticed that the existence of reactions (5) and (6) and their supposed role in the TEA self-ignition process still remain questionable. For example, in [4], on the basis of molecular dynamics simulation of the reaction of TEA (in the condensed phase) with gaseous oxygen at a temperature above 2000 K, it is assumed that the reaction begins with the rapid removal of a hydrogen atom by an oxygen molecule, with the formation of the HO_2 radical in the gas phase, and subsequently H_2 molecules (the latter is associated with air humidity), i.e., the rapid occurrence of reactions in the gas phase is associated with the high reactivity of hydrogen. For the kinetic parameters A and ε in the analogue of formula (7), the values $9.67 \cdot 10^9$ s^{-1} and 0.3 kcal/mol, respectively, were obtained. Another example is Ref. [19], where the priority is given to the channel of TEA decomposition through the breaking of the bond between oxygen atoms in the $(C_2H_5)_2Al–O–O–(C_2H_5)$ molecule with the formation of the C_2H_5O radical. In this case, the total process of oxygen intrusion and $(C_2H_5)_2Al–O–O–(C_2H_5)$ molecule decomposition is exothermic and proceeds with the release of 15.4 kcal/mol, whereas the same process proceeding through the C_2H_5 radical is claimed to be endothermic. The question of which mechanism is actually implemented can only be answered after systematic experiments, which are planned for the future. Note that the model proposed herein, even if it is not implemented in relation to the self-ignition of TEA, TEB, and TEA–TEB mixtures, can be applied to other reactions, the rate constant of which depends on temperature according to the Arrhenius law.

5. Conclusions

A novel scheme of the heterogeneous interaction of gaseous oxygen with liquid TEA–TEB droplets accompanied by the release of light hydrocarbon radicals C_2H_5 and/or C_2H_5O into the gas phase was used for calculating the self-ignition of a spatially homogeneous mixture of TEA–TEB droplets in ambient air at normal pressure and temperature conditions. Calculations were performed with the variation of the activation energy of the rate-limiting reaction intended for comparison with experiments. Calculations showed that the self-ignition delay of a stoichiometric C_2H_5O–air mixture was much longer than that of a stoichiometric C_2H_5–air mixture, indicating the much greater reactivity of the C_2H_5 radical compared to the C_2H_5O radical in air. The proposed kinetic model with the formation of $(C_2H_5)_2Al–O–O–(C_2H_5)$ or $(C_2H_5)_2B–O–O–(C_2H_5)$ molecules seems plausible due to the indisputable experimental fact of the self-ignition of liquid TEA, TEB, and TEA–TEB mixtures upon contact with air. Experiments on the self-ignition of pulsed and continuous TEA–TEB mixture sprays in air at normal pressure and temperature conditions provided the data for estimating the activation energy of the rate-limiting reaction, which appeared to be close to 2 kcal/mol. The ignition delay was shown to decrease with the decrease in the droplet size, both in the model and in the experiment.

Author Contributions: Conceptualization, S.M.F.; methodology, S.M.F., I.O.S. and V.S.A.; investigation, V.Y.B., A.A.B., I.O.S., V.S.A., F.S.F., P.A.S. and S.L.G.; data curation, A.A.B., I.O.S. and V.S.A.; writing—original draft preparation, S.M.F.; writing—review and editing, S.M.F.; visualization, I.O.S. and V.S.A.; supervision, S.M.F.; project administration, S.M.F.; funding acquisition, S.M.F.; resources, S.L.G. All authors have read and agreed to the published version of the manuscript.

Funding: This research was supported by a subsidy given to Semenov Federal Research Center for Chemical Physics of the Russian Academy of Sciences to implement the state assignments with registration numbers 122040500073-4 and 122040500068-0.

Data Availability Statement: Not applicable.

Acknowledgments: The authors offer a tribute of respect to the late Nickolai N. Kuznetsov for his valuable contribution to this study.

Conflicts of Interest: The authors declare no conflict of interest.

Appendix A. Heat Capacities of TEA–Air Mixture

According to reaction (1), there are 10.5 moles of O_2 per 1 mole of TEA. The heat capacities c_v and c_p of 1 cm^3 of air under normal conditions are $\frac{\frac{5}{2}R}{22{,}400} = 9.3 \times 10^{-4}$ J/(cm^{3}K) and $\frac{\frac{7}{2}R}{22{,}400} = 1.3 \times 10^{-3}$ J/(cm^{3}K). Moreover, 1 cm^3 contains $\frac{0.21}{22{,}400 \times 10.5} = 8.9 \cdot 10^{-7}$ moles of TEA. The heat capacity of such an amount of TEA is much less than the heat capacity of air, and it can be taken into account approximately, equating it to the heat capacity of, e.g., oil. The heat capacity of oil is roughly half that of water and is approximately 4.5R per mole. Hence, in terms of the number of moles in 1 cm^3, one obtains $9 \times 4.18 \times 8.9 \times 10^{-7} = 0.33 \times 10^{-4}$ J/(cm^{3}K). The heat capacities of the mixture are

$$c_v = 9.6 \times 10^{-4}\ \mathrm{J}/\left(\mathrm{cm}^3\mathrm{K}\right);$$

$$c_p = 1.33 \times 10^{-3}\ \mathrm{J}/\left(\mathrm{cm}^3\mathrm{K}\right).$$

Appendix B. The Number of TEA Droplets in 1 cm^3

The number of droplets, N, is determined by the ratio

$$N = \frac{n_{\mathrm{TEA}}}{n_{\mathrm{TEA,d}}},\ \mathrm{cm}^{-1}$$

where n_{TEA} is the number of TEA molecules in 1 cm^3; $n_{\mathrm{TEA,d}}$ is the number of such molecules in one droplet. Considering that the number n_{TEA} is 10.5 times less than the number of oxygen molecules in air, $n_{O_2} = 0.21 N_A/22{,}400\ \mathrm{cm}^{-3}$, one obtains $n_{\mathrm{TEA}} = n_{O_2}/10.5$. The number $n_{\mathrm{TEA,d}}$ is equal to the ratio of the droplet mass, $4\pi r_d^3 \rho_d/3$, to the mass of one TEA molecule, equal to W_{TEA}/N_A, where $\rho_d \approx 0.83$ g/cm^3 [1] is the drop density, and W_{TEA} = 114 g is the molecular mass of TEA. Substitution of all indicated quantities and $r_d = 5 \times 10^{-3}$ cm into the equation for N gives N = 210 cm^{-3}.

References

1. Heck, W.B.; Johnson, R.L. Aluminum alkyls—Safe handling. *Ind. Eng. Chem.* **1962**, *54*, 35–38. [CrossRef]
2. Triethylaluminum (CAS 97-93-8)—Chemical & Physical Properties by Cheméo. Available online: https://www.chemeo.com/cid/63-022-7/Aluminum-triethyl (accessed on 22 September 2022).
3. Marsel, J.; Kramer, L. Spontaneous ignition properties of metal alkyls. *Proc. Combust. Inst.* **1958**, *7*, 906–912. [CrossRef]
4. Goncalves, R.F.B.; Iha, K.; Rocco, J.A.F.F. Reactive molecular dynamics simulation and chemical kinetic evaluation of combustion of triethylaluminium (TEA). *Quim. Nova* **2018**, *41*, 507–511. [CrossRef]
5. Sato, Y.; Okada, K.; Akiyoshi, M.; Tokudome, K.; Matsunaga, T. Reaction hazards of triethylaluminum under closed conditions. *J. Loss Prev. Process Ind.* **2011**, *24*, 656–661. [CrossRef]
6. Cotton, F.A.; Wilkinson, G. *Advanced Inorganic Chemistry*, 3rd ed.; John Wiley & Sons: New York, NY, USA, 1972.
7. Streitwieser, A., Jr.; Heathcock, H. *Introduction to Organic Chemistry*, 3rd ed.; Macmillan: New York, NY, USA, 1989.
8. Sydora, O.L. Selective ethylene oligomerization. *Organometallics* **2019**, *38*, 997–1010. [CrossRef]
9. Brotherton, R.J.; Weber, C.J.; Guibert, C.R.; Little, J.L. Boron Compounds. In *Ullmann's Encyclopedia of Industrial Chemistry*; Wiley-VCH GmbH: Weinheim, Germany, 2000. [CrossRef]
10. Allais, F.; Boivin, J.; Nguyen, V.T. Part 2. Mechanistic aspects of the reduction of S-alkyl-thionocarbonates in the presence of triethylborane and air. *Beilstein J. Org. Chem.* **2007**, *3*, 46. [CrossRef] [PubMed]
11. Mirviss, S.B.; Rutkowski, A.J.; Seelbach, C.W.; Oakley, H.T. SAFETY—Pyrophoric Organometallics. *Ind. Eng. Chem.* **1961**, *53*, 58A–62A. [CrossRef]
12. Davis, S.M.; Yilmaz, N. Thermochemical analysis of hypergolic propellants based on triethylaluminum/nitrous oxide. *Int. J. Aerosp. Eng.* **2014**, *2014*, 269836. [CrossRef]
13. You, Z. (Ed.) Chapter 8—Micropropulsion. In *Micro and Nano Technologies, Space Microsystems and Micro/Nano Satellites*; Butterworth-Heinemann: Oxford, UK, 2018; pp. 295–339. [CrossRef]
14. Zhou, Y.; Dai, L.; Jiao, N. Review of Bubble Applications in Microrobotics: Propulsion, Manipulation, and Assembly. *Micromachines* **2022**, *13*, 1068. [CrossRef] [PubMed]
15. Seedhouse, E. *SpaceX: Starship to Mars—The First 20 Years*; Springer: Cham, Switzerland, 2022.

16. Triethylaluminum Market Size is Expected to Expand at a CAGR of 1.8% to 2026. Press Release Published: 18 October 2022. Available online: https://www.marketwatch.com/press-release/triethylaluminum-market-size-is-expected-to-expand-at-a-cagr-of-18-to-2026-new-report-spread-across-96-pages-2022-10-18?tesla=y (accessed on 17 November 2022).
17. Kuznetsov, N.M.; Frolov, S.M.; Storozhenko, P.A.; Shamshin, I.O. Kinetic model of oxidation and self-ignition of triethyl aluminum in air. *Combust. Explos.* **2019**, *12*, 91–97.
18. Kuznetsov, N.M.; Frolov, S.M.; Storozhenko, P.A. Calculation of the standard enthalpy of formation and heat of complete combustion of triethylaluminum in water vapor and in air. *Combust. Explos.* **2019**, *12*, 10–13. [CrossRef]
19. Kroupnov, A.A.; Pogosbekian, M.J. Energy and structural characteristics for initial stage of self-ignition triethylaluminum in air. *Combust. Explos.* **2021**, *14*, 93–97. [CrossRef]
20. Johnson, W.H.; Kilday, M.V.; Prosen, E.J. Heats of combustion and formation of trimethylborane, triethylborane, and tri-*n*-butylborane. *J. Res. Natl. Bur. Stand. A Phys. Chem.* **1961**, *65A*, 215–219. [CrossRef] [PubMed]
21. Kondratiev, V.N.; Nikitin, E.E.; Reznikov, A.O.; Umanskii, S.Y. *Thermal Bimolecular Reactions in Gases*; Nauka: Moscow, Russia, 1976.
22. Ollivier, C.; Renaud, P. Organoboranes as a source of radicals. *Chem. Rev.* **2001**, *101*, 3415–3434. [CrossRef] [PubMed]
23. Curran, D.P.; McFadden, T.R. Understanding initiation with triethylboron and oxygen: The differences between low-oxygen and high-oxygen regimes. *J. Amer. Chem. Soc.* **2016**, *138*, 7741–7752. [CrossRef]
24. Semenov, N.N. *Selected Works, Volume 3: On Some Problems of Chemical Kinetics and Reactivity*; Nauka: Moscow, Russia, 2005.
25. Zel'dovich, Y.B.; Barenblatt, G.I.; Librovich, V.B.; Makhviladze, G.M. *Mathematical Theory of Combustion and Explosion*; Nauka: Moscow, Russia, 1980.
26. Basevich, V.Y.; Belyaev, A.A.; Frolov, S.M. The mechanisms of oxidation and combustion of normal alkane hydrocarbons: The transition from C1–C3 to C4H10. *Russ. J. Phys. Chem. B* **2007**, *2*, 477–484. [CrossRef]
27. Williams, F.A. *Combustion Theory*; Addison-Wesley Publishing Co.: Boston, MA, USA, 1965.
28. Burcat, A. Ideal Gas Thermodynamic Data in Polynomial Form for Combustion and Air Pollution Use. Thermodynamic Data at the Web Site of the Laboratory for Chemical Kinetics. 2005. Available online: http://garfield.chem.elte.hu/Burcat/burcat.html (accessed on 29 September 2022).
29. Elkotb, M.M. Fuel atomization for spray modelling. *Prog. Energy Combust. Sci.* **1982**, *8*, 61–91. [CrossRef]
30. Frolov, S.M.; Basevich, V.Y.; Aksenov, V.S.; Polikhov, S.A. Optimization study of spray detonation initiation by electric discharge. *J. Shock Waves* **2005**, *14*, 175–186. [CrossRef]
31. Azatyan, V.V.; Kogan, A.M.; Neigauz, M.G.; Poroikova, A.I.; Aleksandrov, E.N. The role of self-heating during the combustion of hydrogen near the first ignition limit. *Kinet. Catal.* **1975**, *16*, 577–585.

micromachines

Article

Prediction of Dispersion Rate of Airborne Nanoparticles in a Gas-Liquid Dual-Microchannel Separated by a Porous Membrane: A Numerical Study

Zohreh Sheidaei [1,2], Pooria Akbarzadeh [1,*], Carlotta Guiducci [2] and Navid Kashaninejad [3,*]

1 Faculty of Mechanical and Mechatronics Engineering, Shahrood University of Technology, Shahrood 3619995161, Iran
2 Laboratory of Life Sciences Electronics, École Polytechnique Fédérale de Lausanne, 1015 Lausanne, Switzerland
3 Queensland Micro- and Nanotechnology Centre, Nathan Campus, Griffith University, 170 Kessels Road, Brisbane, QLD 4111, Australia
* Correspondence: akbarzad@ut.ac.ir or p.akbarzadeh@shahroodut.ac.ir (P.A.); n.kashaninejad@griffith.edu.au (N.K.); Tel.: +98-9122022776 (P.A.); +61-737355391 (N.K.)

Abstract: Recently, there has been increasing attention toward inhaled nanoparticles (NPs) to develop inhalation therapies for diseases associated with the pulmonary system and investigate the toxic effects of hazardous environmental particles on human lung health. Taking advantage of microfluidic technology for cell culture applications, lung-on-a-chip devices with great potential in replicating the lung air–blood barrier (ABB) have opened new research insights in preclinical pathology and therapeutic studies associated with aerosol NPs. However, the air interface in such devices has been largely disregarded, leaving a gap in understanding the NPs' dynamics in lung-on-a-chip devices. Here, we develop a numerical parametric study to provide insights into the dynamic behavior of the airborne NPs in a gas–liquid dual-channel lung-on-a-chip device with a porous membrane separating the channels. We develop a finite element multi-physics model to investigate particle tracing in both air and medium phases to replicate the in vivo conditions. Our model considers the impact of fluid flow and geometrical properties on the distribution, deposition, and translocation of NPs with diameters ranging from 10 nm to 900 nm. Our findings suggest that, compared to the aqueous solution of NPs, the aerosol injection of NPs offers more efficient deposition on the substrate of the air channel and higher translocation to the media channel. Comparative studies against accessible data, as well as an experimental study, verify the accuracy of the present numerical analysis. We propose a strategy to optimize the affecting parameters to control the injection and delivery of aerosol particles into the lung-on-chip device depending on the objectives of biomedical investigations and provide optimized values for some specific cases. Therefore, our study can assist scientists and researchers in complementing their experimental investigation in future preclinical studies on pulmonary pathology associated with inhaled hazardous and toxic environmental particles, as well as therapeutic studies for developing inhalation drug delivery.

Keywords: lung-on-a-chip; porous membrane; gas–liquid dual-channel chip; nanoparticle; numerical simulation

Citation: Sheidaei, Z.; Akbarzadeh, P.; Guiducci, C.; Kashaninejad, N. Prediction of Dispersion Rate of Airborne Nanoparticles in a Gas-Liquid Dual-Microchannel Separated by a Porous Membrane: A Numerical Study. *Micromachines* **2022**, *13*, 2220. https://doi.org/10.3390/mi13122220

Academic Editor: Pingan Zhu

Received: 21 September 2022
Accepted: 11 December 2022
Published: 14 December 2022

1. Introduction

As a main component of the lungs, alveoli, with a sizeable capillary interface, enable mass transfer and gas exchange in the human body through their air–blood barrier (ABB) (see Figure 1a) [1]. They are consistently exposed to various inhaled NPs, which due to their small size, can easily transfer throughout the airways and reach them. Some NPs are deposited on the alveolar epithelial cells, while others tend to cross the high-permeable ABB and transfer toward other organs and tissues through the vascular system [2–4]. Over the last few decades, morbidity statistics have demonstrated a significant increase in severe pulmonary disorders, including asthma, emphysema, chronic obstructive pulmonary disease, and lung cancer, and highlighted inhaled exposure of alveoli to ambient toxic NPs as the main cause [5,6]. Moreover, new evidence unexpectedly reveals a growing list of other extrapulmonary diseases associated with the translocation of inhaled toxin NPs from the alveoli to the other organs such as brain, kidney, and liver [7]. On the other hand, this transfer mechanism has been recently taken into account in inhalation therapies to develop inhaled nano-drug delivery systems as a targeted and noninvasive method for the treatment, with fewer side effects, of pulmonary and even other organs' diseases [5]. Therefore, all of this results in a burgeoning demand for pathology, toxicology, and therapeutical studies linked with the inhaled NPs in the alveolar region in order to assess the adverse and beneficial effects of inhaled NPs on the target organ or tissue.

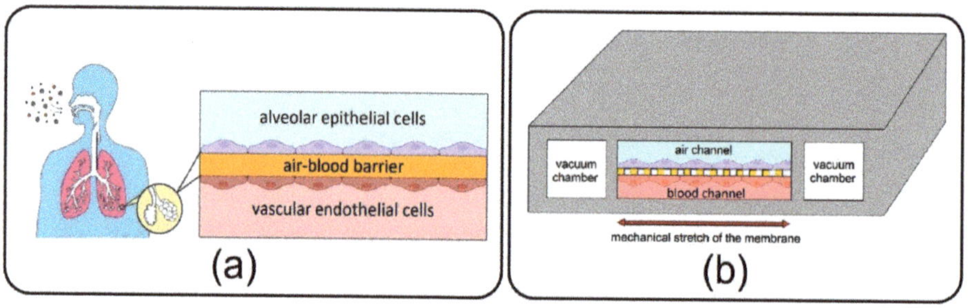

Figure 1. (**a**) Air–blood barrier (ABB) in the alveolar space of human lungs; (**b**) Microfluidic lung-on-a-chip device comprises air/blood channels separated by a thin, porous membrane and two side vacuum channels.

Lung-on-a-chip devices have recently drawn attention in biomedical studies owing to their unique potential in bio-mimicking the physiology of the lung alveolar ABB and providing more accuracy in cellular responses as compared to conventional cell-culture models. First introduced by Huh et al. [8] in 2010, the lung-on-a-chip comprises air/blood (alveolar/vascular) microchannels separated by a thin, porous membrane to enable the co-culturing of epithelial and endothelial cells neighboring on the opposite sides of the membrane and two side vacuum chambers to emulate a breathing lung (see Figure 1b). Adventitiously, this device has opened new research doors in toxico-pharmacological fields, in addition to promising an enormous capacity in recapitulating complex biological pulmonary malfunctions such as edema formation and thrombosis [9,10]. Lung-on-a-chip devices could be conveniently utilized to inject toxic and medical aerosol NPs to replicate inhaled delivery into the alveoli for pathological and therapeutic studies, avoiding unnecessary clinical trials and animal tests. While the air interface in the lung-on-a-chip makes it tremendously unique among the other common liquid-based cellular microsystems, most of the relevant experimental studies aim at streaming an aqueous solution of NPs [8,11,12], and exposure of cultivated cells to the aerosols has been disregarded. Controlled injection of NPs into the lung-on-a-chip will return appropriate sedimentation and distribution in the target area, which is critical for attaining the desired physiochemical efficacy. However, a big research gap exists in the comprehensive understanding of the dynamics of NPs in the lung-on-a-chip device. Accordingly, the distribution and deposition of particles on the endothelial cells in the air channel, as well as their translocation to the media channel, are imperative subjects that should be examined carefully. It is evident that the dynamic behavior of particles in microdevices is determined by many crucial factors, such as the geometry of the device and properties of fluid flow and particles. Nevertheless, optimizing all these parameters is challenging for an enhanced delivery process. Complications associated with fabricating the lung-on-a-chip device and especially the membrane with micro-scale pores, aerosol delivery, and transient monitoring of NPs, which consequently lead to costly and laborious optimization procedures, are the prime suspects responsible for such deficiencies in the field [13–15].

Many efforts have been directed toward employing numerical approaches to overcome challenges associated with experimental analyses of various cell-based microfluidic devices to improve their functionality [16,17]. A few recent investigations are devoted to studying the dynamic behavior of solid particles in the lung-on-a-chip. According to their research objectives, these studies follow two main perspectives, namely, the Eulerian approach, which focuses on the concentration of particles considering the problem governed by typical convection–diffusion phenomena, and the Lagrangian approach, which deals with individual particles, tracing the trajectory of each particle separately. For instance, Frost et al. [18] numerically investigated the effect of a porous membrane, one modeled by molecular diffusivity, on the molecular convection–diffusion transport in a bilayer microfluidic device. Their results showed that the molecular concentration at the outlet of the bottom channel increased by membrane porosity augmentation, while it was invertedly affected by the upper channel height.

Although molecule-sized particles are widely considered in many biomedical studies because of their simple diffusivity, particular attention has also been recently paid to particles with diameters ranging between 10 nm to 10 μm due to some other advantages, such as high uptake capacity by cells [19,20]. Because the Eulerian method investigates the average behavior of particles with less computational cost, it is often favored over its more precise counterpart, i.e., the Lagrangian approach. However, the outcome of these two perspectives would not converge for investigating large particles' dynamics, especially when the total number of released particles is relatively low [21]. Using the Lagrangian approach, Arefi et al. [22] studied the deposition of NPs in an air microchannel under a pulsatile bidirectional flow to simulate breathing patterns. They found that increased airflow rate and breathing frequency raise the deposition rate of particles on the substrate of the channel. In another similar study using finite element simulation, Moghadas et al. [23] reported that airborne delivery of NPs to the cellular region in a microchannel is affected by airflow velocity as well as particle diameter.

The lung-on-a-chip device, which enables particle tracing in air and media phases, allows for simultaneously investigating the particles' deposition and translocation. Thus, it is an essential prerequisite for precise toxicological and pharmaceutical studies. Despite its great potential in mimicking the in vivo conditions, the dynamics of NPs in a typical lung-on-a-chip device with a porous membrane and air/blood channels have never been investigated, to the best of our knowledge. To bridge this research gap, the current paper aims to conduct a comprehensive numerical parametric study to provide a clear insight into the role of various hydrodynamical and geometrical properties on the dynamics of airborne NPs in the lung-on-a-chip system. The impact of fluid-flow velocity, membrane porosity, and particle diameter as dominant factors is examined on the dispersion of the particles in both air and media channels as well as their deposition and transfer rates.

This paper is organized as follows: the numerical modeling of the lung-on-a-chip, the governing equations, and the boundary conditions are introduced in Section 2; comparative studies against accessible data and an experimental study are conducted in Section 3 to establish the accuracy of the present numerical study; a case study and parameter sensitivity are discussed in Section 4; and finally, Section 5 provides the conclusions and summarizes the results.

2. Governing Equations and Boundary Conditions

In the current work, the dynamics of airborne NPs in the gas/liquid (air/blood) channels of a lung-on-a-chip device separated by a thin, porous membrane are investigated against variations in fluid flow velocity, membrane porosity, and particle diameter. For this purpose, a 2D numerical model is developed using Laminar Flow and Particle Tracing for Fluid Flow modules in COMSOL Multiphysics 5.6 finite element software (see Figure 2) [24] to trace the motion of NPs. For simplicity, the system of governing equations is assumed to be one-way coupled, considering only the effect of the fluid regime on particles and ignoring the reactive effect due to the tiny size of particles. Therefore, first, the time-dependent velocity profile of the fluids is calculated in the channels, and then the obtained solutions are employed to solve the particle tracing. Here, it should be noted that small dimensions and low fluid velocity generally result in laminar fluid flow in microchannels and also allow the airflow to be treated as an incompressible fluid. Moreover, with a Knudsen number smaller than the critical value ($Kn \cong 0.0004 < 0.01$), the continuum hypothesis and no-slip velocity condition are valid for the airflow field in the channel [25]. The Knudsen number for the airflow is calculated using the hard-sphere collision model as follows [26]:

$$Kn = \frac{\lambda}{2h} = \frac{k_B}{2\sqrt{2}\pi\sigma^2\rho_a Rh}, \tag{1}$$

where λ is the mean free path and h is the channel height. In this model, σ $(= 346\ \text{pm})$ demonstrates the collision diameter of fluid molecules, R $\left(\cong 290\ \text{J.K}^{-1}/\text{Kg}\right)$ and ρ_a represent the specific gas constant and air density, and k_B is the Boltzmann constant.

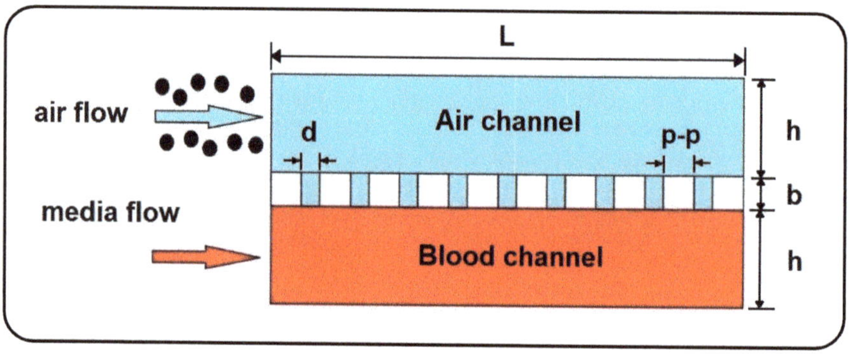

Figure 2. 2D model of airborne delivery of NPs into a gas-liquid dual-channel microdevice with a thin, porous membrane.

In many experimental studies, blood flow is mainly replaced by a medium flow to supply the cells with required nutrition in microfluidic devices, which could be conveniently considered as a Newtonian fluid with similar hydrodynamic properties of water at 37 °C [23]. Consequently, the Newtonian and incompressible air and medium phases are governed by continuity and momentum equations as follows [27]:

$$\nabla \cdot u_i = 0, \tag{2}$$

$$\rho_i\left(\frac{\partial u_i}{\partial t} + u_i \cdot \nabla\ u_i\right) = -\nabla p_i + \mu_i \nabla^2 u_i + \rho_i g, \tag{3}$$

where u and p represent the fluid velocity field and pressure, and t is the time, while the subscript $(i = \text{a}, \text{m})$ stands for air and media, respectively.

The Lagrangian approach is used to model the dynamics of NPs under the action of gravity and hydrodynamic forces arising from the motion of the fluid:

$$m_\text{p}\frac{du_\text{p}^i}{dt} = F_\text{d}^i + F_\text{b}^i + m_p\left(1 - \frac{\rho_i}{\rho_\text{p}}\right)g, \tag{4}$$

where m_p and ρ_p are the particle mass and density, and u_p^i indicates the particle velocity in air and media channels, respectively. The right-hand side of Equation (4) is the summation of drag force (F_d^i), the Brownian force (F_b^i), and the particle's buoyant weight, while the gravity acceleration is considered to be perpendicular to the flow direction. The drag force in Equation (4) is determined in accordance with Stokes law as follows [28]:

$$F_{\rm d}^{i} = 3\pi\mu_i\left(u_i - u_{\rm p}^{i}\right)d_{\rm p}, \tag{5}$$

where $d_{\rm p}$ is the particle diameter. Also, the Brownian force exerted on particles at each time step taken by the numerical solver could be calculated as [29]:

$$F_{\rm b}^{i} = \zeta_i\sqrt{6\pi k_B\mu_i T_i d_{\rm p}/\Delta t}, \tag{6}$$

where T_i is the fluid temperature, which is considered to be at a constant level of 37 °C, and Δt represents the time step. In this calculation, ζ_i, which indicates the direction of Brownian force, refers to a vector whose components are randomly selected with a Gaussian distribution [29]. Inertial lift force contribution is neglected in Equation (4) since the particle Reynolds number is significantly smaller than 1 ($Re_{\rm P} = \rho_i u_i d_{\rm p}^2/\mu_i l$), resulting in viscous forces dominancy [30].

Fluid flow boundary conditions in the air and media channels are considered as constant inlet velocity, zero outlet pressure, and no-slip on all solid-fluid interfaces. Furthermore, the top surface of the membrane is employed as the sticky boundary condition for the injected airborne NPs allowing evaluation of particle deposition in the air channel. On the other hand, the pass-through boundary condition is applied to the air-media interfaces to allow NPs to enter the media channel unimpededly. It is worthwhile to note that the numerical simulation is terminated once all the transferred particles flow out of the media channel.

3. Model Verification

In this section, comparative studies against accessible data as well as an experimental study are conducted to establish the accuracy of the present numerical study. Accordingly, four limiting-case verifications are considered by setting the membrane's porosity equal to zero.

3.1. Verification of Numerical Results with Analytical Data

In the absence of solid particles, the first verification study compares the dimensionless axial fluid velocity in the middle of the air channel with analytical data for a 2D steady laminar flow between two fixed parallel plates in Ref. [31]. For this purpose, fluid velocity and vertical distance from the substrate (y) are normalized with respect to the inlet velocity of 0.3 mm/s and the channel height of 100 µm. As depicted in Figure 3a, a fair agreement is observed between the results of the current study and those of Ref. [31].

3.2. Verification of Numerical Results with Other Numerical Studies

3.2.1. Verification with a 2D Model

Next, the normalized mean concentration of particles with a diameter of 1 µm along the air channel is compared with that obtained by Saidi et al. [21], who developed a customized numerical code to track 2D dynamics of particles under drag and Brownian forces between two parallel plates. Accordingly, a fully developed inlet flow regime with a mean velocity ($\overline{u}$) of 0.05 m/s and channel height (h) of 1 mm is considered. The concentration of particles is calculated by partitioning the computational space into equal bandwidths along the channel length and then dividing the number of particles in each band to the volume of each bandwidth. Finally, distance from the entrance (x) and particle concentration (C) are nondimensionalized, respectively, with respect to a specific length of $3hPe/16$, and initial concentration ($C_0 = 10^5/\text{m}^2$) [21]. $Pe = 2\overline{u}h/D$ stands for the Peclet number, where $D\left(= 2.96 \times 10^{-11}\ \text{m}^2/\text{s}\right)$ is the diffusion coefficient. The outcome, as shown in Figure 3b, demonstrates a good agreement with Ref. [21].

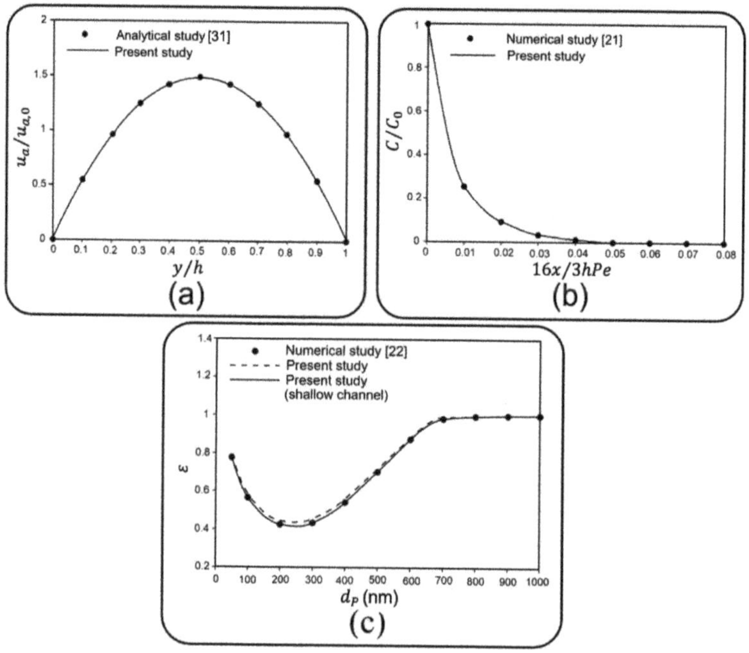

Figure 3. Comparison of the present numerical results with (**a**) analytical data in Ref. [31] for a 2D airflow between two parallel plates. Axial fluid velocity and vertical distance from the substrate are normalized with respect to the inlet velocity of 0.3 mm/s and the channel height of 100 μm. Subsequently, (**b**) numerical data in Ref. [21] is shown for the mean concentration of particles with a diameter of 1 μm along two parallel plates. Distance from the entrance and particle concentration are nondimensionalized, respectively, with respect to a specific length of $3hPe/16$, and initial concentration ($C_0 = 10^5/\text{m}^2$). $Pe = 2\overline{u}h/D$ is the Peclet number, and where $D\ (= 2.96 \times 10^{-11}\ \text{m}^2/\text{s})$ is the diffusion coefficient. Finally, the (**c**) the numerical result of Ref. [22] for deposition rate of NPs with various diameters on the substrate of an air channel is shown.

3.2.2. Verification with a 3D Model

Subsequently, the correctness of the developed finite element model is examined to evaluate the deposition of airborne NPs. With the definition of "the ratio between the number of deposited particles on the substrate and the number of total released particles" for the deposition rate (ε), the third verification study compares the deposition rate obtained through the current 2D model with that reported for a 3D model developed by Arefi et al. [22]. All adjustments for boundary conditions are taken into account, according to Ref. [22]. Here, the effect of channel depth is also surveyed, enabling the "Shallow Channel" interface in the 2D model. Figure 3c compares the results obtained for a 2D channel with finite and infinite depth and the 3D model of Ref. [22], where the independence of deposition rate and the channel depth as well as an excellent agreement between these three is perceivable.

3.3. *Qualitative Verification of the Numerical Results with Experimental Data*

Besides verification against the analytical and numerical data in the literature, an experimental study was accomplished to support the results of the present numerical model. The experiment was carried out with the injection of particles into a microchannel under a water flow. Therefore, a numerical simulation was developed in accordance with the experimental parametric values to compare the obtained results from the two studies. Accordingly, a polydimethylsiloxane (PDMS) microchannel with rectangular cross-section was fabricated using a silicone mold (see Figure 4a). The mold was made with maskless ultraviolet (UV) lithography method, as shown in Figure 4b. A silicon wafer coated with a positive photoresist was exposed to direct UV laser to transfer the channel design on it. Subsequently, the wafer was dry etched to engrave the microchannel mold. Then, a 10:1 mixture of PDMS was cast on the mold and the microchannel was sealed with a glass slide to finalize the fabrication procedure (see Figure 4b). A syringe pump was used to inject the particles into the channel under different fluid velocities (see Figure 5 for the experimental setup). In order to have a similar inlet boundary condition to that of the numerical model, the particles were first accumulated at the entrance of the inlet and injected altogether into the channel. Fluid injection was stopped once the particles reached the channel outlet or deposited entirely.

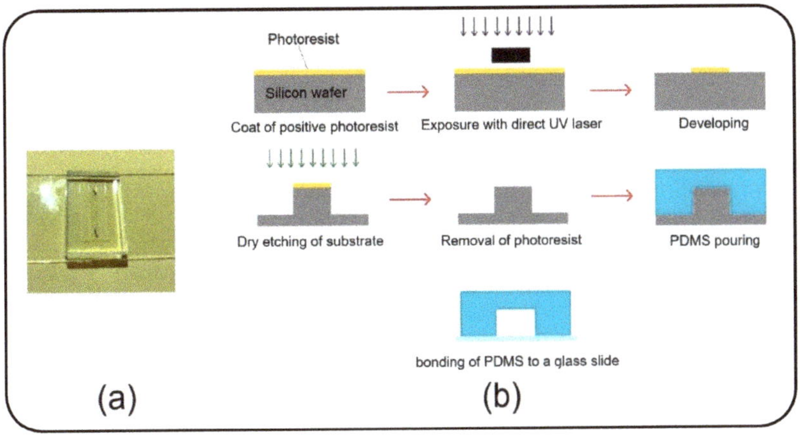

Figure 4. (**a**) Microfluidic device, which is bonded to a glass slide and contains a channel with one inlet and one outlet; (**b**) Schematic diagram of the fabrication process of the microfluidic device.

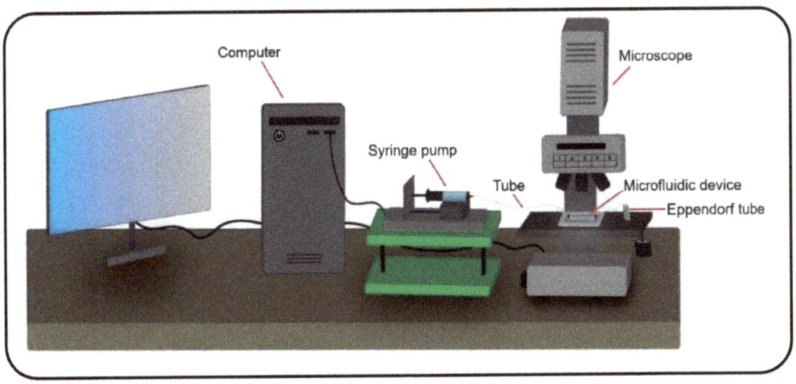

Figure 5. Schematic diagram of the experimental setup for injecting the particles into a microfluidic device using a syringe pump and monitoring their dynamics throughout the device with a microscope.

Magnetic particles of diameter 1 μm and 10 μm were used in the experiments to investigate their sedimentation on the channel substrate under three different water flow velocities. In addition to more deposition efficiency due to their higher inertia and dominant buoyant weight, better control on injection and easier tracing are other main reasons for selecting these particles. The experimental parameter values, including the physical properties of particles, fluid velocities, and the dimension of the microchannel, are listed in Table 1.

Table 1. The values of the parameter used for the experimental test.

Diameter of Particles (d_p) [μm]	Density of Particles (ρ_p) [kg/m^3]	Dimension of Channel ($w \times h \times l$) [μm^3]	Fluid Velocity ($u_{m,0}$) [mm/s]	Flow Rate [μL/min]
1	2200	2000 × 100 × 15,000	0.03 and 0.1	0.36 and 1.2
10	1470	2000 × 300 × 15,000	0.1 and 0.3	3.6 and 10.8

Figure 6 qualitatively compares the trend of particle deposition obtained from the numerical and experimental studies under four different cases. The experimental results are actual images showing the particle deposition rate on the bottom layer of the microchannels. These images were obtained from two different parts of the channel substrate, one from the first half of the channel and the other from its second half.

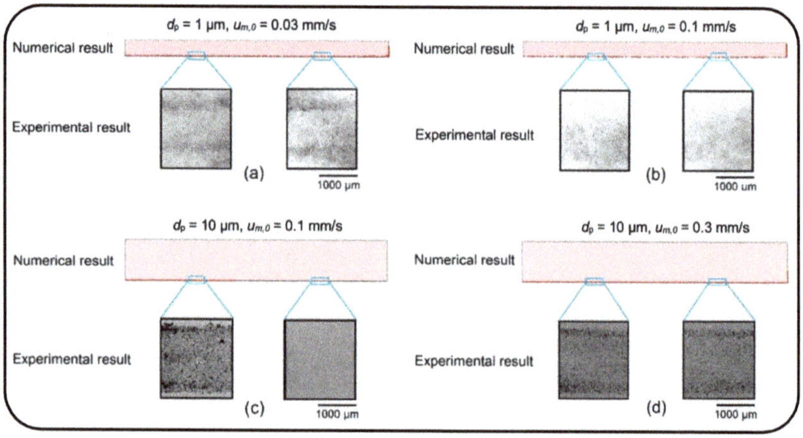

Figure 6. Qualitative comparison of distribution and deposition of the particles on the microchannel substrate obtained from numerical and experimental studies. (**a**) $d_p = 1$ μm, $u_{m,0} = 0.03$ mm/s; (**b**) $d_p = 1$ μm, $u_{m,0} = 0.1$ mm/s; (**c**) $d_p = 10$ μm, $u_{m,0} = 0.1$ mm/s; and (**d**) $d_p = 10$ μm, $u_{m,0} = 0.3$ mm/s.

First, the particles with a diameter of 1 μm were injected into the channel with a height of 100 μm under a fluid velocity of 0.3 mm/s. Due to the dominancy of drag force over the buoyant weight, all the particles were washed out from the channel without any deposition on the substrate (data not shown). Then, the dynamics of particles were examined with a lower fluid velocity of 0.1 mm/s and 0.03 mm/s. Under the fluid velocity of 0.1 mm/s, the particles again streamed quickly toward the microchannel outlet, resulting in a low deposition rate along the channel length (see Figure 6a). The fluid velocity, therefore, was reduced to 0.03 mm/s so that the particles could be uniformly distributed all over the channel substrate and demonstrate a higher deposition rate, Figure 6b.

There is a mild deposition gradient along the channel width due to the parabolic profile of the fluid velocity in the channel. Moreover, on the substrate next to the channel side walls, the deposition rate is very low as a consequence of the no-slip boundary condition. In the second part of the experiment, particles with a diameter of 10 μm were examined under fluid velocities of 0.1 mm/s and 0.3 mm/s (see Figure 6c,d). The height of the microchannel was increased to 300 μm to allow the particles easily to move inside the channel. The distribution of particles on the substrate of the channel reveals that particles with a diameter of 10 μm under the fluid velocity of 0.1 mm/s provide a concentrated deposition at the channel's entrance. However, an increase in the fluid velocity to 0.3 mm/s drives the particles quickly forward, resulting in a uniform distribution on the channel substrate. The experimental results for deposition and distribution of the particles on the substrate of the microchannel show a good agreement with the numerical results.

4. Numerical Results

The current numerical parametric study investigates the dynamics of airborne NPs in a gas–liquid dual-channel lung-on-a-chip device with a thin, porous membrane using Equations (2)–(4). Accordingly, the impact of fluid flow velocity, membrane porosity, and particle diameter, as dominant factors, are analyzed on the dispersion of the particles in both air and media channels as well as their deposition and translocation. Numerical results are confined only to some specific cases due to computational limitations, as well as for the sake of brevity. The diameter of solid particles is assumed to be ranging between 10 to 900 nm, according to their high deposition rate reported in the alveolar region [4]. The effect of membrane porosity is examined for pore diameters (d) of 3 and 10 μm, and pore to pore distances ($p - p$) of 5 and 10 μm. The other numerical values used for simulations are all listed in Table 2.

Table 2. The values of the parameter used for the numerical simulations.

Parameters	Values	Descriptions	Reference
h	100 μm	Height of the channels	[32]
l	2 mm	Length of the channels	[22]
d	10 μm, 3 μm	Diameter of the membrane pore	[32,33]
b	10 μm	Thickness of the membrane	[8]
ρ_p	1180 kg/m^3	Density of particles	[22]
ρ_m	1000 kg/m^3	Density of media	[34]
μ_m	0.718 mPa.s	Viscosity of media	[34]
ρ_a	1.123 kg/m^3	Density of air	[35]
μ_a	0.019 mPa.s	Viscosity of air	[35]
$u_{m,0}$	0.3 mm/s, 1 mm/s	Media velocity at inlet	[23,32]
$u_{a,0}$	0.3 mm/s, 1 mm/s	Air velocity at inlet	[22]

Before presenting the main results, a brief discussion is presented here to investigate the dynamics of NPs injected with an aqueous solution, which substitutes aerosol injection in many relevant experimental studies [8,11,12]. Accordingly, the pore diameter of 10 μm and pore-to-pore distance of 10 μm are taken into consideration, and the deposition rate and transfer rate are compared in Figure 7 for NPs injected via air and media flow into the upper channel of the device. As previously mentioned, the deposition rate is referred to as the proportion of the total released particles that are deposited on the channel substrate. Similarly, the transfer rate defines the ratio between the number of particles transferred to the media channel and the number of released particles. The results obtained demonstrate a noticeable decrease in deposition/transfer rate when media flow is used for the injection of NPs, regardless of their size. To put it clearly, particles follow the fully-developed parabolic profile of media velocity with a lower chance of settling on the substrate or passing through it (data not shown for brevity). Buoyant weight almost fades for submerged particles with a density of the same order as the fluid, which explains the reduction in deposition/transfer rate, especially for large particles. Furthermore, by increasing the fluid viscosity, the drag force dominates the Brownian force, which plays a key role in the deposition of small NPs. Thus, one can draw the conclusion that injecting through an aqueous solution leads to less efficient particle deposition/translocation, and accordingly, air flow injection should be taken into consideration.

4.1. Deposition of Nanoparticles

This section deals with the sedimentation of airborne NPs in the air channel of the lung-on-a-chip device with a focus on the deposition rate and distribution efficiency. Figure 8 depicts the deposition rate against particle diameter for different membrane porosities and air inflow velocities of 0.3 and 1 mm/s. The results show that the deposition rate declines by increasing the particle diameter up to 200 nm regardless of the inlet velocity. The reason behind this decay is related to the dominancy of Brownian diffusion, which is the main particle deposition mechanism for particles of diameter less than 100 nm according to Stokes–Einstein expression [28,36,37]. On the other hand, as could be deduced from Equations (4)–(6), gravity acceleration becomes more dominant by increasing the particle diameter, while the Brownian effect diminishes. This consequently results in higher deposition rates as the particles enlarge, which can be observed in Figure 8 for diameters larger than 200 nm. A comparison within Figure 8a,b implies an overall decrease in deposition rate by intensifying the inlet flow rate, which makes a larger fraction of NPs leave the device through the air channel outlet. However, both cases follow a similar trend except for particles larger than 700 nm, demonstrating a downtrend of deposition rate for inflow velocity of 0.3 mm/s. By comparing the results of Figure 8a and those of Figure 12a, it is observed that more than 95% of these injected big particles tend to either sediment on the air channel substrate or pass through the perforated membrane with a growing transfer rate against the particle size. Therefore, one can conclude that this unexpected behavior is not linked with blowing NPs off the device but rather with a higher transfer rate into the media channel. Indeed, lower inflow velocity magnitude results in slower axial fluid velocity in the boundary layer above the porous area and prolonged regional residence time, which consequently provide more probability for heavier particles to pass through the membrane. The idea is explicitly illustrated through a series of snapshots in Figure 9, where the trajectory of an initially inert 900 nm particle is compared in the porous region for inflow velocities of 0.3 and 1 mm/s. Such behavior also leads to more dependency of larger NPs' deposition on membrane porosity, which is supported by the divergence of deposition rate curves corresponding with different porosities in Figure 8.

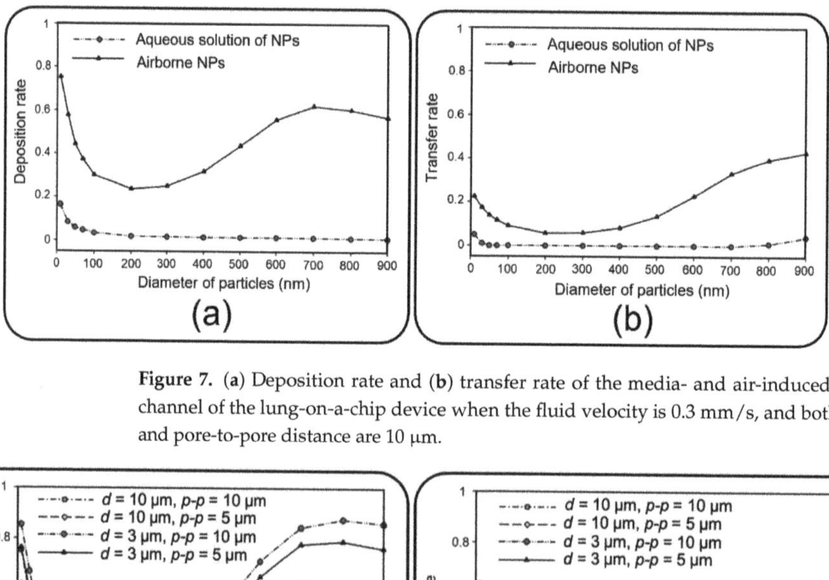

Figure 7. (**a**) Deposition rate and (**b**) transfer rate of the media- and air-induced NPs in the top channel of the lung-on-a-chip device when the fluid velocity is 0.3 mm/s, and both pore diameter and pore-to-pore distance are 10 μm.

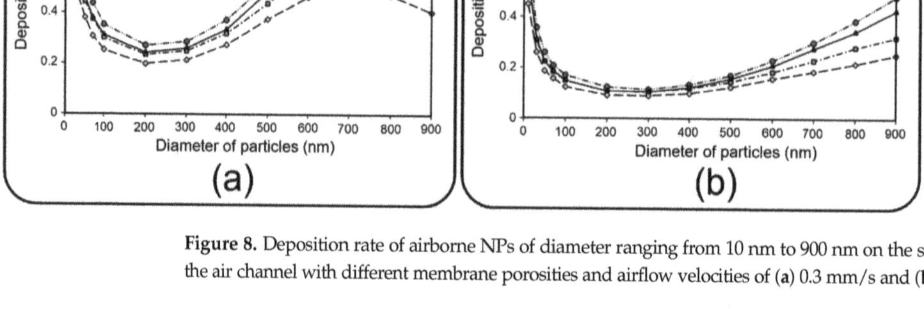

Figure 8. Deposition rate of airborne NPs of diameter ranging from 10 nm to 900 nm on the substrate of the air channel with different membrane porosities and airflow velocities of (**a**) 0.3 mm/s and (**b**) 1 mm/s.

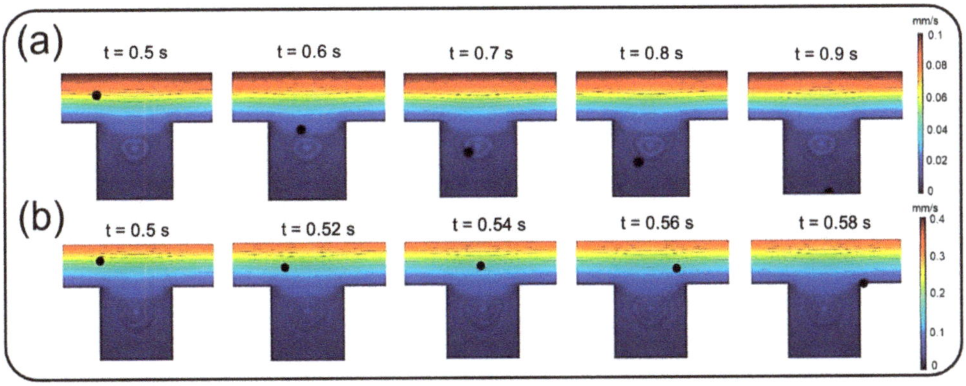

Figure 9. Trajectory of an initially inert 900 nm particle in the porous region for inflow velocities of (**a**) 0.3 mm/s and (**b**) 1 mm/s.

In addition to the high deposition rate of the injected solid NPs, their better distribution on the channel substrate is another significant factor in toxicological and pharmaceutical studies to attain an effective biochemical impact on cultured cells. Here, the distribution of NPs is mathematically described in terms of the mean value and standard deviation of their position [38].

Figure 10 depicts dimensionless distribution indexes of deposited NPs on the air channel substrate under two air inflow velocities, which are normalized with respect to channel length here. Since membrane porosity has no significant effect on statistical characteristics of particle distribution due to the regular alignment of pores, only a single case with a pore diameter of 10 μm and pore-to-pore distance of 10 μm is considered. Distribution efficiency can simply be assessed against an ideal uniform deposition with normalized mean value of 0.5 and a standard deviation of $1/\sqrt{12}$ (shown with a solid red line in Figure 10) [38].

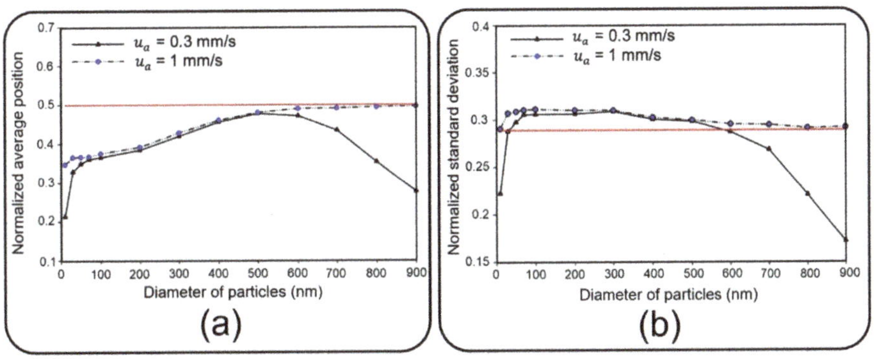

Figure 10. Normalized (**a**) average position and (**b**) standard deviation of deposited NPs of diameter ranging from 10 nm to 900 on the substrate of the air channel with pore diameter and pore to pore distance of 10 μm under two different fluid velocities of 0.3 mm/s and 1 mm/s. The solid red line corresponds with an ideal deposition.

Under an airflow velocity of 0.3 mm/s, both small and large particles demonstrate low values of the mean location and standard deviation, indicating a concentrated deposition at the entrance of the channel. Increasing the airflow velocity would not alter the results unless for particles larger than 600 nm or smaller than 50 nm, giving rise to a significantly widened distribution range. Although the uniform distribution achieved by higher inflow velocity seems to be desirable, it should be noted that the deposition rate would be unfavorably affected in this case. Therefore, depending on the application, a tradeoff between deposition rate and distribution uniformity should be considered to obtain proper particle size and inflow velocity values. To provide a better insight into the problem, snapshots from the distribution of 900 nm particles are provided in Figure 11.

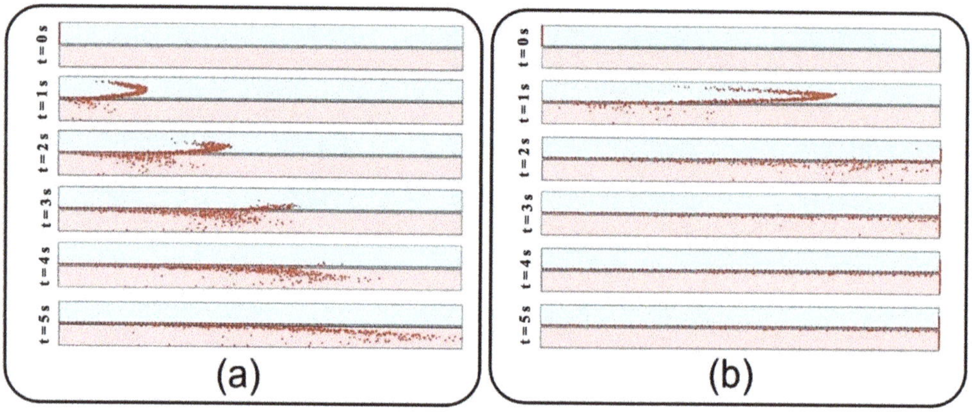

Figure 11. Snapshots from transient distribution of NPs with a diameter of 900 nm in the lung-on-a-chip device with pore diameter and pore-to-pore distance of 10 µm under the fluid velocity of (**a**) 0.3 mm/s, and (**b**) 1 mm/s.

4.2. Translocation of Nanoparticles

This section deals with the translocation of airborne NPs into the media channel of the lung-on-a-chip device, mainly analyzing transfer rate and distribution efficiency. Figure 12 compares the transfer rate of NPs for different membrane porosities and inflow velocities. A higher translocation efficiency seems achievable by increasing the former and decreasing the latter. The obtained results show that for constant membrane porosity and inlet velocity, the transfer rate declines by increasing the particle diameter up to 200 nm, whereafter it starts to go up. Higher transfer rates arise from Brownian diffusivity for small particles and dominancy of gravity force, as well as longer residence time in the porous regions for large particles.

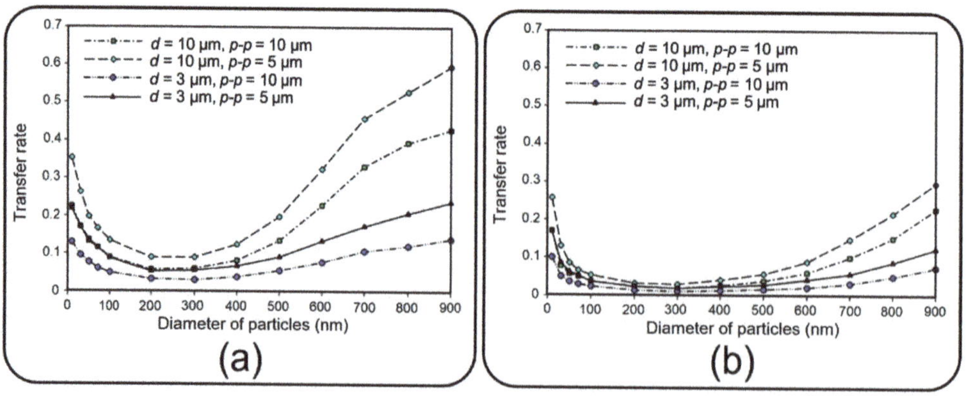

Figure 12. Transfer rate of NPs of diameter ranging from 10 nm to 900 to the media channel of the lung-on-a-chip device with different membrane porosities considering equal inflow velocities of (**a**) 0.3 mm/s and (**b**) 1 mm/s for air and media channels.

Distribution of NPs inside the media channel is also of great importance, especially for targeted particle delivery applications. Since submerged particles with a density close to that of fluid would follow the media velocity profile regardless of inflow velocity, the results are presented only for the inlet velocity of 0.3 mm/s. Accordingly, vertical distribution of transferred NPs are assessed at the outlet of media channel utilizing Equations (4) and (5) with N representing the number of transferred particles and x_n showing the vertical position of each particle. Here, the average position and standard deviation are normalized with respect to media channel height, and, similar to the previous section, are then assessed against an ideal uniform distribution (shown with the solid red line in Figure 13). Higher average position and smaller standard deviation for large particles in Figure 13 indicate their tendency to accumulate close to the media channel's top surface, where the endothelial cells are normally cultured. On the other hand, smaller particles demonstrate a higher penetration power, which is due to their higher vertical velocity magnitude while crossing the membrane as well as their krelatively strong Brownian force in the medium. Moreover, a more uniform distribution is observed for medium-sized particles with a diameter range of 100–300 nm. For a better vision of the impact of size, one can compare the snapshots from the distribution of 10 nm particles in Figure 14 with those obtained from 900 nm particles in Figure 11a.

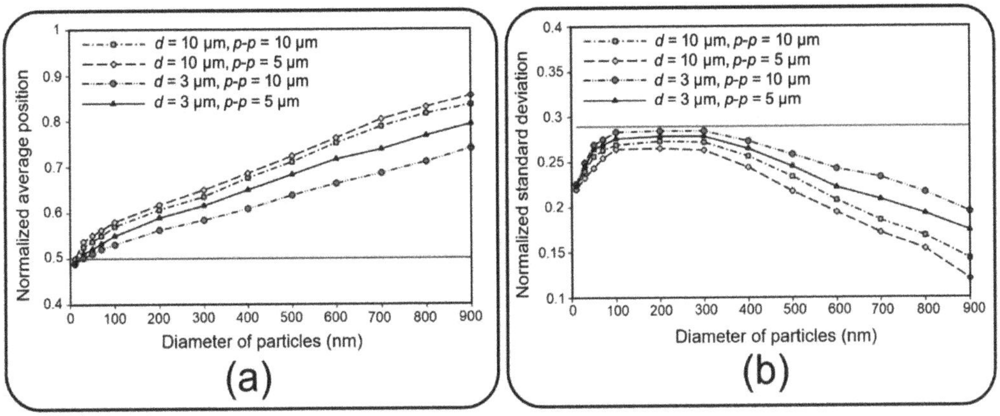

Figure 13. Normalized (**a**) average position and (**b**) standard deviation of transferred NPs of diameters ranging from 10 nm to 900 at the outlet of the media channel in the lung-on-a-chip device with different membrane porosities under the fluid velocity of 0.3 mm/s. The solid red line corresponds with an ideal distribution in the media channel.

In addition to distribution efficiency, the current study investigates the transient evolution of transferred NPs as a significant factor in toxico-pharmacokinetic analyses [39]. First, the relative concentration, which is defined by the ratio of total injected particles that flow out through the outlet of the media channel, is recorded, and then, a numerical differentiation is utilized to obtain the time derivative of the data. After a Gaussian curve fitting, final results are depicted in Figure 15 for different membrane porosities and inlet velocities. Figure 15, which is known as the so-called concentration-time plot [38], provides important details describing the behavior of transferred NPs to the media channel, such as area under each curve, which is equal to the transfer rate (see Figure 12), maximum relative concentration rate (c_{max}), and time to reach this maximal value (t_{max}). For example, it is shown that 10 nm particles demonstrate the highest c_{max}, which is significantly augmented by increasing the inflow velocity. Moreover, it is evident that smaller particles display a faster transmission with a lower t_{max}. Although the overall value of c_{max} grows by increasing the membrane porosity and inflow velocity, it should be noted that this may also considerably change t_{max} depending on the particle size.

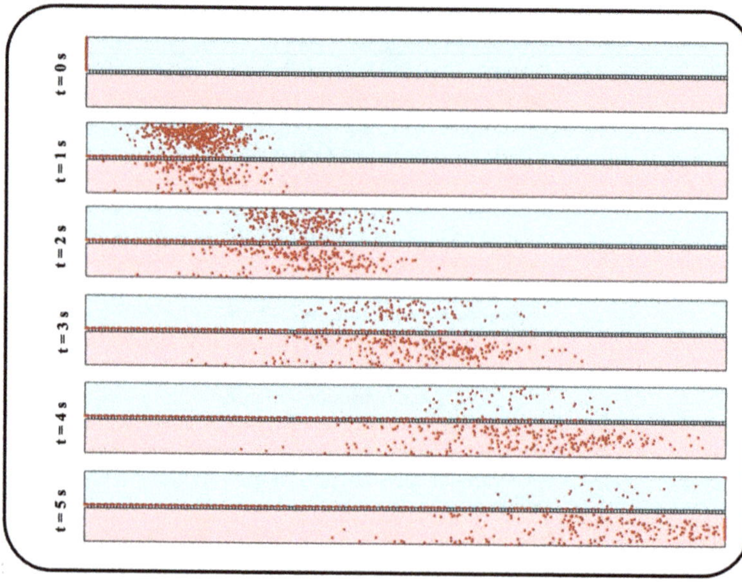

Figure 14. Snapshots from transient distribution of particles with a diameter of 10 nm in the lung-on-a-chip device with pore diameter and pore-to-pore distance of 10 μm and fluid velocity of 0.3 mm/s.

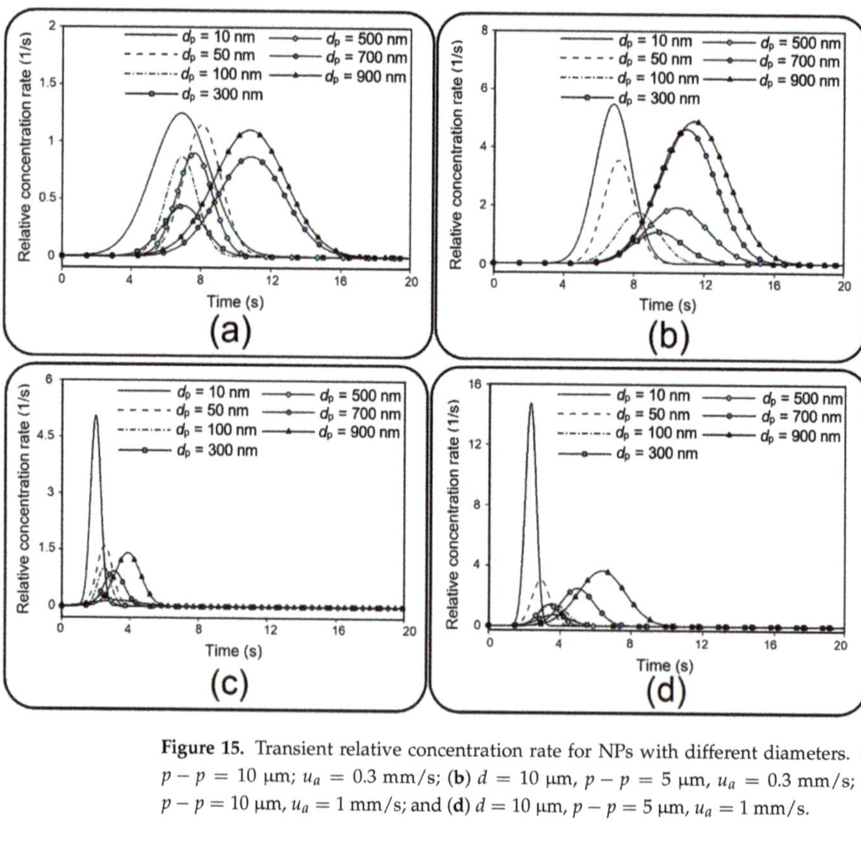

Figure 15. Transient relative concentration rate for NPs with different diameters. (**a**) $d = 3$ μm, $p - p = 10$ μm; $u_a = 0.3$ mm/s; (**b**) $d = 10$ μm, $p - p = 5$ μm, $u_a = 0.3$ mm/s; (**c**) $d = 3$ μm, $p - p = 10$ μm, $u_a = 1$ mm/s; and (**d**) $d = 10$ μm, $p - p = 5$ μm, $u_a = 1$ mm/s.

5. Conclusions

With a promising potential in recapitulating in vivo conditions, lung-on-a-chip technology has recently attracted the attention of researchers. This dual-channel device with a sandwiched porous membrane enables particle tracing in air and media phases simultaneously, which is an essential prerequisite for further toxicological and pharmaceutical studies. Despite this remarkable capability, a detailed investigation into the dynamics of NPs in this device has been overlooked by the literature. Here, a comprehensive numerical parametric study was conducted to study the role of various hydrodynamical and geometrical properties on the dynamics of airborne NPs. Accordingly, the impacts of fluid flow velocity, membrane porosity, and particle diameter were examined on the dispersion of the particles in both air and media channels, as well as their deposition and translocation. Consistent with the experimental observation reported in the literature, we numerically showed that the aerosol injection of NPs provided far more efficient deposition/translocation than the aqueous solution. Although very small particles ($d_p < 50$ nm) as well as very large ones ($d_p > 700$ nm) demonstrated a high deposition rate, they tended to sediment close to the entrance of the channel for low inflow velocities, which could be undesirable for many applications demanding a uniform distribution. On the other hand, a higher inflow velocity provided a more uniform distribution; however, it also resulted in lower deposition and translocation rates. Therefore, it can be concluded that medium-sized particles (500 nm $< d_p <$ 700 nm) are more appropriate for uniform distribution with lower inflow velocities. It was demonstrated that the translocation of NPs into the media channel became more efficient by reducing the particle size as a factor of Brownian diffusivity and also by increasing the particle residence time in the porous region as a consequence of the gravity force's dominancy in large particles. However, smaller particles showed a better distribution in the media channel than did larger ones. Therefore, the results of this paper propose optimizing particle size as well as inflow velocity and membrane porosity, depending on the objectives of biomedical investigations. Even though the current numerical analysis is carried out with some simplifications, such as neglecting the cultured cell layers, it takes the first step towards a perceptive insight into the dynamics of the NPs in the lung-on-a-chip device. As a reliable complementary method seeking to reduce experimental cost, time and efforts, this study can be beneficial for future preclinical studies on pulmonary pathology associated with inhaled hazardous and toxic environmental particles, as well as therapeutic studies for developing inhalation drug delivery.

Author Contributions: Conceptualization, Z.S., P.A. and N.K.; methodology, Z.S.; software, Z.S.; validation, Z.S.; investigation and resources, Z.S.; writing—original draft preparation, Z.S.; writing—review and editing, P.A., N.K. and C.G.; visualization, Z.S.; supervision, P.A., N.K. and C.G.; funding acquisition, Z.S. and P.A. All authors have read and agreed to the published version of the manuscript.

Funding: This research received no external funding.

Data Availability Statement: The data presented in this study are available on request from the corresponding author. The data are not publicly available due to privacy.

Acknowledgments: Z.S. and P.A. would like to acknowledge the Shahrood University of Technology, which supported this study, and École Polytechnique Fédérale de Lausanne for supporting the experimental part of this study and its specific infrastructure for computational resources. Z.S. and P.A. also want to thank Swiss Government Excellence Scholarship. N.K. acknowledges the fellowship from the Australian Research Council (ARC) Discovery Early Career Research Award (DECRA) DE220100205.

Conflicts of Interest: The authors declare no conflict of interest.

References

1. Knudsen, L.; Ochs, M. The Micromechanics of Lung Alveoli: Structure and Function of Surfactant and Tissue Components. *Histochem. Cell Biol.* **2018**, *150*, 661–676. [CrossRef] [PubMed]
2. Li, J.J.; Muralikrishnan, S.; Ng, C.-T.; Yung, L.-Y.L.; Bay, B.-H. Nanoparticle-Induced Pulmonary Toxicity. *Exp. Biol. Med.* **2010**, *235*, 1025–1033. [CrossRef] [PubMed]
3. Nel, A.; Xia, T.; Mädler, L.; Li, N. Toxic Potential of Materials at the Nanolevel. *Science* **2006**, *311*, 622–627. [CrossRef] [PubMed]

4. Amararathna, M.; Goralski, K.; Hoskin, D.W.; Rupasinghe, H.P.V. Pulmonary Nano-Drug Delivery Systems for Lung Cancer: Current Knowledge and Prospects. *J. Lung Health Dis.* **2019**, *3*, 11–28. [CrossRef]
5. Madl, A.K.; Plummer, L.E.; Carosino, C.; Pinkerton, K.E. Nanoparticles, Lung Injury, and the Role of Oxidant Stress. *Annu. Rev. Physiol.* **2014**, *76*, 447–465. [CrossRef]
6. Xia, T.; Li, N.; Nel, A.E. Potential Health Impact of Nanoparticles. *Annu. Rev. Public Health* **2009**, *30*, 137–150. [CrossRef]
7. Raftis, J.B.; Miller, M.R. Nanoparticle Translocation and Multi-Organ Toxicity: A Particularly Small Problem. *Nano Today* **2019**, *26*, 8–12. [CrossRef]
8. Huh, D.; Matthews, B.D.; Mammoto, A.; Montoya-Zavala, M.; Hsin, H.Y.; Ingber, D.E. Reconstituting Organ-Level Lung Functions on a Chip. *Science* **2010**, *328*, 1662–1668. [CrossRef]
9. Huh, D.; Leslie, D.C.; Matthews, B.D.; Fraser, J.P.; Jurek, S.; Hamilton, G.A.; Thorneloe, K.S.; McAlexander, M.A.; Ingber, D.E. A Human Disease Model of Drug Toxicity–Induced Pulmonary Edema in a Lung-on-a-Chip Microdevice. *Sci. Transl. Med.* **2012**, *4*, 1–10. [CrossRef]
10. Jain, A.; Barrile, R.; van der Meer, A.D.; Mammoto, A.; Mammoto, T.; De Ceunynck, K.; Aisiku, O.; Otieno, M.A.; Louden, C.S.; Hamilton, G.A. Primary Human Lung Alveolus-on-a-chip Model of Intravascular Thrombosis for Assessment of Therapeutics. *Clin. Pharmacol. Ther.* **2018**, *103*, 332–340. [CrossRef]
11. Sedláková, V.; Kloučková, M.; Garlíková, Z.; Vašíčková, K.; Jaroš, J.; Kandra, M.; Kotasová, H.; Hampl, A. Options for Modeling the Respiratory System: Inserts, Scaffolds and Microfluidic Chips. *Drug Discov. Today* **2019**, *24*, 971–982. [CrossRef]
12. Zhang, M.; Xu, C.; Jiang, L.; Qin, J. A 3D Human Lung-on-a-Chip Model for Nanotoxicity Testing. *Toxicol. Res.* **2018**, *7*, 1048–1060. [CrossRef]
13. Moghadas, H.; Saidi, M.S.; Kashaninejad, N.; Nguyen, N.T. A High-Performance Polydimethylsiloxane Electrospun Membrane for Cell Culture in Lab-on-a-Chip. *Biomicrofluidics* **2018**, *12*, 24117. [CrossRef]
14. Quirós-Solano, W.F.; Gaio, N.; Stassen, O.M.J.A.; Arik, Y.B.; Silvestri, C.; Van Engeland, N.C.A.; Van der Meer, A.; Passier, R.; Sahlgren, C.M.; Bouten, C.V.C. Microfabricated Tuneable and Transferable Porous PDMS Membranes for Organs-on-Chips. *Sci. Rep.* **2018**, *8*, 13524. [CrossRef]
15. Ari, A. Jet, Ultrasonic, and Mesh Nebulizers: An Evaluation of Nebulizers for Better Clinical Outcomes. *Eurasian J Pulmonol.* **2014**, *16*, 1–7. [CrossRef]
16. Sheidaei, Z.; Akbarzadeh, P.; Kashaninejad, N. Advances in Numerical Approaches for Microfluidic Cell Analysis Platforms. *J. Sci. Adv. Mater. Devices* **2020**, *5*, 295–307. [CrossRef]
17. Sheidaei, Z.; Akbarzadeh, P.; Nguyen, N.T.; Kashaninejad, N. A New Insight into a Thermoplastic Microfluidic Device Aimed at Improvement of Oxygenation Process and Avoidance of Shear Stress during Cell Culture. *Biomed. Microdevices* **2022**, *24*, 15. [CrossRef]
18. Frost, T.S.; Estrada, V.; Jiang, L.; Zohar, Y. Convection–Diffusion Molecular Transport in a Microfluidic Bilayer Device with a Porous Membrane. *Microfluid. Nanofluid.* **2019**, *23*, 114. [CrossRef]
19. Yacobi, N.R.; Fazllolahi, F.; Kim, Y.H.; Sipos, A.; Borok, Z.; Kim, K.-J.; Crandall, E.D. Nanomaterial Interactions with and Trafficking across the Lung Alveolar Epithelial Barrier: Implications for Health Effects of Air-Pollution Particles. *Air Qual. Atmos. Health* **2011**, *4*, 65–78. [CrossRef]
20. Mosquera, J.; García, I.; Liz-Marzán, L.M. Cellular Uptake of Nanoparticles versus Small Molecules: A Matter of Size. *Acc. Chem. Res.* **2018**, *51*, 2305–2313. [CrossRef]
21. Saidi, M.S.; Rismanian, M.; Monjezi, M.; Zendehbad, M.; Fatehiboroujeni, S. Comparison between Lagrangian and Eulerian Approaches in Predicting Motion of Micron-Sized Particles in Laminar Flows. *Atmos. Environ.* **2014**, *89*, 199–206. [CrossRef]
22. Amin Arefi, S.M.; Tony Yang, C.W.; Sin, D.D.; Feng, J.J. Simulation of Nanoparticle Transport and Adsorption in a Microfluidic Lung-on-a-Chip Device. *Biomicrofluidics* **2020**, *14*, 44117. [CrossRef] [PubMed]
23. Moghadas, H.; Saidi, M.S.; Kashaninejad, N.; Nguyen, N.-T. Challenge in Particle Delivery to Cells in a Microfluidic Device. *Drug Deliv. Transl. Res.* **2018**, *8*, 830–842. [CrossRef] [PubMed]
24. *COMSOL Multiphysics 5.6*; COMSOL AB: Stockholm, Sweden, 2020. Available online: https://www.comsol.com/support/knowledgebase/1223 (accessed on 1 September 2021).
25. Liu, C.; Zhou, G.; Shyy, W.; Xu, K. Limitation Principle for Computational Fluid Dynamics. *Shock Waves* **2019**, *29*, 1083–1102. [CrossRef]
26. Barber, R.W.; Emerson, D.R. The influence of Knudsen number on the hydrodynamic development length within parallel plate micro-channels. *WIT Tran. Eng. Sci.* **2002**, *36*, 12. [CrossRef]
27. Barisam, M.; Saidi, M.S.; Kashaninejad, N.; Vadivelu, R.; Nguyen, N.-T. Numerical Simulation of the Behavior of Toroidal and Spheroidal Multicellular Aggregates in Microfluidic Devices with Microwell and U-Shaped Barrier. *Micromachines* **2017**, *8*, 358. [CrossRef]
28. Hinds, W.C. *Aerosol Technology: Properties, Behavior, and Measurement of Airborne Particles*; John Wiley & Sons: Hoboken, NJ, USA, 1999; ISBN 0471194107.
29. Zhang, H.; Ahmadi, G. Aerosol Particle Transport and Deposition in Vertical and Horizontal Turbulent Duct Flows. *J. Fluid Mech.* **2000**, *406*, 55–80. [CrossRef]
30. Zhang, J.; Yan, S.; Yuan, D.; Alici, G.; Nguyen, N.-T.; Warkiani, M.E.; Li, W. Fundamentals and Applications of Inertial Microfluidics: A Review. *Lab Chip* **2016**, *16*, 10–34. [CrossRef]

31. White, F.M.; Majdalani, J. *Viscous Fluid Flow*; McGraw-Hill: New York, NY, USA, 2006; Volume 3, pp. 433–434.
32. Huh, D.; Kim, H.J.; Fraser, J.P.; Shea, D.E.; Khan, M.; Bahinski, A.; Hamilton, G.A.; Ingber, D.E. Microfabrication of Human Organs-on-Chips. *Nat. Protoc.* **2013**, *8*, 2135–2157. [CrossRef]
33. Stucki, A.O.; Stucki, J.D.; Hall, S.R.R.; Felder, M.; Mermoud, Y.; Schmid, R.A.; Geiser, T.; Guenat, O.T. A Lung-on-a-Chip Array with an Integrated Bio-Inspired Respiration Mechanism. *Lab Chip* **2015**, *15*, 1302–1310. [CrossRef]
34. Ochs, C.J.; Kasuya, J.; Pavesi, A.; Kamm, R.D. Oxygen Levels in Thermoplastic Microfluidic Devices during Cell Culture. *Lab Chip* **2014**, *14*, 459–462. [CrossRef]
35. Augusto, L.L.X.; Gonçalves, J.A.S.; Lopes, G.C. CFD Evaluation of the Influence of Physical Mechanisms, Particle Size, and Breathing Condition on the Deposition of Particulates in a Triple Bifurcation Airway. *Water Air Soil Pollut.* **2016**, *227*, 56. [CrossRef]
36. Friedlander, S.K. *Smoke, Dust, Haze: Fundamentals of Aerosol Dynamics*; Oxford University Press: Oxford, UK, 2000.
37. Knotter, D.M.; Wali, F. Developments in Surface Contamination and Cleaning. In *Particle Deposition, Control and Removal*, 1st ed.; William Andrew: Norwich, NY, USA, 2010; pp. 81–120. [CrossRef]
38. Hayter, A.J. *Probability and Statistics for Engineers and Scientists*; Cengage Learning: Boston, MA, USA, 2012; ISBN 1133712762.
39. Frost, T.S.; Jiang, L.; Zohar, Y. Pharmacokinetic Analysis of Epithelial/Endothelial Cell Barriers in Microfluidic Bilayer Devices with an Air–Liquid Interface. *Micromachines* **2020**, *11*, 536. [CrossRef]

micromachines

Article

Influence of Filler Materials on Wettability and Mechanical Properties of Basalt/E-Glass Woven Fabric–Reinforced Composites for Microfluidics

Ayyappa Atmakuri [1,*], Lalitnarayan Kolli [2], Arvydas Palevicius [1], Sigita Urbaite [1,*] and Giedrius Janusas [1]

[1] Faculty of Mechanical Engineering and Design, Kaunas University of Technology, Studentu 56, 51424 Kaunas, Lithuania
[2] Faculty of Mechanical Engineering, Sir CRR College of Engineering, Eluru 534007, Andhra Pradesh, India
* Correspondence: ayyappa.atmakuri@ktu.lt (A.A.); sigita.urbaite@ktu.lt (S.U.)

Abstract: This paper presents the development of novel hybrid composites in the presence of filler particles and manufactured using a proposed new fabrication technique. The hybrid composites were fabricated using a basalt and E-glass woven fabric–reinforced epoxy resin matrix combined with graphite powder nanoparticles. Six sets of samples were fabricated using the vacuum-assisted free lamination compression molding technique. After the fabrication, wettability, mechanical properties (tensile, flexural and impact properties) and moisture properties were evaluated. Surface morphology and chemical composition of the composite samples were examined using a scanning electron microscope (SEM) and spectroscopy. The obtained results showed that the use of filler materials in hybrid composites improves the properties of hybrid composites. Basalt/E-glass hybrid composites with 10% graphite material exhibited superior mechanical properties over the other composites, with high-quality, improved adhesion and surface morphology. Thus, novel composites with the combination of exceptional properties may be integrated in the design of flexible electronics and microfluidics devices as a structural layer of the system. High flexibility and good surface tension of the designed composites makes them attractive for using the thermal imprint technique for microfluidics channel design.

Keywords: basalt; E-glass; epoxy resin; graphite particles; wettability; mechanical properties

Citation: Atmakuri, A.; Kolli, L.; Palevicius, A.; Urbaite, S.; Janusas, G. Influence of Filler Materials on Wettability and Mechanical Properties of Basalt/E-Glass Woven Fabric–Reinforced Composites for Microfluidics. *Micromachines* **2022**, *13*, 1875. https://doi.org/10.3390/mi13111875

Academic Editor: Pingan Zhu

Received: 4 October 2022
Accepted: 27 October 2022
Published: 31 October 2022

1. Introduction

The importance of natural fibers in the development of fiber-reinforced hybrid polymer composites is increasing day by day in the modern era because it is directly associated with environmental concerns as well as with a circular economy [1,2]. The combination of natural fibers with synthetic fibers in reinforced hybrid polymer composites offers several attractive features such as light weight, condensed lifecycle costs, high adhesion properties, flexibility and superior mechanical properties [3]. Nature is a decent source for the creation of composite materials using hemp, flax, sisal, palm [4], okra, banana, wood, bamboo, jute, sugarcane, or cowhide. Thus, all these natural materials can be a portion of the regular fiber composites, providing exceptional properties and expanding the application areas in various fields [5].

In recent years, synthetic fiber-based composites have become the leading progressive composite materials for automotive, construction, marine, sporting goods, biomedical, microfluidics and other engineering applications due to their ease of availability, high strength and modulus and also good corrosion and fatigue resistance behavior [6]. Despite the advantages to the presence of natural fibers in composite materials, there are some disadvantages, such as fiber shrinkage, poor adhesion and moisture, which may have a high effect on the mechanical properties. At the same time, most synthetic fibers are brittle in nature, which may cause a catastrophic mode of failure in the composite material [7]. Synthetic fibers such as carbon, glass, mica and nylon also play a prominent role in the composite industry. Fiber-reinforced composite materials have a better strength and modulus than numerous conventional metallic materials. These composites exhibit low explicit gravity, high strength-to-weight ratio and higher modulus–weight proportions [8,9]. Moreover, natural fibers are lower in weight compared with metallic fibers and may be applied in many weight-basic applications such as those in the aviation, civil industry and automotive areas [10].

Many researchers found interest in developing hybrid composites. The reason can be attributed to the adverse effects of polymer composites on nature (environment), their high cost and other unfavorable mechanical properties [11]. The hybridization of polymer composites is known for the improvement of mechanical properties over individual fiber composites. It is obtained by joining two or more reinforcement materials under the same matrix (thermoset or thermoplastic) material. The possible combinations are natural fiber–natural fiber, natural fiber–synthetic fiber and synthetic fiber–synthetic fiber hybrid composites [12,13]. These hybrid composites have been established to overcome the drawbacks of individual fiber composites. Thus, properly configured hybrid composites produce high-quality composites and offer good mechanical properties such as strength, stiffness, flame retardance and corrosive resistance properties [14].

Among all the prominent methods for understanding the wettability properties of composite material, various droplet based techniques are used [15], such as gravitational [16,17], acoustic [18] and magnetic [19] driving methods. Among these droplet driving methods, contact angle measurement is widely popular due to its accuracy, lower operational cost and low complex procedure. The wettability of composite material is used to understand the material's surface properties, such as its hydrophilic or hydrophobic nature, and is also used to understand the surface tension of driven droplets. Thus, the mechanical properties of thermoset and thermoplastic resin-based hybrid polymer composites with synthetic fibers, such as carbon, nylon, glass, rayon, spandex, acrylic fibers and natural fibers, such as hemp, flax, sisal, jute, bamboo and palm fibers, have been broadly investigated by many researchers. Various researchers stated that hybrid composites have superior properties over individual fiber composites. Static and dynamic mechanical properties of basalt fiber–reinforced polymer composites showed their sensitivity to the strain rate. The mechanical properties such as tensile strength, elastic modulus and failure strain were increasing quickly with strain rate when the strain rate was over 120 s^{-1} [20]. Thus, replacing glass fiber with basalt fiber in the composites enhances mechanical properties by 35–42%, i.e., compressive strength, impact energy and Young's modulus [21], and replacing glass fiber with E-glass fiber composites gives superior tensile strength [21,22]. Moreover, when applying the plasma treatment to basalt fibers during the fabrication, the adhesion properties and surface roughness may be improved compared to those of commercial fibers [22]. The influence of waste basalt powder on rigid polyurethane foams had some influence, too. It was observed that the addition of basalt powder triggered an increase in the reactivity of the polyurethane system during the foaming process, but it caused a reduction of mechanical properties [23]. Additionally, the basalt powder can be used as an eco-friendly modifier for the epoxy resin, improving the thermo-mechanical properties for an increase in filler content [24]. The mechanical properties of composites reinforced with fabrics having the same areal density confirmed that basalt fiber composites possess an elastic modulus higher than the corresponding glass fiber composites, while their tensile strength approaches that of equivalent carbon fiber composites, and the results of fatigue behavior confirmed the better performance of basalt fiber composites than the corresponding glass fiber composites, with higher stiffness retention at low fatigue loads and better damping properties [25]. Thus, basalt and hybrid laminates with an added configuration exhibited higher impact energy absorption capacity than glass fiber composites applying hybridization with reinforcement material [26–32]. However, the lack of filler material in the composite fabrication may improve adhesion but significantly reduces its mechanical properties. To overcome these problems, an attempt was made in the present paper.

In this paper, basalt-fiber (BF) and E-glass-fiber (EF)-reinforced epoxy polymer composites including graphite filler material with various weight fractions (5%, 10% and 15%) were fabricated using the vacuum-assisted free lamination compression molding technique. After, the samples were allowed the heat treatment process and then tested for mechanical properties. The mechanical characterization was carried out by conducting tensile, flexural, impact, hardness and moisture analyses. The contact angle measurement technique was used to evaluate the wettability of composite materials and to understand the material's surface properties, such as its hydrophilic or hydrophobic nature and the surface tension of driven droplets. Surface morphology and chemical composition were examined using scanning electron microscopy (SEM) and spectroscopy methods. Thus, the obtained results proved the relevance of properly configured hybrid composites possessing high quality and good mechanical and adhesion properties together with qualitative surface morphology. The obtained results proved the relevance to use novel composites in the design of microelectromechanical devices, flexible electronics and microfluidics.

2. Materials and Methods

2.1. Materials and Fabrication

Basalt (B) and E-glass (E) woven fabrics were used as a reinforcement, and graphite (G) powder nanoparticles were used as a filler material. Epoxy resin, along with hardener, was used as a matrix material. Both basalt and E-glass fabrics have the same areal density of 198 g/m^2, and the fabrics were taken in a square shape (300 mm × 300 mm) to fabricate the composites. The mechanical properties and chemical characteristics of the materials were considered before the experimentation in order to understand the nature and behavior during the process (Table 1).

Table 1. The mechanical properties of materials [33–36].

Property	Basalt (B)	E-Glass (E)	Epoxy Resin
Density (g/cm^3)	2.40–2.72	2.48–2.62	1.15–1.18
Diameter (μm)	15	6	-
Tensile Strength (MPa)	4800	3200	72
Young's Modulus (GPa)	90.0	70.0	2.9
Elongation (%)	3.15	4.00	5.30
Maximum Temp. (°C)	645	450	-
Color	Brown	White	Pale

A total of six sets of hybrid composites were fabricated using the vacuum-assisted free lamination compression molding technique. In each set of composites, a total of six sheets were taken to fabricate 3 mm thickness composite plates. To make the hybrid composites with filler material, the graphite powder nanoparticles with varying weight fractions (5%, 10% and 15%) were mixed in epoxy resin for 20 min and then applied to the composite panels. The ply sequence and orientation for each composite are described in Table 2.

Table 2. Ply sequence and orientation of basalt and E-glass hybrid composites.

Composite	No. of Plys		% Of Graphite Filler Material	Ply Sequence	Orientation (°)
	B	E			
B	6	0	0	B-B-B-B-B-B	0-90-0-90-0-90
E	0	6	0	E-E-E-E-E-E	0-90-0-90-0-90
B/E	3	3	0	B-E-B-E-B-E	0-90-0-90-0-90
B/E/G5	3	3	5	B-E-B-E-B-E	0-90-0-90-0-90
B/E/G10	3	3	10	B-E-B-B-E-E	0-90-0-90-0-90
B/E/G15	3	3	15	B-E-B-E-B-E	0-90-0-90-0-90

After degassing, a nominal pressure of 1.5 MPa was applied, and the composite panels were cured under the same pressure and vacuum for 8 h at room temperature. Then, the composite panels were taken for a heat treatment process for eliminating the moisture content, which also serves as a post-curing process for composites. The heat treatment process was carried out for 8 h at 90 °C. Total of six sets and in each five samples were developed using different materials and filler concentrations (in total 30 samples manufactured). Thus, the composite notations are considered as pure basalt (B), pure E-glass (E), basalt and E-glass (B/E), basalt and E-glass with 5% filler (B/E/G5), basalt and E-glass with 10% filler (B/E/G10) and basalt and E-glass with 15% filler material (B/E/G15) (Table 2).

2.2. Analytical Equipment

After the fabrication process, composites were allowed wettability testing and mechanical characterization by conducting various tests. In each set of composite groups, samples were tested according to ASTM standards (30 measurements in total). Thus, performed mechanical tests were: the flexural test, tensile test, impact test and moisture analysis. The morphological studies were carried using the scanning electron microscope (SEM) and the chemical compositions of the composite samples were tested using the SEM-EDX spectrometer.

Since fabricated hybrid composites contain the filler material throughout the surface, and in future research are intended for use in flexible electronics and microfluidics devices, the wettability property of the fabricated composite surface plays a crucial role. The same test was conducted on composites without filler material for comparison purposes as well. The wettability properties of novel composite materials were measured using the contact angle measurement technique (Figure 1).

Figure 1. A contact angle measurement setup.

The experiment was carried out in a dim laboratory space with ambient lighting. It was vital for the measurement accuracy and image analysis that the liquid drop look black, and thus the light settings were set up accordingly. The experiment was set up to eliminate any light reflections that would have tampered with the measurement (Figure 1). Additionally, precautions were taken to keep airborne contaminants such as dust and particulates from contaminating the drips. Additionally, the critical distances between the parts were kept constant during the measurements (Figure 2).

Further, flexural tests were performed on fabricated samples to analyze the flexural strength and modulus, i.e., three-point bowing tests were performed on each set of composites according to ASTM D-790M-86 testing standards (total of 30 measurements). The dimensions of the samples were considered as 100 mm × 25 mm × 3 mm (length, width, thickness). The three-point bowing tests were conducted on an electronic tensiometer (METM 2000 ER-1 model (Plate II-18), provided by M/S Microtech Pune, Maharashtra, India), with outside rollers in diameter of 70 mm. Thus, composites were loaded at a strain rate of 0.2 mm/min (Figure 3).

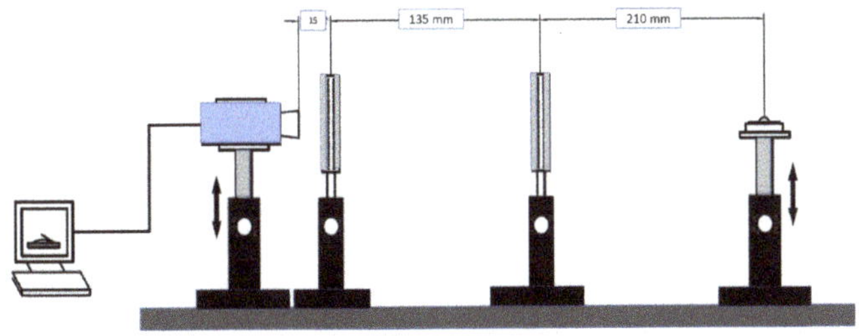

Figure 2. Critical distances between contact angle testing device.

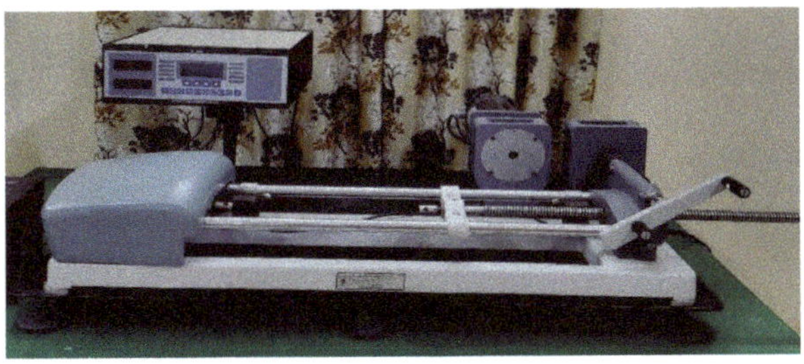

Figure 3. Electronic tensiometer for flexural and tensile properties testing.

Tensile tests were performed on the same electronic tensiometer which was used for flexural tests, at a crosshead speed of 0.2 mm/min. An electronic micrometer was utilized to quantify the necessary thickness and width of composite examples. The check length, width and thickness were estimated with a 0.001 mm negligible count mechanized micrometer. This electronic tensiometer was fixed with weight and expansion pointers, which had an insignificant count of 0.01 kg and 0.01 mm, independently. The tensile tests were performed according to ASTM-D3039 testing standards. The specimen dimensions were considered 165 mm × 20 mm × 3 mm (length, width, thickness). In each set of composites, a total of five samples were tested (total 30 measurements).

Impact tests were performed on a falling dart impact testing machine (Impact analyzer M/S International equipment, Mumbai, India). The composite samples were tested as per ASTM D256-97 testing standards, and the specimen dimensions were 63.5 mm long, 12.36 mm wide and 3 mm in thickness. Designed composites were fixed in the vertical position and a pendulum was used to break the composite. The pendulum sled was delivered from the locking position, which is at a state of 1500 mm and hits the sample with a striking rate of 2.46 m/s. The sample was stripped and energy was exhibited in joules by the pointer on the specific scale.

The amount of moisture absorbed by the composite samples when exposed to moisture was analyzed using the moisture analysis test. The tests were performed according to ASTM-D 570 testing standards. The sample specifications were considered as 60 mm long, 20 mm wide and 3 mm in thickness. The samples were placed in a container full of water at room temperature and the amount of moisture gained by the composites was measured for six weeks' duration (readings taken after every 7 days).

Finally, morphological studies such as surface morphology, porosity content and fracture mechanism were investigated using SEM and the chemical composition of the composite samples was analyzed by using SEM-EDX spectroscopy.

3. Results and Discussion

The basalt and E-glass fiber hybrid composites (B, E, B/E, B/E/G5, B/E/G10 and B/E/G15) in the presence of graphite filler (with varying weight fractions 5%, 10% and 15%) were tested to evaluate their mechanical characteristics, wettability and surface morphology.

3.1. Wettability Analysis

The wettability properties of composite materials were investigated using the contact angle measurement technique imaging water droplet images on the composite surface (Figure 4). The volume of the water droplet was kept constant and the values are noted from the left lower endpoint to the right lower endpoint.

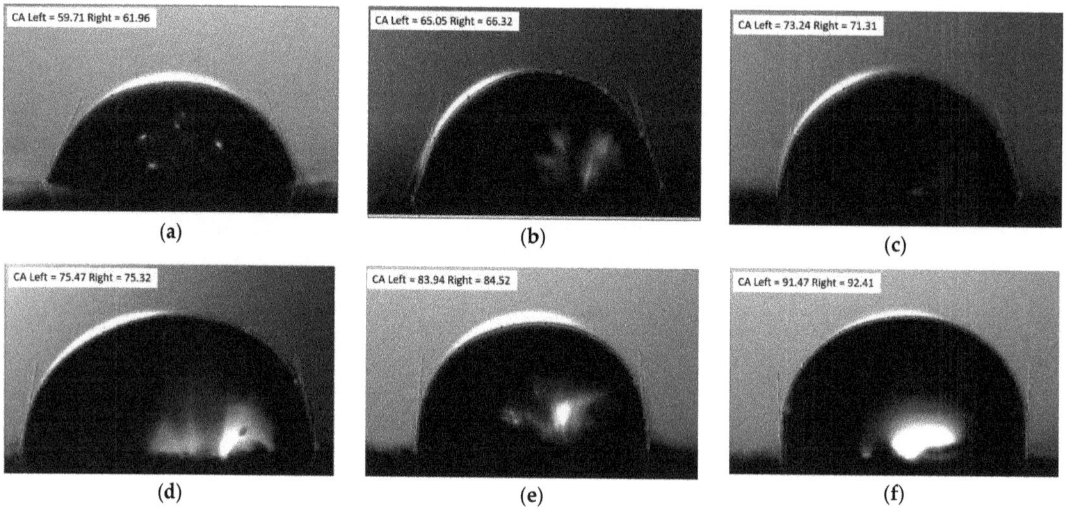

Figure 4. The droplet profile's polynomial fit was achieved by using knots that go from the left to the right while accounting for the contact angle (**a**) B, (**b**) E, (**c**) B/E, (**d**) B/E/G5, (**e**) B/E/G10 and (**f**) B/E/G15.

Results (Figure 5) showed the assessment of contact angle measurement of basalt and E-glass woven-fabric hybrid composites and that all sample surfaces exhibited hydrophilic properties. The measured contact angle formed by the water droplet on the B/E/G15 hybrid composite was 86.45° with an error of ±2.36°, being the maximum among all, whereas for individual basalt (B) composites, the least contact angle of 61.76° with an error of ±1.84° was exhibited. Thus, if the contact angle is less than 90°, then it is considered hydrophilic, and if it is more than 90°, then it is hydrophobic [37–39]. Moreover, hybrid composites with graphite filler material tended to exhibit hydrophobic surface properties. To extend the analysis, the same tests were conducted varying the measurements in time, and the results showed that the contact angle decreased with an increase in time (at the continuous intervals of time such as 10 s, 20 s, 30 s) in all the cases due to the seeping of a water droplet on the composite surface. The hydrophilic surface properties of the composite material play a prominent role during the formation of micro channels on solid surfaces (composite material).

3.2. Flexural Properties

The flexural properties, such as the flexural strength and flexural modulus, of basalt and E-glass woven-fabric reinforced with epoxy matrix composites, including filler material (graphite powder particles) were investigated (Figure 6). Results showed that, the strength and modulus were superior in hybrid composites compared to pure composites. The flexural properties increased with filler material and the impact of hybridization was observed, i.e., there was a slight increment in both strength and flexural modulus in basalt and E-glass hybrid composites over pure basalt and pure E-glass woven-fabric composites. The epoxy composites compared to pure basalt (B) composites exhibited the least flexural strength and modulus at 27.82 MPa and 2.06 GPa, respectively. Thus, basalt and E-glass

(B/E/G15) hybrid composite with 15% graphite filler material showed the highest values of 43.81 MPa flexural strength and 3.57 GPa flexural modulus (Figure 7). Many researchers found that the incorporation of filler materials into the composite material increased both flexural strength and flexural modulus [40,41]. The reason is attributed to the uniform distribution of filler material and the strong adhesion properties between polymer matrix and filler material. Hence, the composites with graphite filler material exhibited superior flexural strength and modulus properties.

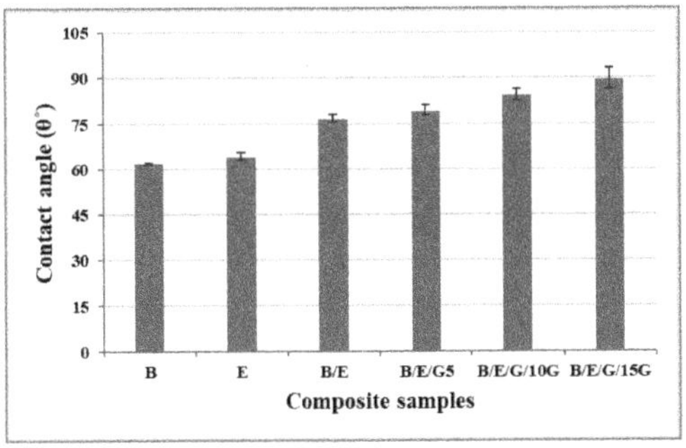

Figure 5. Contact angle measurement for composite samples.

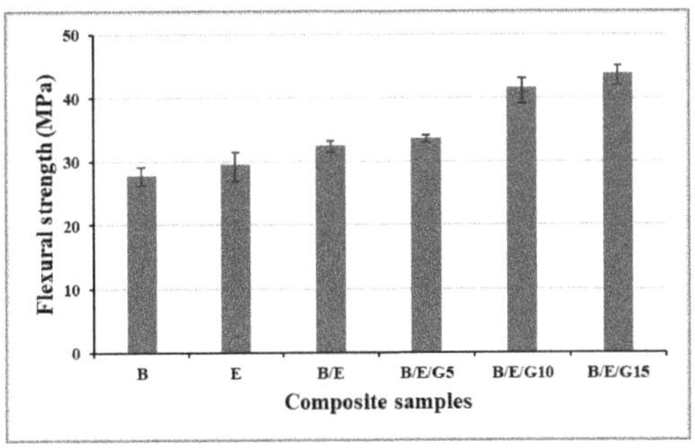

(**a**)

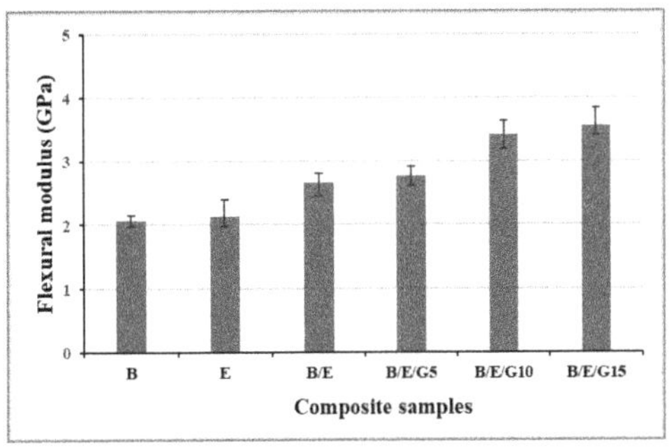

(**b**)

Figure 6. Flexural strength (**a**) and flexural modulus (**b**) of basalt and E-glass composite samples.

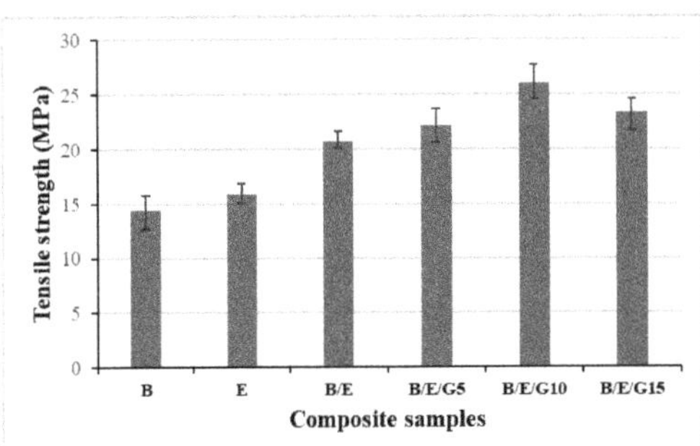

(**a**)

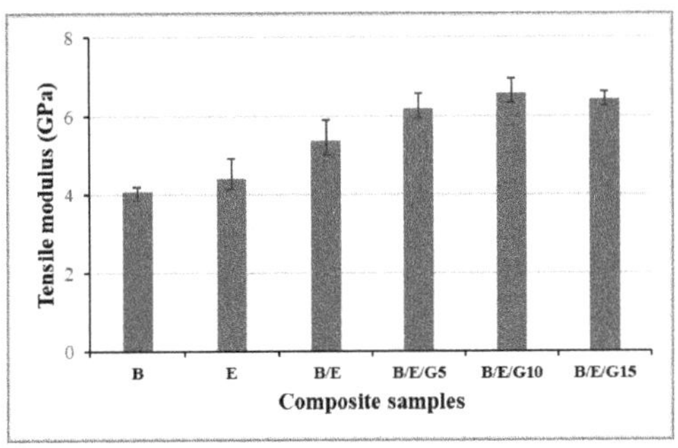

(**b**)

Figure 7. Tensile strength (**a**) and tensile modulus (**b**) of basalt and E-glass composite samples.

Experimental investigations showed that incorporation of the filler in to different matrices played a crucial role in improving the mechanical performance of the designed novel composite materials.

3.3. Tensile Properties

The experiments with basalt and E-glass woven-fabric composites were conducted by performing tensile testing (Figure 7). Thus, the obtained results showed, that the impact of filler material in the hybrid composites was quite noticeable and effective. Unlike the flexural properties, the tensile strength and tensile modulus were increased up to 10% with a graphite filler material. Results for basalt and E-glass (B/E) hybrid composites and basalt and E-glass hybrid composites with 5% graphite particles (B/E/G5) were almost identical. The epoxy-based basalt fiber composite showed the least tensile strength and modulus of 14.68 MPa and 4.07 GPa, respectively (Figure 7a,b). Thus, epoxy-based basalt and E-glass hybrid composites with 10% filler material showed the highest results, i.e., tensile strength of 27.68 MPa (Figure 7a) and tensile modulus of 6.25 GPa (Figure 7b). Hence, the composition in B/E/G10 could act as a very good reinforcing filler in the hybrid composites.

The presence of a filler material in the composites may improve the adhesion properties and also reduce the porosity content. The hybrid composites in the presence of graphite filler material possessed great strength and modulus over the individual fiber composites. Hence, these exceptional properties lead to a potential replacement of other conventional materials used for microfluidics, in the formation of various dimension groves and channels or in absorbent-coated microchannels.

3.4. Impact Properties

Further, investigations were performed on the impact energy absorbed by the composite samples (Figure 8). It was defined that the amount of energy was absorbed by an object when an external load (impact load or sudden load) was applied to it. Tests were conducted using a Charpy impact tester at room temperature with a relative humidity of 30%. The results followed previously obtained results, i.e., the impact energy increased with an increase in filler material. The pure basalt composites showed the least impact energy at 3.1 kJ/m^2, and B/E/G15 hybrid composites showed the highest at 4.9 kJ/m^2. Furthermore, the B/E hybrid composites and B/E/G5 hybrid composites had almost the same impact energy. Composite B/E/G15 exhibited the highest impact energy, and the pure composites showed the lowest impact energy values (Figure 9). This rise in hybrid composites is the result of energy-dissipating fiber-related phenomena such as fiber debonding, pull-out, bridging and fracture, which cause the resin matrix to flex plastically prior to failure. A set of glass fibers with a length greater than the crucial value for effective reinforcement is likely to result in bridging and fiber fractures, while a set of glass fibers with a length less than the critical value is likely to result in debonding and fiber pull-out. Similar to these results, many researchers reported that the hybridization of composites improves the impact strength [42,43]. The same was observed in this research, stating that the hybridization of composites improved the impact strength by two times.

3.5. Moisture Analysis

To measure the amount of moisture absorbed by the composites, a moisture analysis test was conducted (Figure 9). After fabrication, the composite samples were placed in an electric oven to eliminate the moisture in them. To analyze the moisture properties, the composite samples were placed in a jar containing distilled water. The readings were noted by differentiating the initial weight and the final weight (after moisture gain) of the composite samples for 6 weeks. After 6 weeks, results showed that the presence of the filler material in a composite sample absorbed less moisture (Figure 9) because the filler material occupied the porosity content in a composite. Thus, pure samples absorbed more moisture than hybrid composites, meaning that the hybrid composites with filler particles have the property of repelling the water when it comes into contact with it. This occurrence was due to unbalanced molecular forces in the water and the solid interface bringing out surface tension. It was observed that the increase in moisture increased over time in all the composites, and after about 5 weeks it reached a steady state.

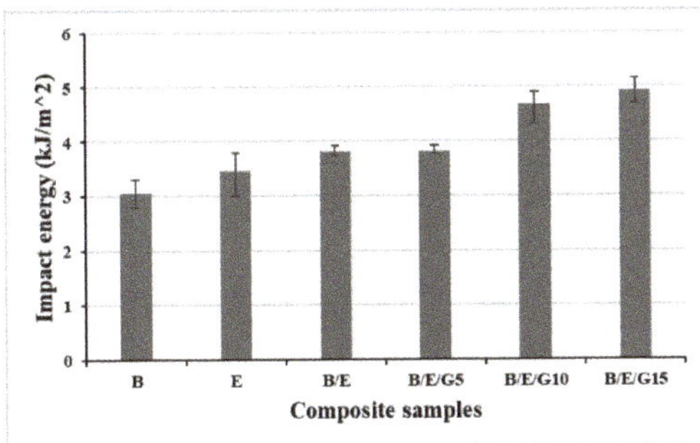

Figure 8. Impact energy of basalt and E-glass composite samples.

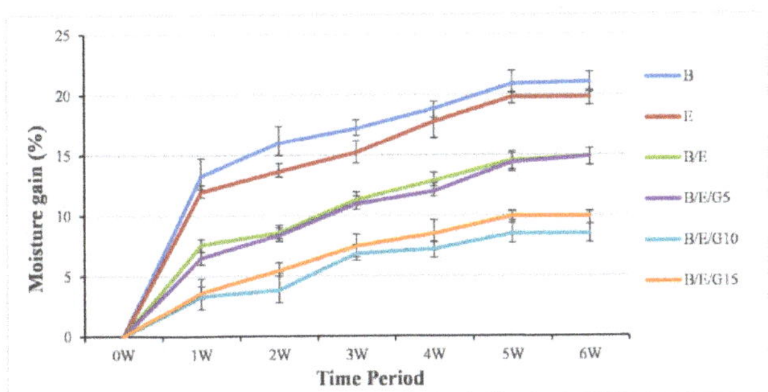

Figure 9. Moisture absorption of basalt and E-glass hybrid composites.

3.6. Morphological Studies

The surface morphology and porosity formation in basalt and E-glass woven-fiber hybrid composites were studied using the SEM. Thus, the presence of pores in a composite may affect the mechanical performance of a composite material. Previous researchers focused on evacuating the air trapped in voids by keeping the volatiles in solution and resin intrusion of the evacuated pore spaces. However, these methods can drive additional problems such as shrinkage and are not suitable for all composites [44,45]. Hence, the graphite powder particles were introduced into the composite material to avoid the microbial effect in it. The SEM results showed that the presence of pores in the form of spherical, cylindrical and helical shapes were observed in the hybrid composites without filler particles (Figure 10a–c). The percentage of porosity in hybrid composites with graphite filler material was smaller when compared to the pure composites (Figure 10d–f). Most of the pores were replaced with nanoparticles in the B/E/G hybrid composites and that can be attributed to improving the mechanical performance and adhesion properties.

The morphological studies were conducted on fractured hybrid samples to understand the chemical composition of the composites. It was observed from SEM-EDX results that hybrid composites exhibited good adhesion properties as well as the surface roughness on composite materials (Figure 11). The map data of chemical composition allowed to evaluate the dominant materials in designed composites: the color notations are red for carbon (C), green for oxygen (O), blue for chlorine (Ci), yellow for calcium (Ca), light blue for sodium (Na) and orange for aluminum (Al). The dominance of carbon in the B/E/G10 hybrid composite was observed clearly, i.e., it was more at fracture over the surface position. In the B/E hybrid case, the presence of oxygen on the surface position was observed. Thus, a detailed description of atomic and weight concentrations is given in the Table 3.

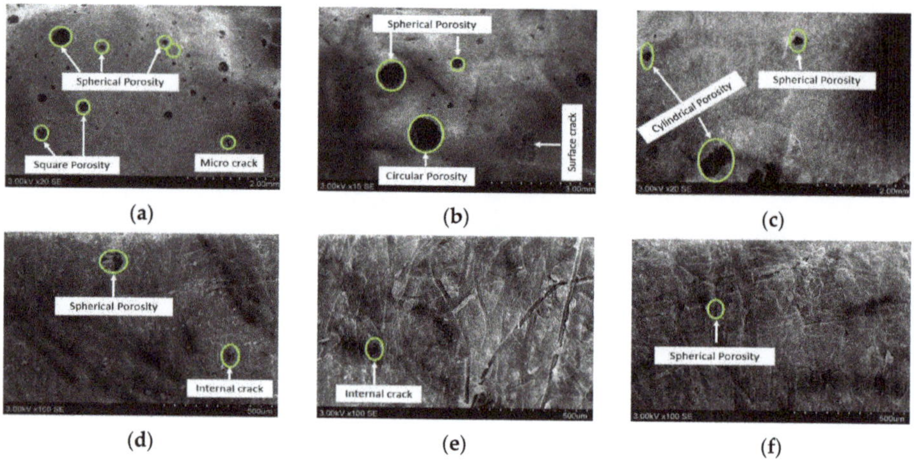

Figure 10. Porosity content in basalt and E-glass woven-fiber hybrid composites (**a**) B, (**b**) E, (**c**) B/E, (**d**) B/E/G5, (**e**) B/E/G10 and (**f**) B/E/G15.

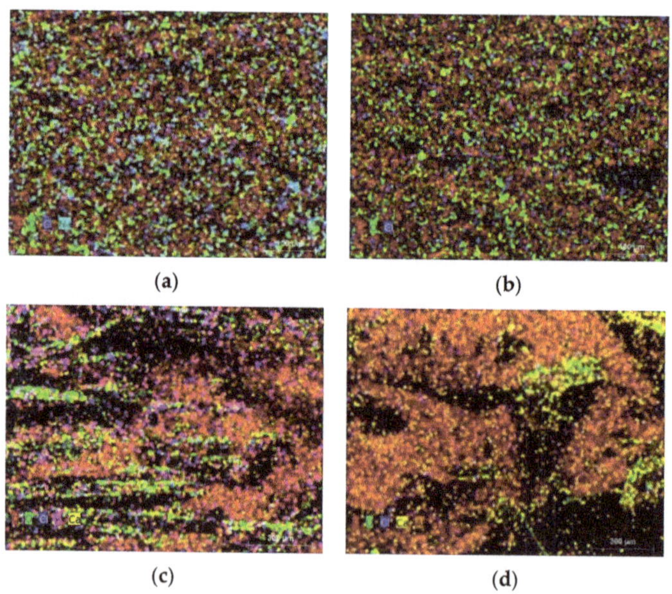

Figure 11. SEM images for map data of hybrid composites (**a**) B/E at surface, (**b**) B/E/G10 at surface, (**c**) B/E at fracture and (**d**) B/E/G10 at fracture.

The chemical composition of basalt and E-glass woven-fabric hybrid composites with and without filler materials was analyzed by using SEM-EDX spectroscopy, too. The chemical composition of a hybrid composite at surface and fracture position was observed for comparison purposes. The highest peak was observed for carbon followed by oxygen. The results stated that the carbon percentage was higher at a fractured place than the surface position (Figure 12).

Table 3. The chemical composition of basalt and E-glass hybrid composites at surface and fracture position.

Element	B/E Surface		B/E Fracture		B/E/G10 Surface		B/E/G10 Fracture	
	Weight (%)	Atom. (%)	Weight (%)	Atom. (%)	Weight (%)	Atom. (%)	Weight (%)	Atom. (%)
Carbon (C)	39.85	48.31	53.56	58.98	67.62	73.24	78.09	81.86
Oxygen (O)	58.77	51.10	45.18	40.57	31.89	26.51	10.85	16.74
Calcium (Ca)	0.00	0.00	0.42	0.14	0.24	0.11	0.08	0.10
Chlorine (Cl)	0.94	0.39	0.00	0.00	0.17	0.08	1.92	1.28
Sodium (Na)	0.44	0.20	0.28	0.13	0.08	0.06	0.00	0.00
Silicon (Si)	0.00	0.00	0.56	0.18	0.00	0.00	0.06	0.02

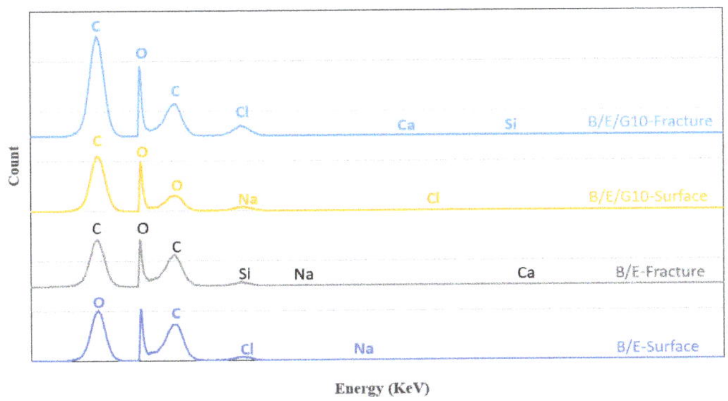

Figure 12. SEM-EDX spectrum analysis on hybrid composites at surface and fracture position.

Due to the presence of graphite particles in B/E/10G, the percentage of carbon reached 78.09% at the fracture position, whereas at the surface position it was 67.72% (Table 3).

Overall, the results proved that the mechanical performance of basalt and E-glass woven-fabric hybrid composites increased with the addition of filler material. The effect of the hybridization concept on woven-fiber laminates was quite impressive. Thus, the designed novel hybrid composites in the presence of filler material can be useful for various engineering applications such as automobiles, construction, biomedical and microfluidics applications. The deeper research into microchannel formation was left for future research.

4. Discussion

Novel composite materials with graphite filler were designed for the future design of composite microfluidics chips. Since microfluidics channels are usually made by layering, these channels need to be perfectly aligned on top of each layer. Moreover, the behaviors of fluidic flow mainly dominate because of the interfaces between the fluids and the microchannels' surface walls, playing critical roles in the regulation of the microflow in the channel. The pressure in microfluidics channel is an important factor, too. Too much pressure in the channel could cause delamination or fracture, depending on the strength of the bonding. Thus, evaluation of surface morphology, wettability and mechanical properties, chemical bonding, etc. are crucial in the design of composites used as the base material for the formation of channels in microfluidics.

Implementation of novel composite materials opens the opportunity to integrate high-complexity physical and chemical surface modification strategies in the design of microchannels in microfluidics devices. The future work will be focused on the formation of such microchannels using designed composite materials as the basic material, evaluating these materials from the perspective of interfacial interactions between the inner surface and the fluid and determining the behavior of the microflow.

5. Conclusions

Basalt and E-glass woven-fiber-based epoxy hybrid composites in the presence of graphite filler material were fabricated using a vacuum-assisted free lamination compression molding fabrication technique.

Novel hybrid composites showed superior mechanical properties over individual basalt and E-glass fiber composites. The addition of filler material to the composite material improved its mechanical performance. It was observed that all the composites exhibited hydrophilic surface properties with a contact angle of less than 90°, and the contact angle increased with an increase in filler material in the composite. From the flexural results, the individual basalt fiber composites exhibited the least flexural strength of 27.82 MPa and flexural modulus of 2.06 GPa. Thus, the B/E/G15 composite showed the highest flexural strength of 43.81 MPa and flexural modulus of 3.57 GPa. From the tensile results, the epoxy-based B/E/10G composites exhibited the highest tensile properties, and epoxy-based basalt composites showed the lower tensile properties. The impact energy values observed in individual basalt and E-glass were 3.1 KJ/m^2 and 3.5 KJ/m^2, respectively. The amount of moisture gained by the composites was investigated using moisture analysis. It was observed that all the composites followed the same trend for up to 5 weeks and remained the same even after a 6-week period.

Moreover, the porosity content in hybrid composites with filler material was less than that of composites without filler material. The presence of filler material in hybrid composites showed a sudden peak in carbon weight concentration, proving that incorporation of filler material and the application of the hybridization concept gave superior mechanical performance. Thus, the obtained exceptional composite properties enable future research in the application of microfluidics and microchannels.

Author Contributions: Conceptualization, A.A.; methodology, A.A. and A.P.; investigation, A.A. and L.K.; calculation and analysis, S.U. and L.K.; writing—original draft preparation, A.A.; writing—review and editing, G.J. and A.P.; visualization, A.A.; validation, S.U. and G.J. All authors have read and agreed to the published version of the manuscript.

Funding: This research received no external funding.

Data Availability Statement: Not applicable.

Conflicts of Interest: The authors declare no conflict of interest.

References

1. Saba, N.; Jawaid, M.; Alothman, O.Y.; Paridah, M.T.; Hassan, A. Recent advances in epoxy resin, natural fiber-reinforced epoxy composites and their applications. *J. Reinf. Plast. Compos.* **2016**, *33*, 447–470. [CrossRef]
2. Barua, A.; Jeet, S.; Bagal, D.K.; Satapathy, p.; Agrawal, P.K. Evaluation of Mechanical Behavior of Hybrid Natural Fiber Reinforced Nano Sic Particles Composite using Hybrid Taguchi-Cocoso Method. *Int. J. Innov. Technol. Explor. Eng.* **2019**, *8*, 3341–3345. [CrossRef]
3. Bharath, K.N.; Basavarajappa, S. Applications of biocomposite materials based on natural fibers from renewable resources: A review. *Sci. Eng. Compos. Mater.* **2016**, *23*, 123–133. [CrossRef]
4. Summerscales, J.; Virk, A.; Hall, W. A review of bast fibres and their composites: Part 3—Modelling. *Compos. Part A Appl. Sci. Manuf.* **2013**, *44*, 132–139. [CrossRef]
5. Sanjay, M.R.; Arpitha, G.R.; Naik, L.L.; Gopalakrishna, K.; Yogesha, B. Applications of Natural Fibers and Its Composites: An Overview. *Nat. Resour.* **2016**, *7*, 108–114. [CrossRef]
6. Nisini, E.; Santulli, C.; Liverani, A. Mechanical and impact characterization of hybrid composite laminates with carbon, basalt and flax fibres. *Compos. Part B Eng.* **2017**, *127*, 92–99. [CrossRef]
7. Moudood, A.; Rahman, A.; Huq, N.M.L.; Ochsner, A.; Islam, M.M.; Francucci, G. Mechanical properties of flax fiber-reinforced composites at different relative humidities: Experimental, geometric, and displacement potential function approaches. *Polym. Compos.* **2020**, *41*, 4963–4973. [CrossRef]
8. Nurazzi, N.M.; Asyraf, M.R.R.; Fatimah Athiyah, S.; Shazleen, S.S.; Ayu Rafiqah, S.; Harussani, M.M.; Kamarudin, S.H.; Razman, M.R.; Rahmah, M.; Zainudin, E.S.; et al. A Review on Mechanical Performance of Hybrid Natural Fiber Polymer Composites for Structural Applications. *Polymers* **2021**, *13*, 2170. [CrossRef]
9. Rana, A.; Mandal, A.; Bandyopadhyay, S. Short jute fiber reinforced polypropylene composites: Effect of compatibiliser, impact modifier and fiber loading. *Compos. Sci. Technol.* **2003**, *63*, 801–806. [CrossRef]

10. Jamshaid, H.; Mishra, R.; Basra, S.; Rajput, A.W.; Hassan, T.; Petru, M. Lignocellulosic Natural Fiber Reinforced Bisphenol F Epoxy Based Bio-composites: Characterization of Mechanical Electrical Performance. *J. Nat. Fibers* **2022**, *19*, 3317–3332. [CrossRef]
11. Ramamoorthy, S.K.; Skrifvars, M.; Persson, A. A Review of Natural Fibers Used in Biocomposites: Plant, Animal and Regenerated Cellulose Fibers. *Polym. Rev.* **2015**, *55*, 107–162. [CrossRef]
12. Ramadan, N.; Taha, I.; Hammouda, R.; Abdellatif, M.H. Behaviour of Hybrid SiC/Jute Epoxy Composites Manufactured by Vacuum Assisted Resin Infusion. *Polym. Polym. Compos.* **2017**, *25*, 333–344. [CrossRef]
13. Tang, J.; Swolfs, Y.; Longana, M.L.; Yu, H.; Wisnom, M.R.; Lomov, S.V.; Gorbatikh, L. Hybrid composites of aligned discontinuous carbon fibers and self-reinforced polypropylene under tensile loading. *Compos. Part A Appl. Sci. Manuf.* **2019**, *123*, 97–107. [CrossRef]
14. Aji, I.S.; Zainudin, E.; Abdan, K.; Sapuan, S.M.; Khairul, M.D. Mechanical properties and water absorption behavior of hybridized kenaf/pineapple leaf fibre-reinforced high-density polyethylene composite. *J. Compos. Mater.* **2013**, *47*, 979–990. [CrossRef]
15. Patel, Y.; Janusas, G.; Palevicius, A.; Vilkauskas, A. Development of nanoporous AAO membrane for nano filtration using the acoustophoresis method. *Sensors* **2020**, *20*, 3833. [CrossRef]
16. Yoon, J.Y. Open-Surface Digital Microfluidics. *Open Biotechnol. J.* **2008**, *2*, 94–100. [CrossRef]
17. Abdelgawad, M.; Park, P.; Wheeler, A.R. Optimization of device geometry in single-plate digital microfluidics. *J. Appl. Phys.* **2009**, *105*, 1–9. [CrossRef]
18. Zida, S.I.; Lin, Y.; Khung, Y.L. Current Trends on Surface Acoustic Wave Biosensors. *Adv. Mater. Technol.* **2021**, *6*, 1–12. [CrossRef]
19. Guo, J.; Lin, L.; Zhao, K.; Song, Y.; Huang, M.; Zhu, Z.; Zhou, L.; Yang, C. Auto-affitech: An automated ligand binding affinity evaluation platform using digital microfluidics with a bidirectional magnetic separation method. *Lab A Chip* **2020**, *20*, 1577–1585. [CrossRef]
20. Chen, W.; Hao, H.; Jong, M.; Cui, J.; Shi, Y.; Chen, L.; Pham, T.M. Quasi-static and dynamic tensile properties of basalt fibre reinforced polymer. *Compos. Part B Eng.* **2017**, *125*, 123–133. [CrossRef]
21. Lopresto, V.; Leone, C.; De Iorio, I. Mechanical characterisation of basalt fibre reinforced plastic. *Compos. Part B Eng.* **2011**, *42*, 717–723. [CrossRef]
22. Seghini, M.C.; Touchard, F.; Sarasini, F.; Cech, V.; Chocinski-Arnault, L.; Mellier, D.; Tirllo, J.; Bracciale, M.P.; Zvonek, M. Engineering the interfacial adhesion in basalt/epoxy composites by plasma polymerization. *Compos. Part A Appl. Sci. Manuf.* **2019**, *122*, 67–76. [CrossRef]
23. Kuranska, M.; Barczewski, M.; Uram, K.; Lewandowski, K.; Prociak, A.; Michatowski, S. Basalt waste management in the production of highly effective porous polyurethane composites for thermal insulating applications. *Polym. Test.* **2019**, *76*, 90–100. [CrossRef]
24. Matykiewicz, D.; Barczewski, M.; Michalowski, S. Basalt powder as an eco-friendly filler for epoxy composites: Thermal and thermo-mechanical properties assessment. *Compos. Part B Eng.* **2019**, *164*, 272–279. [CrossRef]
25. Dorigato, A.; Pegoretti, A. Flexural and impact behaviour of carbon/basalt fibers hybrid laminates. *J. Compos. Mater.* **2014**, *48*, 1121–1130. [CrossRef]
26. Sakthivel, M.; Vijayakumar, S.; Prasad, N.K. Drilling analysis on basalt/sisal reinforced polymer composites using ANOVA and regression model. *Appl. Math. Sci.* **2015**, *9*, 3285–3290. [CrossRef]
27. Sabet, S.M.M.; Akhlaghi, F.; Eslami-Farsani, R. The effect of thermal treatment on tensile properties of basalt fibers. *J. Ceram Sci. Technol.* **2015**, *6*, 245–248.
28. Sarasini, F.; Tirillo, J.; Valente, M.; Valente, T.; Cioffi, S.; Iannace, S.; Sorrentino, L. Effect of basalt fiber hybridization on the impact behavior under low impact velocity of glass/basalt woven fabric/epoxy resin composites. *Compos. Part A Appl. Sci. Manuf.* **2013**, *47*, 109–123. [CrossRef]
29. Amaro, A.M.; Reis, P.N.B.; de Moura, M.F.S.F. Residual Strength after Low Velocity Impact in Carbon-Epoxy Laminates. *Mater. Sci. Forum* **2006**, *514*, 624–628. [CrossRef]
30. Chenghong, H.; Yubin, L.; Zuoguang, Z.; Zhijie, S. Impact Damage Modes and Residual Flexural Properties of Composites Beam. *J. Reinf. Plast. Compos.* **2008**, *27*, 1163–1175. [CrossRef]
31. Santiuste, C.; Sanchez-Saez, S.; Barbero, E. Residual flexural strength after low-velocity impact in glass/polyester composite beams. *Compos. Struct.* **2010**, *92*, 25–30. [CrossRef]
32. Shim, V.P.W.; Yang, L.M. Characterization of the residual mechanical properties of woven fabric reinforced composites after low-velocity impact. *Int. J. Mech. Sci.* **2005**, *47*, 647–665. [CrossRef]
33. Valentino, P.; Sgambitterra, E.; Furgiuele, F.; Romano, M.; Ehrlich, I.; Gebbeken, N. Mechanical characterization of basalt woven fabric composites: Numerical and experimental investigation. *Frat. Ed Integrita. Strutt.* **2014**, *28*, 1–10. [CrossRef]
34. King, F.L.; Arul Jeya Kumar, A.; Vijayaragahavan, S. Mechanical characterization of polylactic acid reinforced bagasse/basalt hybrid fiber composites. *J. Compos. Mater.* **2019**, *53*, 33–43. [CrossRef]
35. Harsha Vardhan, V.S.S.; Ravi Prakash Babu, K.; Raghu Kumar, B.; Rama Krishna, G. Mechanical Characterization of E-Glass Fibre Reinforced Composite with Graphene Nano Particles as Filler. *Int. J. Emerg. Trends Eng. Res.* **2020**, *8*, 5043–5047.
36. Arun Prakash, V.R.; Rajadurai, A. Thermo-mechanical characterization of siliconized E-glass fiber/hematite particles reinforced epoxy resin hybrid composite. *Appl. Surf. Sci.* **2016**, *384*, 99–106.
37. Caylak, S.; Ertas, M.; Donmez Cavdar, A.; Angin, N. Mechanical characteristics and hydrophobicity of alkyl ketene dimer compatibilized hybrid biopolymer composites. *Polym. Compos.* **2021**, *42*, 2324–2333. [CrossRef]

38. Meng, F.; Zhang, Y.; Xiong, Z.; Wang, G.; Li, F.; Zhang, L. Mechanical, hydrophobic and thermal properties of an organic-inorganic hybrid carrageenan-polyvinyl alcohol composite film. *Compos. Part B Eng.* **2018**, *143*, 1–8. [CrossRef]
39. Wang, F.; Wu, Y.; Huang, Y. Novel application of graphene oxide to improve hydrophilicity and mechanical strength of aramid nanofiber hybrid membrane. *Compos. Part A Appl. Sci. Manuf.* **2018**, *110*, 126–132. [CrossRef]
40. Tarawneh, M.A.; Ahmad, S.H.; Yahya, S.Y.; Rasid, R.; Noum, S.Y.E. Mechanical properties of thermoplastic natural rubber reinforced with multi-walled carbon nanotubes. *J. Reinf. Plast. Compos.* **2011**, *30*, 363. [CrossRef]
41. Sreekanth, M.S.; Joseph, S.; Mhaske, S.T.; Mahanwar, P.A. Effects of Mica and Fly Ash Concentration on the Properties of Polyester Thermoplastic Elastomer Composites. *J. Thermoplast. Compos. Mater.* **2011**, *24*, 317–330. [CrossRef]
42. Atmakuri, A.; Palevicius, A.; Kolli, L.; Vilkauskas, A.; Janusas, G. Development and Analysis of Mechanical Properties of Caryota and Sisal Natural Fibers Reinforced Epoxy Hybrid Composites. *Polymers* **2021**, *13*, 864. [CrossRef] [PubMed]
43. Abd El-Baky, M.A.; Attia, M.A.; Abdelhaleem, M.M.; Hassan, M.A. Flax/basalt/E-glass Fibers Reinforced Epoxy Composites with Enhanced Mechanical Properties. *J. Nat. Fibers* **2022**, *19*, 954–968. [CrossRef]
44. Malysheva, G.V.; Guzeva, T.A. Technological support for porosity reduction of polymer composite material components. *Technol. Met.* **2021**, *13*, 1692–1695.
45. Pollard, M.; Tran, P.; Dickens, T. Porosity reducing processing stages of additive manufactured molding (Amm) for closed-mold composite fabrication. *Materials* **2020**, *13*, 5328. [CrossRef]

micromachines

MDPI

Article

Variable Stiffness Conductive Composites by 4D Printing Dual Materials Alternately

Fei Long [1,2,3], Gaojie Xu [1], Jing Wang [4], Yong Ren [2,3,5,*] and Yuchuan Cheng [1,6,*]

1 Zhejiang Key Laboratory of Additive Manufacturing Materials, Ningbo Institute of Materials Technology and Engineering, Chinese Academy of Sciences, Ningbo 315201, China
2 Department of Mechanical, Materials and Manufacturing Engineering, University of Nottingham Ningbo China, Ningbo 315100, China
3 Research Group for Fluids and Thermal Engineering, University of Nottingham Ningbo China, Ningbo 315100, China
4 Department of Electrical and Electronic Engineering, University of Nottingham Ningbo China, Ningbo 315100, China
5 Key Laboratory of Carbonaceous Wastes Processing and Process Intensification Research of Zhejiang Province, University of Nottingham Ningbo China, Ningbo 315100, China
6 Center of Materials Science and Optoelectronics Engineering, University of Chinese Academy of Sciences, Beijing 100049, China
* Correspondence: yong.ren@nottingham.edu.cn (Y.R.); yccheng@nimte.ac.cn (Y.C.)

Abstract: Materials that can be designed with programmable properties and which change in response to external stimuli are of great importance in numerous fields of soft actuators, involving robotics, drug delivery and aerospace applications. In order to improve the interaction of human and robots, materials with variable stiffness are introduced to develop their compliance. A variable stiffness composite has been investigated in this paper, which is composed of liquid metals (LMs) and silicone elastomers. The phase changing materials (LMs) have been encapsulated into silicone elastomer by printing the dual materials alternately with three-dimensional direct ink writing. Such composites enable the control over their own stiffness between soft and rigid states through LM effective phase transition. The tested splines demonstrated that the stiffness changes approximately exceeded 1900%, and the storage modulus is 4.75 MPa and 0.2 MPa when LM is rigid and soft, respectively. In the process of heating up, the stretching strain can be enlarged by at least three times, but the load capacity is weakened. At a high temperature, the resistance of the conductive composites changes with the deformation degree, which is expected to be applied in the field of soft sensing actuators.

Keywords: liquid metal; variable stiffness; 4D printing; thermal response; phase change

Citation: Long, F.; Xu, G.; Wang, J.; Ren, Y.; Cheng, Y. Variable Stiffness Conductive Composites by 4D Printing Dual Materials Alternately. *Micromachines* **2022**, *13*, 1343. https://doi.org/10.3390/mi13081343

Academic Editor: Pingan Zhu

Received: 25 July 2022
Accepted: 11 August 2022
Published: 19 August 2022

1. Introduction

In contrast to the uniaxial or torsional displacements of traditional actuators, soft elastomers can be programmed to undergo changes throughout the whole structure with the significant advantages in low density, high mechanical flexibility and multidimensional movement. As an aspect that has received a lot of attention surrounding soft actuators recently, tunable stiffness refers to the ability of materials or systems to transform between a compliant and rigid load-bearing state after applying external stimulus. Until now, such controllable stiffness materials have been applied in a wide range of fields, such as more efficient catheters and endoscopes to perform non-invasive procedures [1] in medicine, adaptive wings that improve aircraft performance [2] in aeronautics and building materials that lower the damages from wind and earthquakes [3] in architecture. To avoid time and material consuming with the tedious fabrication process, a suitable approach with additive manufacturing (AM) is required for effective fabrication of 3D structure with complex topographical feature. Taking a particular instance, various 3D printing techniques have been applied for designing versatile microfluidic systems [4] to detect different analytes [5] and different clinically relevant diseases [6].

The additive manufacturing technique, also known as three-dimensional printing (3D printing), can enable one-step patterning of multi-materials, such as plastics [7], metals [8], ceramics [9] and woods [10]. Traditional processing techniques such as milling, molding and engraving result in high ambient temperatures, which have been a hindrance to pattern metal and polymeric materials directly. The conventional 3D printing objects are considered as statistic, whose dynamic evolutions were restricted in potential applications. Recently, the significant efforts on 3D printing have yielded four-dimensional (4D) printing structures that the structure or properties change over time; in other words, it is an updated version of 3D printing technologies with stimuli-materials involving heat [11], light [12], magnetic field [13], electrical field [14], etc. Until now, 4D printing has attached worldwide attention from micro- to macroscale as a result of various functional applications in soft robotics, flexible electronics, and biomedicine [15].

The materials with significant variable stiffness can be primarily classified into three groups that are shape memory polymers (SMP), smart materials and low melting point materials. Shape memory materials can experience the irreversible transitions from metastable to globally stable states [16] or response to environmental conditions to undergo reversible changes (such as LCEs [17]). The phase transition of major shape memory materials operates through temperature change, in detail that SMPs must be heated indirectly using external heaters [18]. For smart materials (for example, electrorheological fluids (ERF) [19] and magnetorheological fluids (MRF)) [20], the principle of phase transitions depends on the order of the inner nanometer or micrometer to change own viscosity. Low melting point materials represented by metal alloys [21] can be heated directly due to the thermal conductivity, and its heating rate is almost 100 times of SMPs with 18 W/mK [22]. Since both smart fluids and low melting point materials will exist as flowing state in a compliant state, they must be encapsulated to perform stiffness-changing functions without the loss of fluids.

As one of the low melting point materials, liquid metals (LMs) exhibit the properties of traditional rigid metal in a solid state, while it can be dispensed, stretched and deformed easily after melting. With the inherent soft state, LMs are suitable for applications in the devices where the desired materials need to endure varying degrees of stress, such as soft actuators and soft robotics. In addition, the features in high electrical and thermal conductivity can be applied in cases where design stretchable circuits and strain sensors [23], such as heat dissipation [24]. However, the surface tension of LMs is high with approximately 700 mN/m [25], that is poor wettability with solid substrates, resulting in the leakage or spillage problems when utilizing devices [26]. Cheng et al. [27] reported that the oxidized LMs have ability to reduce surface tension and improve wettability, but its lower thermal conductivity with around 1 Wm^{-1} K^{-1} will deteriorate severely. Until now, to balance wettability and thermal performance, Cu particles have been mixed by various shapes and sizes [26]. Due to LMs fluidity in the liquid state and inherent brittleness in the stiff state, it is difficult to pattern in the solid form for further processing [28]. It is possible to introduce direct ink printing (DIW) that is a common approach in the AM field in spite of existing challenges to print LM directly [29], so the precise control of distance between nozzle and surface enables the direct printing of non-spherical shapes by shearing metal from the nozzle. As the printed structures maintain a liquid state at room temperature, further encapsulations are required for most applications.

In this paper, a controllable stiffness composite has been proposed, which consisted of LM as functional materials and silicone elastomer as the encapsulation layer, respectively. Here, to explore the absolute stiffness and hardness, three types of silicone have been selected, including Ecoflex00-30, PDMS Sylgard 184 and SE 1700. Additionally, different mixing ratios of silicone have been adjusted to satisfy the necessary rheological conditions for DIW. Through printing dual materials alternately, simple splines, hollow flower patterns and Poisson structures can be fabricated. The stiffness change of LM composites can be controlled by the thermal response of LMs, which results in large changes in stiffness after LM melting completely, while the silicone guarantees that the melted LMs retain the pre-molten shape. The DIW of LMs has broken through the difficulty of filling approach and can achieve a larger volume content of LM composites that were promoted a wider range of stiffness change. At a temperature above the melting point, the addition of LM will greatly increase the tensile deformation capacity, but the load-carrying ability will be weakened to a certain extent. Owing to the electrical conductivity of LMs, their resistance can change during stretching and the recyclability has certain advantages compared to previous studies, which is expected to be applied into soft sensing actuators.

2. Experimental

2.1. Dual Phase Direct Write Printing

A challenge for LM elastomers is that filling LM into the cured silicone mold is difficult at the room temperature due to the high surface tension of LM. A multi-material alternate printing approach, previously used for water-soluble support in 3D bioprinting systems has been proposed to stabilize the surface contact between materials. Bodaghi et al. [30] has fabricated dual-material lattice-based meta-structures by fused deposition modelling technology. In this study, the progress of dual phase direct write printing (Figure 1a) has been divided into three layers as a 'sandwich' structure, shown as in Figure 1b. In the DIW printing process, the relevant typical parameters can refer to the 3D printing of stretchable elastomers by Zhou et al. [31]. With a suitable pressure P = 0.1 MPa, the slurries are extruded from the narrow nozzle with a diameter D of 450 mm at a speed c which is pressure dependent. Since the extruded slurries will experience die-swelling [32], the diameter of the filament can be defined as αD, in which α is die-swelling ratio. Meanwhile, the deposited slurries are extruded from the moving nozzle at a speed V of 8 mm/s and a height H of 0.05 mm which is the distance between the printer layer and the nozzle. As the surface tension between the LM in solid state and the silicone elastomer in semi-solid state is relatively small [33,34], cooling LMs to solid can alleviate such tough problems. In detail, after completing the first layer, the carrier platform needs to be cooled down to −10 °C for printing the second layer, which ensures that LMs can be cured quickly after printing to contact with the first layer in the solid state. Finally, the third layer is printed to encapsulate the LMs. This one-step additive manufacturing technology can complete complex structures in a short time, effectively reducing preparation time and saving costs. The process of extruding the slurries from the narrow nozzle can be regarded as preparing the tiny droplets under the microfluidic control, so the rheological parameters are extremely significant.

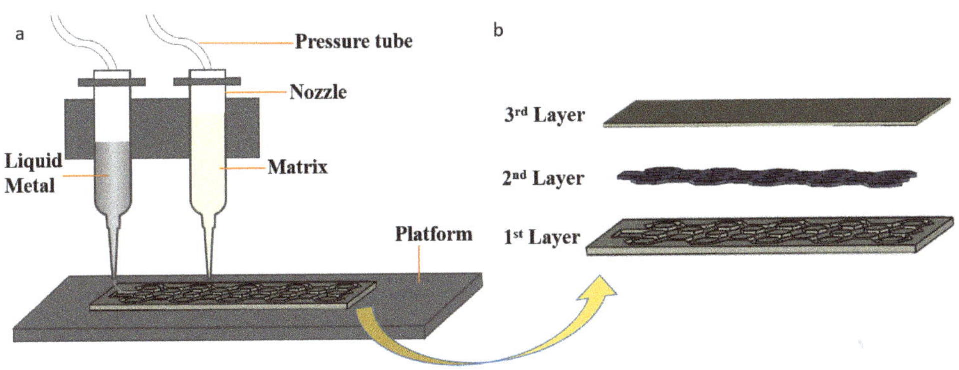

Figure 1. (**a**) The schematic diagram of dual phase direct write printing. (**b**) The design structure in a 'sandwich' structure.

2.2. Materials

The liquid metal applied in this research is composed of 75.5% gallium (Ga) and 24.5% indium (In), purchased from Jintai Alloy Corporation, Guangdong, China. The selected LMs start to melt from 20 °C to 40 °C approximately, which has been verified by DSC measurements, as shown in Figure S1. At room temperature, LMs remain liquid state of high surface tension with around 0.624 N/m and low viscosity with 0.0024 Pas; the above data were provided by supplier, which are consistent basically with previous study by Koster [35]. Here, there are three types of silicon elastomers selecting in this study, containing EcoflexTM 0030 (Smooth-On, Macungie, PA, USA), PDMS 184 (DOWSILTM, Sylgard, Midland, MI, USA) and PDMS 1700 (DOWSILTM, SE, Midland, MI, USA). Ecoflex 0030 was prepared in a 1:1 base to curing agent weight ratios, while two kinds of PDMS were mixed at 1:10. In order to obtain the printable slurries, Ecofelx 0030 and SE1700 would be mixed at 1:1 in weight, while Sylgard 184 and SE 1700 should be mixed at 1:2. The printable substrates have been mixed with a planetary mixer (VM300SA3, Miantangshinuo Corporation, Jiangsu, China).

2.3. Measurements

2.3.1. Thermal Characterization

The experiments on the melting point of LMs were used a DCS 214 (NETZSCH, Selb, German) with high purity alumina ceramic crucible that can withstand 100 °C and bear the corrosion of Ga. The measurement temperature range was from −40 °C to 40 °C, and the heating rate was 10 K/min.

2.3.2. Rheological Characterization

The experiments on rheology were conducted by Discovery HR-20 (TA, New Castle, DE, USA), equipped with a 20 mm parallel plate geometry. To minimize the effect by measuring, all samples were pre-sheared and tested for three times. Additionally, the specific steps have been discussed in further detail in the ESI.

2.3.3. Mechanical Characterization

Mechanical experiments were conducted using Dynamic Mechanical Analyzer DMA Q800 (TA, New Castle, DE, USA) and universal testing machine UTM Roell Z030 (Zwick, Ulm, German). Both instruments have been equipped with the heating function, and the detail settings have been shown in the ESI.

2.3.4. Electrical Characterization

A high-precision LCR digital bridge TH2827c (Tonghui, Changzhou, China) was used to confirm whether the elastomer is conductive. The change of resistance during the stretch progress was monitored, and related electrical data has been synchronized directly to the computer.

3. Results and Discussions

3.1. Rheological Properties of Matrix

In the progress of DIW, the slurries pass through the narrow constriction of a needle which generates high shear force; once extruded, such shear force disappears instantly. Thus, it is required that the printed slurries can be smoothly extruded from the nozzle and perform well self-supporting after extrusion. In other words, the proportioning material should meet the characterization in both shear thinning [36] that viscosity decreases with shear strain, and viscoelastic inversion [37] that the changes of storage modulus and viscoelastic modulus show an intersection with the increase in shear strain. Inspired by Sangchul et al. [38], the above characterization can be obtained by adding other polymers. Here, SE 1700 was mixed with Ecoflex 00-30 and Sylgard 184, respectively, at a ratio of 1:1 and 1:1.5 in weight. Prior to combining, rheological tests have been carried out on these three silicones, all of which occurred shear thinning, but merely SE 1700 has the characteristic of viscoelastic inversion (Figure 2a–c). The plateau value of storage modulus of SE 1700 is two orders of magnitude larger than that of the other two silicone rubbers, while its loss modulus is larger as well (Figure 2d,e). Furthermore, two types of mixed silica gels were also conducted systematic rheological investigations that reflect direct write printability (Figure 2f). The combination of SE 1700 and Sylgard 184 at a weight ratio of 1:1.5 (as shown in Figure S2) has both the properties of shear thinning and viscoelastic inversion. Additionally, the above materials have been mixed in a 1:1 ratio (consistent with Ecoflex 0030), and G′ and G″ have no intersecting trajectories (Figure S3), that is, they cannot be applied to DIW printing.

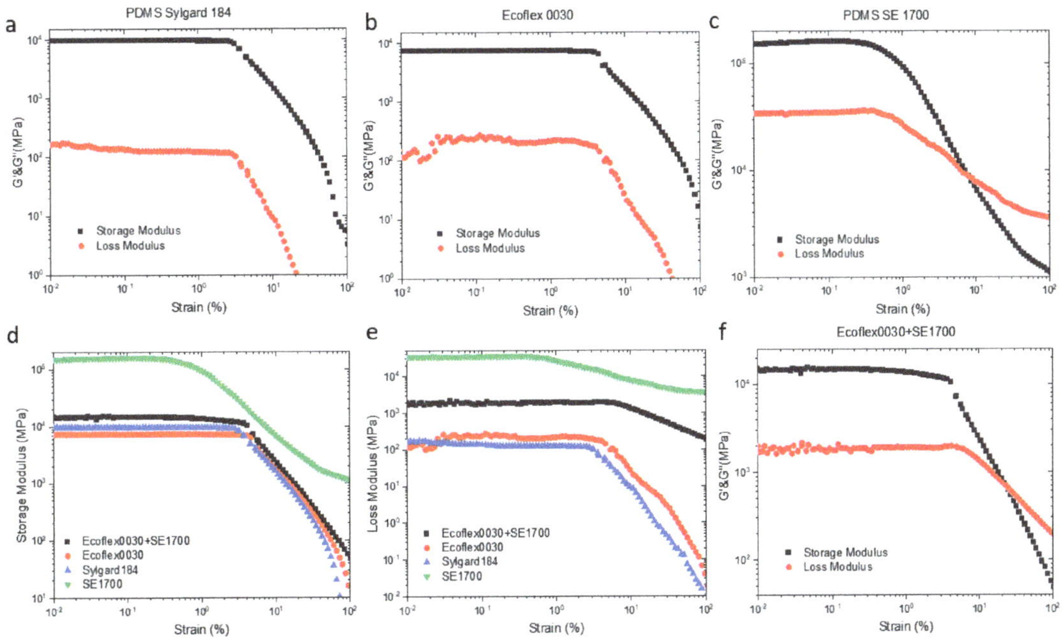

Figure 2. Viscoelastic inversion measurement of (**a**) PDMS Sylgard 184, (**b**) Ecoflex 0030 and (**c**) PDMS SE 1700. (**d**) The storage modulus and (**e**) loss modulus versus strain for silicone elastomer. (**f**) The viscoelastic inversion characteristics of the combination of Ecoflex 0030 and PDMS SE 1700.

The shear thinning behavior can be derived from a power−law variant of Herschel and Bulkley model [39], in detail that if the value of n is between 0 and 1 in the fitted linear function relationship based on the Equation (1), the material possesses the characteristics of shear thinning.

$$\tau = \tau_0 + k\dot{\gamma}^n \quad (1)$$

where τ is shear stress, $\tau_0 > 0$ is the yield stress, $k > 0$ is the consistency parameter, and $n > 0$ is the power index. Figure 3b can be obtained by fitting a linear function by taking a logarithmic relationship to Equation (1), and all value of n below 1. As the increasing of n, the phenomenon of shear thinning becomes more obvious. So, the joint of a certain amount of SE 1700 can effectively make the substrate printable.

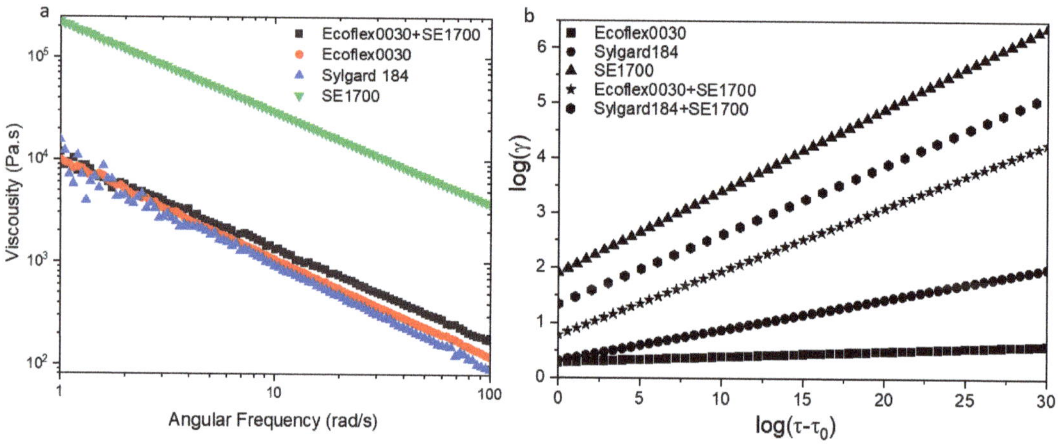

Figure 3. (**a**) Shear thinning of the selected silicone slurries (**b**) The fitting linear function of selected silicone slurries by Herschel and Bulkley model.

3.2. Mechanical Properties of Matrix

The stiffness is a vital metric for defining the forces changing that the composites can support. After the printed splines are cured (Figure 4a, and the detail dimensions has been labeled in the engineering drawing), a series of DMA measurements have been established that all the tested LM composites are elastomeric in nature. As shown in Figure S4a, the storage modulus of silicone elastomer mixed with Ecoflex 0030 and SE 1700 is smaller than that of Sylgard 184 but greater than that of SE 1700, which is more conductive to highlight the effect of LM additive. So, the former mixed matrix has been focused on, while the latter analysis can be referred to in the Supporting Information section. Despite that LMs possess fluidity in the liquid state at room temperature, the storage modulus of LM composites increases with LM volume fraction (Figure 5a), compared to the unfilled LM composites with 0.429 ± 0.01 MPa, storage modulus at 60 vol% experienced increasing by a factor of around 11 to 4.71 ± 0.1 MPa. The incorporation of liquid inclusions enables to improve the stiffness of polymer composites to a certain extent, which has been demonstrated [40,41]. In addition, the stiffness is also influenced by the interfacial tension between LM and silicone matrix [42]. Taking example of the composites with 60 vol%, its storage modulus decreases gradually as temperature rises, but a plunge has been occurred when the temperature reaches the melting point T_m of LM. Such a dip becomes more pronounced as the LM volume content increases, and the change of storage modulus can be over 200% during the transformation between the rigid and soft state. Moreover, the relationship between deformation strain and load bearing of LM composites has been explored by UTM, as shown in Figure 5b. As a result of the solid-liquid transformation taking place below 60 °C, it can be stretched much more than that keeping at 0 °C, particularly, up to 3.5 times for LM (60 vol%) composites mixed with Ecoflex 0030 and SE 1700. Meanwhile, the corresponding carrying load has been weakened due to the soft state of LM. Similar changes occurred in the LM composites with Sylgard 184 and SE 1700 as the carrier (Figure S5).

Apart from the simple splines, the flower-shaped and Poisson structure have been designed (shown in Figure 4b–e). The modeling and fabrication of relatively complex structures in a short period of time further confirm the high efficiency of additive manufacturing. As the temperature rises, the changes on tensile strength of the above designed structures have been recorded in Movies S1 and S2. In terms of Poisson structures, the connection between bearing capacity and deformation degree at high temperature has been displayed in Figure S6.

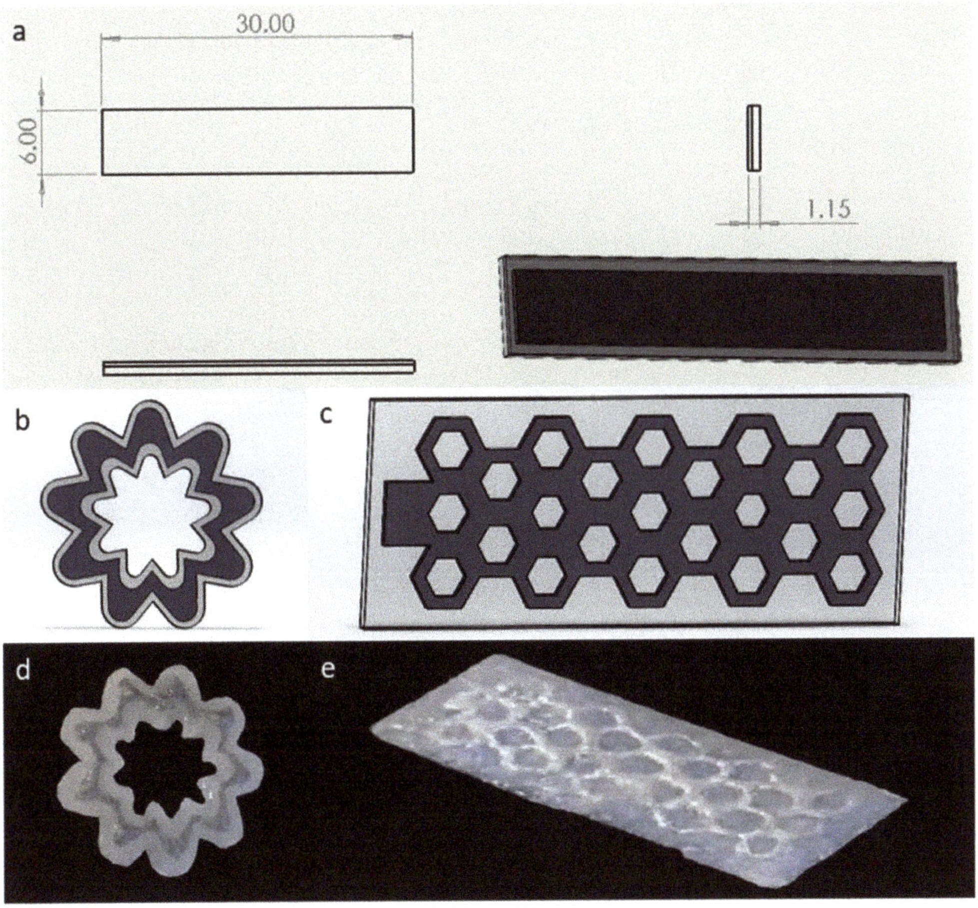

Figure 4. (**a**) The detail engineering diagram of sample splines, (**b**) the design structure, (**c**) the printed real part, (**d**) hollow flower structure and (**e**) Poisson structure.

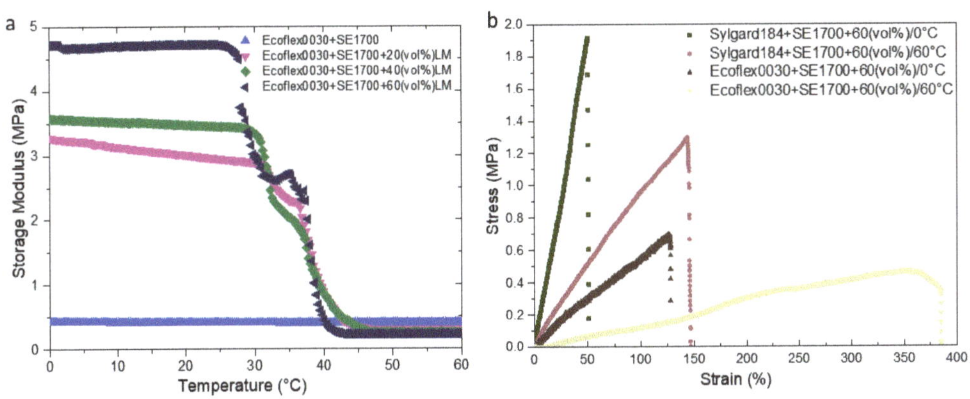

Figure 5. (**a**) The storage modulus with LM composite with the increasing volume fraction of LM. (**b**) The load capacity versus the stretchable strain at low and high temperature.

3.3. Resistance Changes in the Process of Stretching

As the typical types of PCMs, LMs possess good electrical conductivity. The state of LM can be judged by monitoring changes in resistance, thereby determining the softness of composites. In theoretical, the resistance of LM composites can be referred as the standard equation for the wire:

$$R = \frac{\rho L}{A} \tag{2}$$

in which ρ is the electrical resistivity, L is the length and A is the cross-sectional area of conductor. It can be seen that the resistance varies with the geometry for a certain material. During the stretching process at high temperature, the samples will not only undergo thermal expansion but also length elongation, and the length increases faster than the cross-sectional area.

To characterize the change in resistance, the samples have been clamped on UTM, connecting with LCR digital bridge simultaneously. The relative resistance change has been introduced:

$$\frac{\Delta R}{R_0} = \frac{R - R_0}{R_0} \tag{3}$$

R and R_0 are corresponding to the resistance values with and without deformation, respectively. The $\Delta R/R_0$ increases with stretching deformation, which proves that the external strain has a certain influence on the relative resistance change. To investigate the sensing performance in terms of tensile strain, its sensitivity can be defined by gauge factor (GF):

$$GF = \frac{\Delta R/R_0}{\Delta L/L_0} \tag{4}$$

where, L_0 represents the initial size of the splines, and ΔL indicates the size change. As shown in Figure 6a, with the good corresponding consistency, the highest strain can reach 9400% approximately, and GF value is around 60 at this point. Moreover, the relationship between resistance changes and deformation of LM composites with Sylgard 184 and SE 1700 has been explored (Figure S8a), which can be up to around 120% with the GF value of 1.28 approximately.

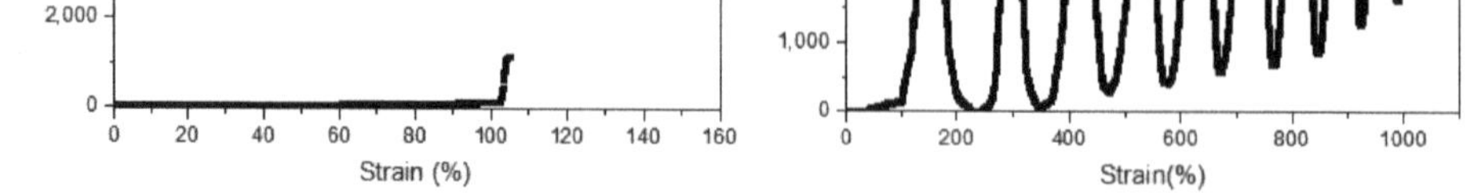

Figure 6. (**a**) At high temperature (60 °C), the relative resistance changes with the stretchable strain, and (**b**) the stretch repeatability over time.

According to the maximum strain obtained from the measurement, the sample of LM composites mixed with Ecoflex 0030 and SE 1700 have been applied to the strain from 0% to 120% at a constant rate of 10 mm/min, and then released until they return to the initial state. The relative resistance change versus time has been achieved in Figure 6b by applying and releasing pressure several times repeatedly, and the spline takes place fracture when the number of cycles is about 10 times. With the increase in strain, $\Delta R/R_0$ improves gradually; and after reaching the peak value, $\Delta R/R_0$ deceases as a result that the external force is released. The sample can basically return to its original shape when the external force disappears. However, the stretching process will cause a certain degree of permanent loss for no contract of internal LMs fracture due to elastomer stretching, which will make the value of resistance become larger when returning to the origin point. At the tenth stretch, the sample has broken. Moreover, the reproducibility test graph for LM composites with Sylgard 184 and SE 1700 have been demonstrated in the Supporting Information (Figure S8b).

4. Conclusions

We have developed a variable stiffness composite that consists of LM and silica gel with different mixing ratios, which can change properties in response to the thermal stimuli. In DIW printing, the process of slurry extrusion can be regarded as material prepared by a microfluidic channel, so the related rheological properties are necessary for the combined slurries. With a certain printability, dual material printing alternately has been applied to fabricate LM composites in one step for the relative complex structures. The samples presented here illustrate the stiffness change of greater than 1900% from a stiff to soft state, while the storage modulus decrease from 4.75 MPa and 0.2 MPa after heating up. Furthermore, by changing the inner structure design or volume fraction between LMs and silicon elastomer, different stiffness values for these two steady states can be achieved. Owing to the electrical conductivity of LMs, the composite exhibits electrical resistance that changes with stretching. However, each stretch will lead to irreversible damage in the elastomer to a certain extent; the spline fracture is generated after about ten repetitions of the tensile test. Overall, this work has demonstrated the LM composites undergo the changes in mechanical and electrical properties under temperature stimuli. With the tuning capability, LM composites are expected to be used in the field of soft sensing actuators, even towards artificial muscle applications after enhancing adhesion.

Supplementary Materials: The following supporting information can be downloaded at: https://www.mdpi.com/article/10.3390/mi13081343/s1; Figure S1: The thermal properties of LM by DSC measurements; Figure S2: a. Shear thinning and b. Viscoelastic inversion measurement of the combination of PDMS Sylgard 184 and PDMS SE1700 with 1:1.5 in weight ratio; Figure S3: a. Shear thinning and b. Viscoelastic inversion measurement of the combination of PDMS Sylgard 184 and PDMS SE1700 with 1:1 in weight ratio; Figure S4: The storage modulus of splines composed by different silicone elastomer a. Comparison with Sylgard 184, SE 1700 and their mixture based on 1:2 in volume fraction. b. Comparison with Ecoflex 0030, SE 1700 and their mixture on the basis of volume fraction with 1:1; Figure S5: Th storage modulus of LM composites that subtracted by Sylgard184 and SE1700 with the LM increasing volume fraction; Figure S6: The load capacity versus the stretchable strain at low and high temperature for the Poisson structure with 60 vol% in LM; Figure S7: The Schematic diagram of real-time monitoring resistance measurements; Figure S8: At high temperature (60 °C), the relative resistance of LM composites (that based on the mixture of Sylgard184 and SE 1700) changes with the stretchable strain, and b. the stretch repeatability over time; Movie S1: Flower; Movie S2: Poisson Structure.

Author Contributions: Conceptualization, F.L., Y.R. and Y.C.; Data curation, F.L. and Y.C.; Formal analysis, F.L. and Y.C.; Funding acquisition, G.X., J.W. and Y.R.; Investigation, F.L. and Y.R.; Methodology, F.L., Y.R. and Y.C.; Project administration, G.X. and Y.C.; Resources, G.X., J.W., Y.R. and Y.C.; Supervision, Y.R. and Y.C.; Validation, F.L.; Writing—original draft, F.L.; Writing—review & editing, G.X., J.W., Y.R. and Y.C. All authors have read and agreed to the published version of the manuscript.

Funding: This research was funded by Zhejiang Provincial Natural Science Foundation of China under grant No. LZ22E030003, LY19E060001 and LQ19F050003, Ningbo Science and Technology Bureau under Service Industry Science & Technology Programme with project code 2019F1030, and Zhejiang Provincial Department of Science and Technology under its Provincial Key Laboratory Programme (2020E10018). F.L. acknowledges the Ph.D. scholarship of Doctor Training Program between University of Nottingham Ningbo China and Ningbo Institute of Materials Technology and Engineering, Chinese Academy of Sciences.

Data Availability Statement: The data that support the findings of this study are available from the corresponding author upon reasonable request.

Acknowledgments: This work was financially supported by Zhejiang Provincial Natural Science Foundation of China under grant No. LZ22E030003, LY19E060001 and LQ19F050003, Ningbo Science and Technology Bureau under Service Industry Science & Technology Programme with project code 2019F1030. The Zhejiang Provincial Department of Science and Technology is also acknowledged for this research under its Provincial Key Laboratory Programme (2020E10018).

Conflicts of Interest: The authors declare no conflict of interest.

References

1. Loeve, A.; Breedveld, P.; Dankelman, J. Scopes Too Flexible ... and Too Stiff. *IEEE Pulse* **2010**, *1*, 26–41. [CrossRef] [PubMed]
2. Sofla, A.; Meguid, S.; Tan, K.; Yeo, W. Shape morphing of aircraft wing: Status and challenges. *Mater. Des.* **2010**, *31*, 1284–1292. [CrossRef]
3. Janke, L.; Czaderski, C.; Motavalli, M.; Ruth, J. Applications of shape memory alloys in civil engineering structures—Overview, limits and new ideas. *Mater. Struct.* **2005**, *38*, 578–592. [CrossRef]
4. Prabhakar, P.; Sen, R.K.; Dwivedi, N.; Khan, R.; Solanki, P.R.; Srivastava, A.K.; Dhand, C. 3D-Printed Microfluidics and Potential Biomedical Applications. *Front. Nanotechnol.* **2021**, *3*, 609355. [CrossRef]
5. Bazaz, S.R.; Rouhi, O.; Raoufi, M.A.; Ejeian, F.; Asadnia, M.; Jin, D.; Warkiani, M.E. 3D Printing of Inertial Microfluidic Devices. *Sci. Rep.* **2020**, *10*, 5929. [CrossRef]
6. Romanov, V.; Samuel, R.; Chaharlang, M.; Jafek, A.R.; Frost, A.; Gale, B.K. FDM 3D Printing of High-Pressure, Heat-Resistant, Transparent Microfluidic Devices. *Anal. Chem.* **2018**, *90*, 10450–10456. [CrossRef]
7. Zmarzły, P.; Gogolewski, D.; Kozior, T. Design guidelines for plastic casting using 3D printing. *J. Eng. Fibers Fabr.* **2020**, *15*, 1558925020916037. [CrossRef]
8. Buchanan, C.; Gardner, L. Metal 3D printing in construction: A review of methods, research, applications, opportunities and challenges. *Eng. Struct.* **2019**, *180*, 332–348. [CrossRef]
9. Chen, Z.; Li, Z.; Li, J.; Liu, C.; Lao, C.; Fu, Y.; Liu, C.; Li, Y.; Wang, P.; He, Y. 3D printing of ceramics: A review. *J. Eur. Ceram. Soc.* **2019**, *39*, 661–687. [CrossRef]
10. Das, A.K.; Agar, D.A.; Rudolfsson, M.; Larsson, S.H. A review on wood powders in 3D printing: Processes, properties and potential applications. *J. Mater. Res. Technol.* **2021**, *15*, 241–255. [CrossRef]
11. Lei, Z.; Wang, Q.; Wu, P. A multifunctional skin-like sensor based on a 3D printed thermo-responsive hydrogel. *Mater. Horizons* **2017**, *4*, 694–700. [CrossRef]
12. Yu, C.; Schimelman, J.; Wang, P.; Miller, K.L.; Ma, X.; You, S.; Guan, J.; Sun, B.; Zhu, W.; Chen, S. Photopolymerizable Biomaterials and Light-Based 3D Printing Strategies for Biomedical Applications. *Chem. Rev.* **2020**, *120*, 10695–10743. [CrossRef] [PubMed]
13. Kania, A.; Berent, K.; Mazur, T.; Sikora, M. 3D printed composites with uniform distribution of Fe_3O_4 nanoparticles and magnetic shape anisotropy. *Addit. Manuf.* **2021**, *46*, 102149. [CrossRef]
14. Valentine, A.D.; Busbee, T.A.; Boley, J.W.; Raney, J.R.; Chortos, A.; Kotikian, A.; Berrigan, J.D.; Durstock, M.F.; Lewis, J.A. Hybrid 3D Printing of Soft Electronics. *Adv. Mater.* **2017**, *29*, 1703817. [CrossRef] [PubMed]
15. Kuang, X.; Roach, D.J.; Wu, J.; Hamel, C.M.; Ding, Z.; Wang, T.; Dunn, M.L.; Qi, H.J. Advances in 4D Printing: Materials and Applications. *Adv. Funct. Mater.* **2019**, *29*, 1805290. [CrossRef]
16. Wei, H.; Zhang, Q.; Yao, Y.; Liu, L.; Liu, Y.; Leng, J. Direct-Write Fabrication of 4D Active Shape-Changing Structures Based on a Shape Memory Polymer and Its Nanocomposite. *ACS Appl. Mater. Interfaces* **2017**, *9*, 876–883. [CrossRef]
17. Xiao, Y.; Jiang, Z.; Hou, J.; Chen, X.; Zhao, Y. Electrically driven liquid crystal network actuators. *Soft Matter* **2022**, *18*, 4850–4867. [CrossRef]
18. Long, F.; Cheng, Y.; Ren, Y.; Wang, J.; Li, Z.; Sun, A.; Xu, G. Latest Advances in Development of Smart Phase Change Material for Soft Actuators. *Adv. Eng. Mater.* **2022**, *24*, 2100863. [CrossRef]
19. Cao, C.; Zhao, X. Tunable stiffness of electrorheological elastomers by designing mesostructures. *Appl. Phys. Lett.* **2013**, *103*, 041901. [CrossRef]
20. Kallio, M.; Lindroos, T.; Aalto, S.; Järvinen, E.; Kärnä, T.; Meinander, T. Dynamic compression testing of a tunable spring element consisting of a magnetorheological elastomer. *Smart Mater. Struct.* **2007**, *16*, 506–514. [CrossRef]
21. Shintake, J.; Schubert, B.; Rosset, S.; Shea, H.; Floreano, D. Variable stiffness actuator for soft robotics using dielectric elastomer and low-melting-point alloy. In Proceedings of the 2015 IEEE/RSJ International Conference on Intelligent Robots and Systems (IROS), Hamburg, Germany, 28 September–3 October 2015; pp. 1097–1102. [CrossRef]
22. Liu, C.; Qin, H.; Mather, P.T. Review of progress in shape-memory polymers. *J. Mater. Chem.* **2004**, *17*, 1543–1558. [CrossRef]
23. Lin, Y.; Genzer, J.; Dickey, M.D. Attributes, Fabrication, and Applications of Gallium-Based Liquid Metal Particles. *Adv. Sci.* **2020**, *7*, 2000192. [CrossRef] [PubMed]
24. Jia, L.-C.; Jin, Y.-F.; Ren, J.-W.; Zhao, L.-H.; Yan, D.-X.; Li, Z.-M. Highly thermally conductive liquid metal-based composites with superior thermostability for thermal management. *J. Mater. Chem. C* **2021**, *9*, 2904–2911. [CrossRef]

25. Zhao, X.; Xu, S.; Liu, J. Surface tension of liquid metal: Role, mechanism and application. *Front. Energy* **2017**, *11*, 535–567. [CrossRef]
26. Xing, W.; Wang, H.; Chen, S.; Tao, P.; Shang, W.; Fu, B.; Song, C.; Deng, T. Gallium-Based Liquid Metal Composites with Enhanced Thermal and Electrical Performance Enabled by Structural Engineering of Filler. *Adv. Eng. Mater.* **2019**, *29*, 1805290. [CrossRef]
27. Cheng, Z.; Wheeler, V.D.; Bai, T.; Shi, J.; Tadjer, M.J.; Feygelson, T.; Hobart, K.D.; Goorsky, M.S.; Graham, S. Integration of polycrystalline Ga_2O_3 on diamond for thermal management. *Appl. Phys. Lett.* **2020**, *116*, 062105. [CrossRef]
28. Neumann, T.V.; Facchine, E.G.; Leonardo, B.; Khan, S.; Dickey, M.D. Direct write printing of a self-encapsulating liquid metal–silicone composite. *Soft Matter* **2020**, *16*, 6608–6618. [CrossRef]
29. Cook, A.; Parekh, D.P.; Ladd, C.; Kotwal, G.; Panich, L.; Durstock, M.; Dickey, M.D.; Tabor, C.E. Shear-Driven Direct-Write Printing of Room-Temperature Gallium-Based Liquid Metal Alloys. *Adv. Eng. Mater.* **2019**, *21*, 1900400. [CrossRef]
30. Bodaghi, M.; Serjouei, A.; Zolfagharian, A.; Fotouhi, M.; Rahman, H.; Durand, D. Reversible energy absorbing meta-sandwiches by FDM 4D printing. *Int. J. Mech. Sci.* **2020**, *173*, 105451. [CrossRef]
31. Zhou, L.-Y.; Gao, Q.; Fu, J.-Z.; Chen, Q.-Y.; Zhu, J.-P.; Sun, Y.; He, Y. Multimaterial 3D Printing of Highly Stretchable Silicone Elastomers. *ACS Appl. Mater. Interfaces* **2019**, *11*, 23573–23583. [CrossRef]
32. Wang, K. Die Swell of Complex Polymeric Systems. *Viscoelast. Theory Biol. Appl.* **2012**, *1*, 77–96. [CrossRef]
33. Kumikov, V.K.; Khokonov, K.B. On the measurement of surface free energy and surface tension of solid metals. *J. Appl. Phys.* **1983**, *54*, 1346–1350. [CrossRef]
34. Lu, H.M.; Jiang, Q. Surface Tension and Its Temperature Coefficient for Liquid Metals. *J. Phys. Chem. B* **2005**, *109*, 15463–15468. [CrossRef]
35. Koster, J.N. Directional solidification and melting of eutectic GaIn. *Cryst. Res. Technol. J. Exp. Ind. Crystallogr.* **1999**, *34*, 1129–1140. [CrossRef]
36. Poslinski, A.J.; Ryan, M.E.; Gupta, R.K.; Seshadri, S.G.; Frechette, F.J. Rheological Behavior of Filled Polymeric Systems I. Yield Stress and Shear-Thinning Effects. *J. Rheol.* **1988**, *32*, 703–735. [CrossRef]
37. Li, L.; Lin, Q.; Tang, M.; Duncan, A.J.E.; Ke, C. Advanced Polymer Designs for Direct-Ink-Write 3D Printing. *Chem.—Eur. J.* **2019**, *25*, 10768–10781. [CrossRef] [PubMed]
38. Roh, S.; Parekh, D.P.; Bharti, B.; Stoyanov, S.D.; Velev, O.D. 3D Printing by Multiphase Silicone/Water Capillary Inks. *Adv. Mater.* **2017**, *29*, 1701554. [CrossRef]
39. Saramito, P. A new elastoviscoplastic model based on the Herschel–Bulkley viscoplastic model. *J. Non-Newton. Fluid Mech.* **2019**, *158*, 154–161. [CrossRef]
40. Ford, M.J.; Palaniswamy, M.; Ambulo, C.P.; Ware, T.H.; Majidi, C. Size of liquid metal particles influences actuation properties of a liquid crystal elastomer composite. *Soft Matter* **2020**, *16*, 5878–5885. [CrossRef]
41. Ambulo, C.P.; Ford, M.J.; Searles, K.; Majidi, C.; Ware, T.H. 4D-Printable Liquid Metal–Liquid Crystal Elastomer Composites. *ACS Appl. Mater. Interfaces* **2020**, *13*, 12805–12813. [CrossRef]
42. Ladd, C.; So, J.-H.; Muth, J.; Dickey, M.D. 3D Printing of Free Standing Liquid Metal Microstructures. *Adv. Mater.* **2013**, *25*, 5081–5085. [CrossRef] [PubMed]

micromachines

Article

Development of Finite Element Models of PP, PETG, PVC and SAN Polymers for Thermal Imprint Prediction of High-Aspect-Ratio Microfluidics

Justas Ciganas, Paulius Griskevicius, Arvydas Palevicius, Sigita Urbaite * and Giedrius Janusas

Department of Mechanical Engineering, Kaunas University of Technology, Studentu 56, 51424 Kaunas, Lithuania
* Correspondence: sigita.urbaite@ktu.lt

Abstract: Polymeric microstructures and microchannels are widely used in biomedical devices, optics, microfluidics and fiber optics. The quality, the shape, the spacing and the curvature of microstructure gratings are influenced by different mechanisms and fabrication techniques used. This paper demonstrates a cost-effective way for patterning high-aspect-ratio thermoplastic microstructures using thermal imprint technology and finite element modeling. Polymeric materials polypropylene (PP), polyethylene terephthalate glycol (PETG), polyvinyl chloride (PVC) and styrene-acrylonitrile (SAN) were chosen for the experimental investigations. A finite element model was constructed to define the most suitable parameters (time, heating temperature, pressure, etc.) for the formation of microstructures using the thermal imprint procedure. To confirm the relevance of the finite element model, different types of PP, PETG, PVC and SAN microstructures were fabricated using theoretically defined parameters. Experimental investigations of imprinted microstructures' morphological and optical properties were performed using scanning electron microscopy, atomic force microscopy and a diffractometer. Obtained results confirmed the relevance of the created finite element model which was applied in the formation of high-aspect-ratio microstructures. Application of this model in thermal imprint would not only reduce the fabrication time, but also would highly increase the surface quality and optical properties of the formed structures.

Keywords: microstructure; finite element simulation; thermal imprint; polymer

Citation: Ciganas, J.; Griskevicius, P.; Palevicius, A.; Urbaite, S.; Janusas, G. Development of Finite Element Models of PP, PETG, PVC and SAN Polymers for Thermal Imprint Prediction of High-Aspect-Ratio Microfluidics. *Micromachines* **2022**, *13*, 1655. https://doi.org/10.3390/mi13101655

Academic Editor: Pingan Zhu

Received: 8 September 2022
Accepted: 27 September 2022
Published: 30 September 2022

1. Introduction

In recent years, novel micro- and nanomanufacturing technologies have been intensively developed for fabrication of microstructures [1]. Different applications require different grating parameters and materials to be used (shape, size and structure) in microfluidics. The developed microstructures are used in different applications, such as biosensors, wavelength division multiplexing devices, optical devices, nano- and microfluidics, electronics, sample separation or single-molecule analysis [2–5], to transport particles or materials, separate, dispense or mix liquids [6].

Because of the precision required for fabrication of microchannels, the lack of suitable and effective conventional technologies is hindering the development of microstructure-based devices. Processes such as photolithography, laser processing or etching are conventional, but mostly time-consuming grating manufacturing technologies. A thermal imprint technology is one of the possible choices for forming microstructures in thermoplastics at a relatively high speed and low cost. Basic advantages of this technique include low material flow and low flaw rates, which allows to avoid internal stress resulting in more delicate gratings [7]. The mostly used thermoplastics for fabrication of microchannels are: polypropylene (PP), polyethylene terephthalate glycol (PETG), polyvinyl chloride (PVC) and styrene-acrylonitrile (SAN). PP thermoplastic is distinguished by being biocompatible and is mostly used in MEMS devices such as thermal microactuators [8]. PETG plastic has a higher glass transition temperature (T_g), meaning that a formed microstructure can encounter higher temperatures. In some cases, PETG thermoplastic is used as a template for embossing in plastics with low glass transition temperature [9]. PVC microstructural film can be used as an anti-reflection layer on solar cells with a high efficiency [10]. The glass transition temperature of SAN thermoplastic is higher than 100 °C, which makes the material resistant to boiling water [11]. Thus, selected plastics are widely used in thermal printing and have different applicability in MEMS devices.

However, to create high-aspect-ratio microstructures and microchannels using the thermal imprint technique, it is necessary to study the behavior of the materials in the plastic deformation stage. High-aspect-ratio microchannels need to be designed in order to avoid uneven flow of the liquid, to achieve a larger active surface area, to create massive microdevices with parallelization functionality or to achieve higher system throughput. Thus, finite element modeling was used in order to optimize and better understand structural changes in thermoplastics and analysis of the thermomechanical changes throughout the process. It allowed to define the optimal thermal imprint parameters of the plastics (PP, PETG, PVC and SAN) used in the formation of gratings on the surface. This paper covers the numerical simulation and experimental investigations used to find the basic parameters for formation of microstructures on different types of thermoplastics, PP, PETG, PVC and SAN. Evaluation of thermoplastic properties and tensile experiment showed the behavior of PP, PETG, PVC and SAN under stress. Obtained results were used in the interpolation of temperature, stress and strain, i.e., the properties of the plastics were determined by multilinear isotropic hardening using finite element modeling. Further, determined parameters were used in thermal imprint procedures to form qualitative microstructures of four different thermoplastics. Surface properties were evaluated using scanning electron microscopy (SEM) and atomic force microscopy (AFM), and optical properties were analyzed using a He-Ne laser diffractometer.

2. Materials and Methods

Materials. Four thermoplastics, polypropylene (PP), polyethylene terephthalate glycol (PETG), polyvinyl chloride (PVC) and styrene-acrylonitrile (SAN), were used for the investigations. Theoretical material characteristics of PP, PETG, PVC and SAN are given in Table 1, including Young's Modulus, Poisson's Ratio, Bulk Modulus, etc.

Table 1. Properties of PP, PETG, PVC and SAN plastics [12–14].

	PP	PETG	PVC	SAN
Young's Modulus (MPa)	1325.0	2950.0	3275.0	3650.0
Poisson's Ratio	0.43	0.4	0.4	0.4
Bulk Modulus (MPa)	3888.9	4868.0	5458.3	4055.6
Shear Modulus (MPa)	1296.3	1054.3	1169.6	1351.9
Isotropic Secant Coefficient of Thermal Expansion (1/°C)	6.8×10^{-5}	4.3×10^{-5}	3×10^{-5}	6.8×10^{-5}
Tensile Ultimate Strength (MPa)	32.94	67.4	52.0	85.0
Tensile Yield Strength (MPa)	26.1	58.7	54.8	83.4

For analysis of geometrical behavior of formed microstructures using the thermal imprint technique, it is necessary to define additional properties of investigated thermoplastics. Deformations of the elastic part can be described by Young's modulus, but additional data are required at the onset of plastic deformation, i.e., multilinear isotropic hardening.

Tensile testing. Behavior of thermoplastics in the plastic deformation stage can be described in several ways. One of the most popular, the Mooney–Rivlin model, was used in these investigations. So, first of all, experiments on tensile tests of four thermoplastics were performed at room temperature, i.e., analysis of materials' behavior as the temperature reaches the glass transition temperature in order to observe the direct multilinear isotropic hardening.

The tensile experiment was performed with an Instron E10000 test apparatus (Figure 1c). The test machine consists of standard tensile components and an additional thermal chamber (Figure 1c), which was used to create the temperature medium. Standard ISO 527-2 dog bone specimens were made from PP, PETG, PVC and SAN laser-cut sheet plastic. Each thermoplastic was prepared in five blanks that were stretched at different temperatures (Figure 1a,b). All the samples were stretched out (Figure 1d), except the specimens made from SAN (numbered as 3.7–3.10). Since SAN is brittle, after reaching the glass transition temperature, the thermoplastic changed its properties.

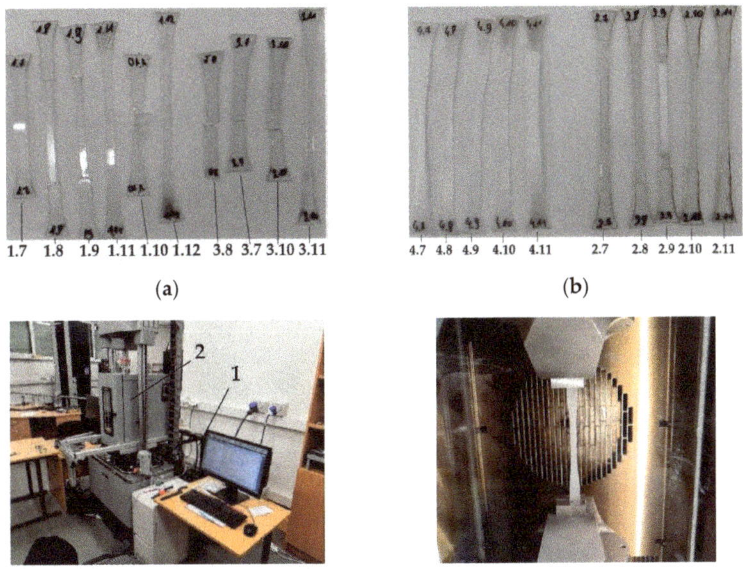

(a) (b) (c) (d)

Figure 1. Tensile experiment. Specimens after the experiment: (**a**) 1.7–1.12—PVC, 3.7–3.11—SAN; (**b**) 4.7–4.11—PP, 2.7–2.11—PETG; (**c**) Testing machine Instron E10000: 1—control computer, 2—heating chamber and testing machine; (**d**) The test sample in the grippers placed in the heating chamber.

Thermal imprint. A master grating for formation of microstructures on analyzed thermoplastics was fabricated by lithography and reactive ion etching technologies, using crystalline silicon material deposited on the nickel base. Thus, a fabricated microstructure consisted of a series of two-dimensional grooves, with defined parameters of 2 μm width, 4 μm periodicity and 1 μm depth (Figure 2a).

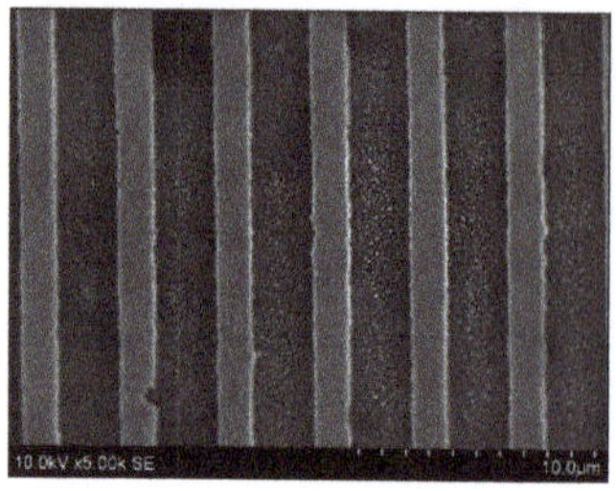

(a)

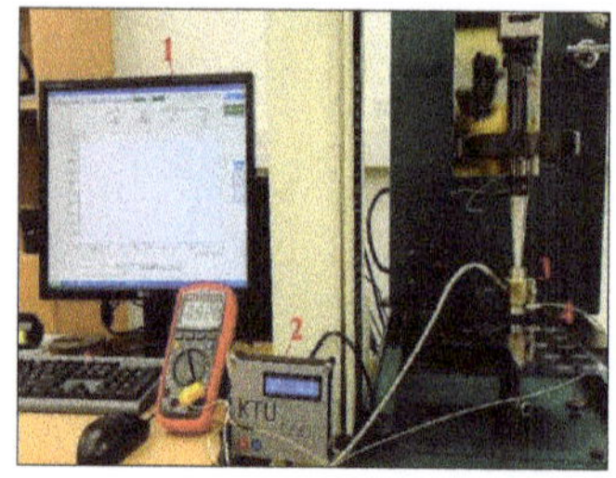

(b)

Figure 2. (**a**) SEM image of the master matrices. (**b**) Tensile testing experimental equipment: 1—a computer that controls the test machine, 2—temperature controller, 3—cathode element, 4—sample.

A 'Tinius Olsen' test machine, together with a punch (in this case, the basis for the microstructure), the heating element and the controller were used for the fabrication of different thermoplastics' microstructures (Figure 2b). Before starting the embossing process, a mold with a thermoplastic plate was first heated to a defined temperature and allowed to stabilize. The force, temperature and time acting on the structure were varied in response to the resulting structure view. Accuracy of the machine was within ±0.5% of the indicated load from 0.2% to 100% of capacity.

Atomic Force Microscope. Surface morphology was investigated using Atomic Force Microscope NT-206 in the static/dynamic mode at 10 μm/s.

Diffractometer. Diffraction efficiency measurements at different peaks (Figure 3a) were performed using a He-Ne laser diffractometer system (Figure 3b) to measure the efficiency at all peaks of the microstructure. Thus, samples were mounted on a rack (Figures 3b and 2) and illuminated with a green laser light of wavelength 532 nm. An Extreme Low Power Laser Detector 11XLP12-3S-H2 (Figure 3b, element 3) was used for the low power μW regime with very low thermal drift and a repeatability of ±0.5%, calibration uncertainty ±2.5% and sensitivity of 200 mV/W. Thus, transmitted light peaks were collected by a photodiode and measured with a Maestro energy monitor (Figures 3b and 4).

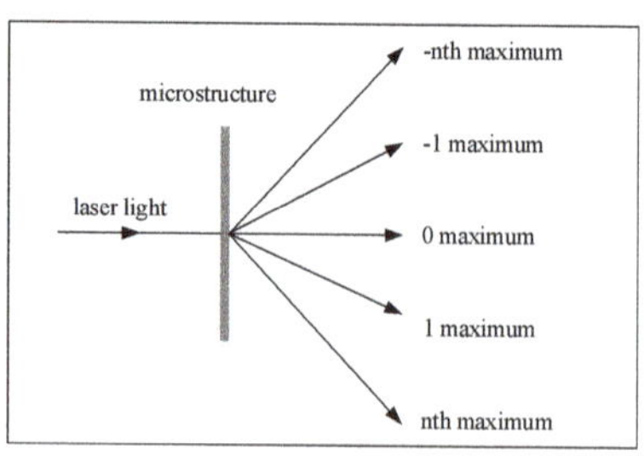

(a)

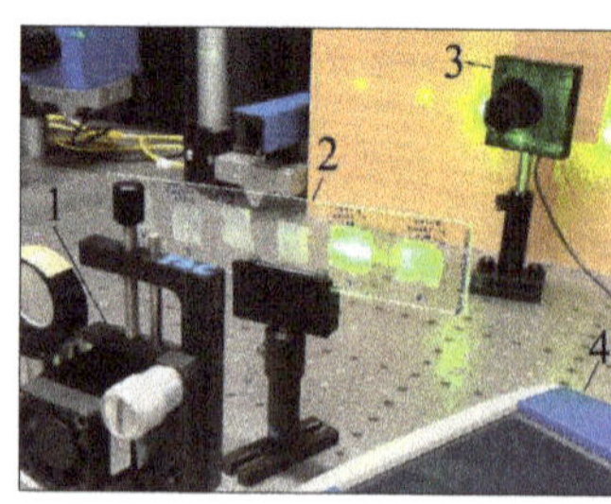

(b)

Figure 3. Diffraction efficiency measurements: (**a**) structural image of the diffraction peaks; (**b**) measurement equipment: 1—laser, 2—grid, 3—photodiode, 4—a visual display of results.

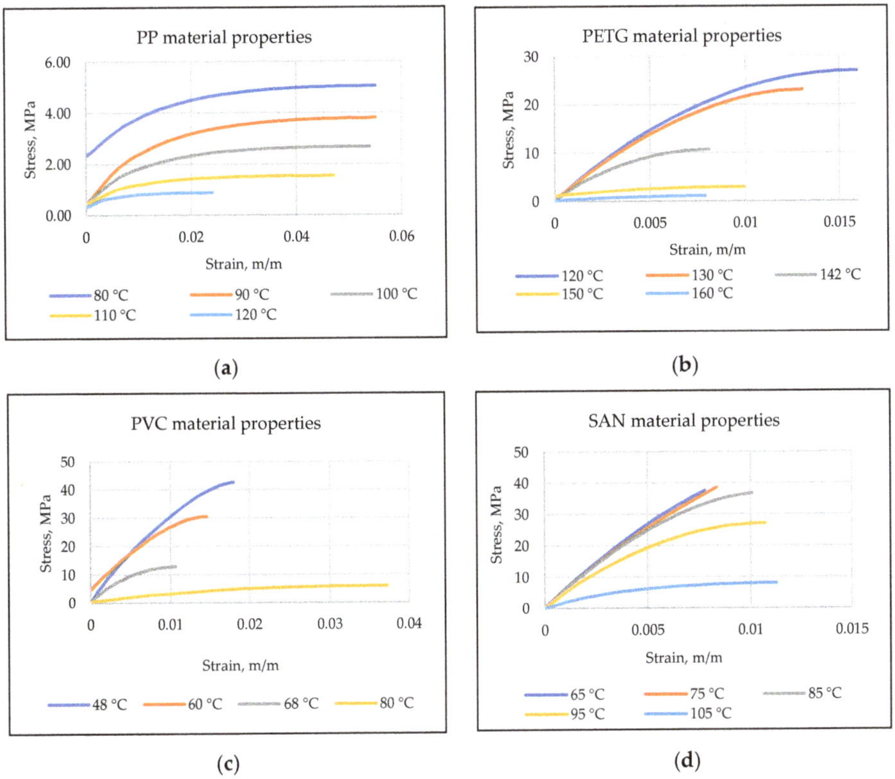

Figure 4. Stress and strain curves of (**a**) PP, (**b**) PETG, (**c**) PVC and (**d**) SAN thermoplastics.

Basically, the efficiency of the microstructure depends on the geometry and the quality of the formed gratings. One of the main parameters evaluating the optical properties of a structure is relative diffraction efficiency (RDE), which can be calculated by the equation:

$$RDE_{i,j} = \frac{P_{i,j}}{\sum_i P_{i,j}} \tag{1}$$

where $RDE_{i,j}$ is the relative diffraction efficiency and $P_{i,j}$ is the light power intensity of the maximum.

3. Results

3.1. *Tensile Testing of Termoplastics*

The specimens were stretched until they were broken or the maximum limits of the test machine were reached. One of the grippers was moving at a constant speed of 20 mm/min. The thermal integrity of the samples was ensured by heating, i.e., the samples were allowed to heat up and then to stabilize as the temperature in the chamber changed. Registered experimental stress–strain curves for different thermoplastics are given in Figure 4.

Using obtained stress–strain curves and the tangent lines, passing through the slopes, Young's modulus for each thermoplastic was calculated defining the behavior of the material under the stress and results are given in Table 2. Thus, increasing the temperature lead to decreasing Young's modulus for different plastics.

Table 2. Young's modulus of PP, PETG, PVC and SAN thermoplastics at different temperatures.

PP	**Temperature, °C**	**80**	**90**	**100**	**110**	**120**
	Young's modulus, MPa	420.0	320.0	240.0	152.0	113.3
PETG	**Temperature, °C**	**120**	**130**	**142**	**150**	**160**
	Young's modulus, MPa	3125.0	3000.0	2666.7	333.3	250.0
PVC	**Temperature, °C**	**48**	**60**	**68**	**80**	**90**
	Young's modulus, MPa	3750.0	2916.7	2333.3	187.5	-
SAN	**Temperature, °C**	**65**	**75**	**85**	**95**	**105**
	Young's modulus, MPa	6000.0	5692.3	5142.9	4222.2	1846.1

Experimental results, given in Table 2, showed that, for PP thermoplastic, the temperature range was chosen from 80 to 120 °C and Young's modulus decreased, respectively, with increasing temperature. During the experiment with PETG thermoplastic, because of the temperature change, PETG suddenly changed its color and became matte (matting appeared at 142 °C when the thermoplastic began to stretch). Temperature limits for PVC thermoplastic were from 48 °C to 90 °C, however, at 90 °C, it did not have enough acceleration to exceed the stress limit, i.e., PVC thermoplastic relaxed more quickly due to temperature than stresses formed. Finally, for SAN, the temperature range was chosen from 65 °C to 105 °C, thus, Young's modulus decreased with increasing temperature until the sample broke. Working temperature ranges were chosen according to the thermoplastics' theoretical thermal properties.

Thus, experimental results show that temperature changes have a significant influence on Young's modulus, i.e., decreasing with increasing temperature. These results can be interpolated to find intermediate Young's modulus values.

3.2. Numerical Simulation of Thermal Imprint

Finite element modeling was used to investigate the geometrical behavior during the thermal imprint of PP, PETG, PVC and SAN. The designed model analyzed the hot stamping technology at the micro-level using the parameters defined during the tensile testing (given in Table 2).

In the numerical simulation, parameters of the fabricated nickel master grating with two-dimensional grooves of 2 μm width, 4 μm periodicity and 1 μm depth (Figure 5) were used.

(a)

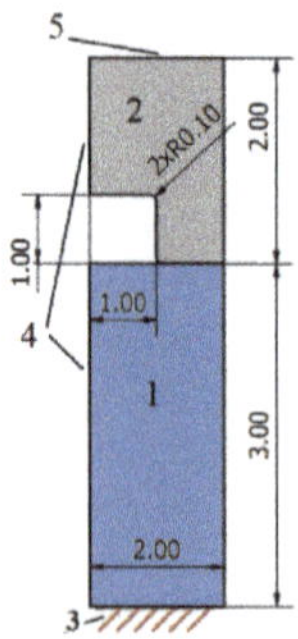

(b)

Figure 5. (**a**) Template of nickel master grating plate; (**b**) drawing of one master-grating element with boundary conditions: 1—substrate, 2—stamp, 3—fixed support, 4—symmetry region and 5—an external force.

Further, ANSYS software was used for the finite element model and the geometry of the microstructure with a mold (or master grating) (Figure 6a) was simplified to half of single element of master grating, and defined under boundary conditions (Figure 6b). Application of the Mooney–Rivlin model, when the glass transition temperature (T_g) was reached or exceeded, was used to analyze and simulate the thermoplastic properties of PP, PETG, PVC and SAN obtained during the experimental research. Defined properties were used in the interpolation of the temperature, stress and strain. Thus, in the finite element model, the properties of the materials were determined by multilinear isotropic hardening.

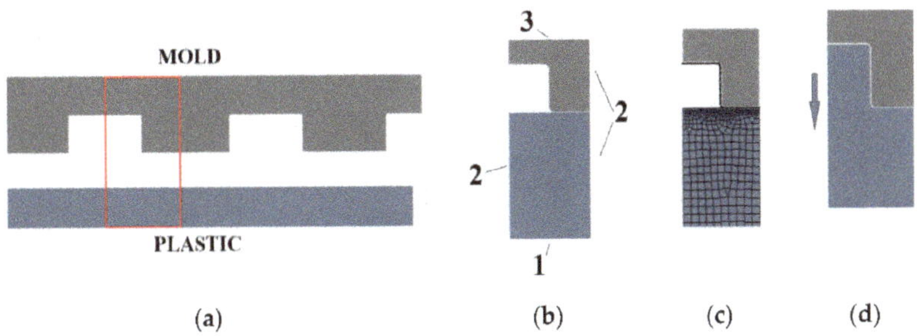

Figure 6. Simplified microstructure. (**a**) Cross-sectional shape of the mold and thermoplastic substrate; (**b**) analyzed model: 1—fixed support, 2—frictionless support and 3—remote displacement; (**c**) meshing of the model; (**d**) deformed analyzed model.

The created mathematical model consists of a flexible thermoplastic (in stiffness) and a non-deformable mold. It was described as two-dimensional and the parameters were defined as 1 mm long. Frictional contact with a 0.2 friction coefficient was used between thermoplastic and metal. Thus, the coefficient of friction, due to plastic deformations and constraints, is not significant in these investigations [15]. The model was divided into 0.2 μm elements, but in the contact areas the scale was reduced to 0.05 μm (Figure 6c). In the simulation, the plastic base was rigidly fixed. Frictionless support was used to assess the integrity of the model. The mold moved toward the plastic in automatic steps up to 2.25 μm and then retracted (Figure 6d). A nonlinear adaptive region was used to modify the grid during formation in order to obtain more accurate results. Output data for total deformation, equivalent elastic strain, equivalent stress and reaction were selected to obtain the plot of stress and strain (Figure 7).

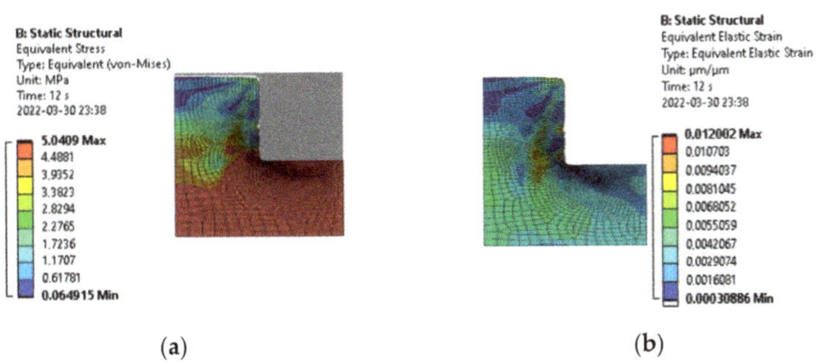

Figure 7. Stress and strain distribution: (**a**) Equivalent stress legend and stresses at maximum strain; (**b**) residual stress image with the equivalent elastic strain legend.

The grid-independent verification study showed that the influence of the grid is not significant in these calculations. The results of strain, stress and reaction did not significantly change after grid compaction. The mesh can be of coarse quality to optimize the calculations (Table 3).

Table 3. The grid-independent verification.

	Elements	Strain, μm/μm	Stress, MPa	Reaction, μN
Coarse	1533	1.2107×10^{-2}	5.0409	16,655
Medium	2873	1.2383×10^{-2}	5.2205	16,701
Fine	5129	1.2089×10^{-2}	5.2511	16,647

After all simulations, the maximum stresses, strains and reaction forces were found (Table 4). The reaction force was calculated for an area of 2000 μm^2. To maintain the same pressure, the total force required for a full mold embossing must be increased, respectively.

Table 4. Simulation results of thermoplastics' behavior during thermal imprint.

PP	**Temperature, °C**	**80**	**90**	**100**	**110**	**120**
	Stress, MPa	5.0409	3.793	2.668	1.5368	0.8648
	Strain, μm/μm	0.012	0.011	0.011	0.010	0.0076
	Reaction force, μN MAX	16,655	12,458	8736	5024.9	2804.8
PETG	**Temperature, °C**	**120**	**130**	**142**	**150**	**160**
	Stress, MPa	26.923	22.966	10.703	2.927	1.0588
	Strain, μm/μm	0.0086	0.00765	0.00401	0.00878	0.00423
	Reaction force, μN MAX	87,723	74,711	35,215	9585	3454.8
PVC	**Temperature, °C**	**48**	**60**	**68**	**80**	**90**
	Stress, MPa	42.387	30.351	12.75	5.793	-
	Strain, μm/μm	0.0113	0.0104	0.00546	0.0308	-
	Reaction force, μN MAX	138,500	99,140	41,886	18,974	-
SAN	**Temperature, °C**	**65**	**75**	**85**	**95**	**105**
	Stress, MPa	37.457	38.379	36.66	26.987	8.0495
	Strain, μm/μm	0.0062	0.0067	0.0071	0.00639	0.00436
	Reaction force, μN MAX	125,080	126,100	119,700	88,337	26,436

Thus, the simulation result (Table 4) proved the relevance of the finite element model because the maximum stresses reached the maximum described limits in all cases. Thus, the reaction force decreased steadily with decreasing stresses. Further, a created finite element model will be applied in the formation of high-aspect-ratio microstructures, i.e., applying theoretically determined forces during the thermal imprint process would not only reduce the fabrication time but also would highly increase the quality of the formed structures. Knowing the properties of the material helps to find out what reaction force is needed to perform thermal imprint. So, the simplification of the model optimizes the calculation time and allows to obtain the highest quality and repeatability of the deformable plastics during the thermal imprint procedure.

3.3. *Thermal Imprint Process Based on Finite Element Modeling Data*

To prove the relevance of the created finite element model, fabrication of PP, PETG, PVC and SAN microstructures, using defined theoretical parameters, was performed. A numerical simulation allowed to determine the exact value required for the formation of a microstructure during thermal imprint technology. In the experiment, a 'Tinius Olsen' test machine (Figure 2b) with a nickel (Ni) master grating (Figure 2a) was used. By varying the force, temperature and time during the process, microstructures in thermoplastics were imprinted (Figure 8a–d).

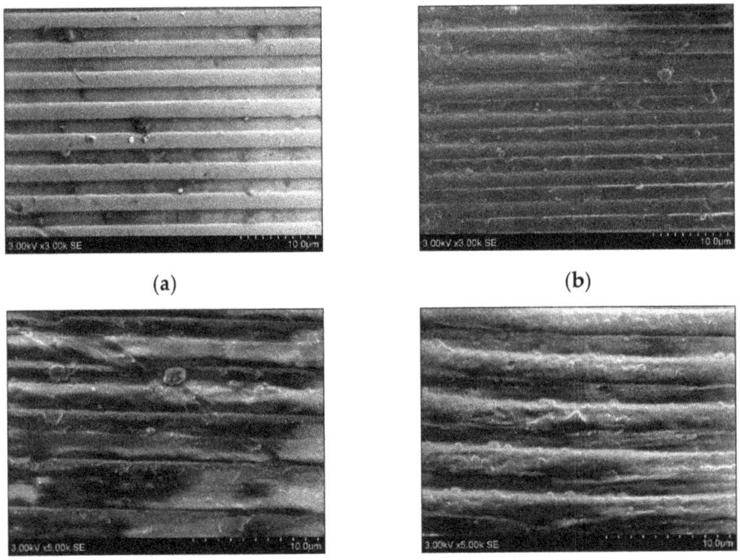

(a) (b) (c) (d)

Figure 8. Formation of microstructures: (**a**) microstructure formed in PETG; (**b**) microstructure formed in PP; (**c**) microstructure formed in PVC; (**d**) microstructure formed in SAN.

Thus, microstructures were embossed in all four thermoplastics—PP, PETG, PVC and SAN—and their 3D views were made using a scanning electron microscope (SEM) (Figure 8). Thus, PETG and PP showed higher-quality reproduction and smoother surfaces than PVC and SAN thermoplastic microstructures.

3.3.1. Surface Morphology of Imprinted Microstructures

Surface view and geometry of the microstructures' topographic profiles were examined using an atomic force microscope. Best results were obtained of the microstructure formed in PETG thermoplastic (Figure 9a,b): a rather smooth surface with the average surface roughness Ra = 174.6 nm and an average grating depth of 400 $\pm$ 20 nm. The microstructure from PP thermoplastic had a smooth surface relief of average surface roughness Ra = 340.2 nm with a geometry similar to master grating with an average grating depth of 330 $\pm$ 30 nm (Figure 9c,d). PVC and SAN microstructures had many defects on the surface with an average roughness of Ra = 437.6 nm and Ra = 298.7 nm, respectively (Figure 9e,g). From profile views, it is seen that the geometry of the gratings is uneven in depth, width and the form itself (Figure 9f,h), compared to parameters of nickel master grating (Figure 9i,j).

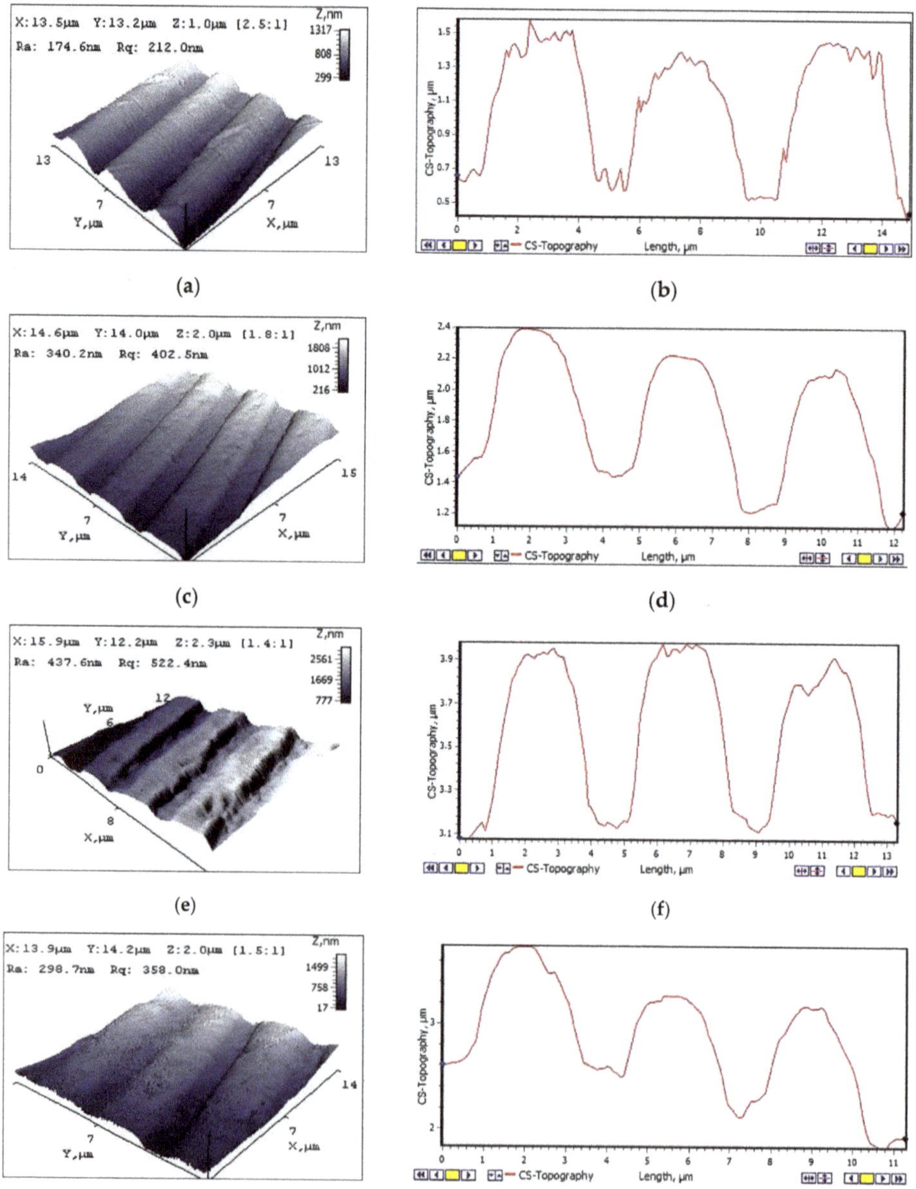

(a) (b)

(c) (d)

(e) (f)

(g) (h)

Figure 9. *Cont.*

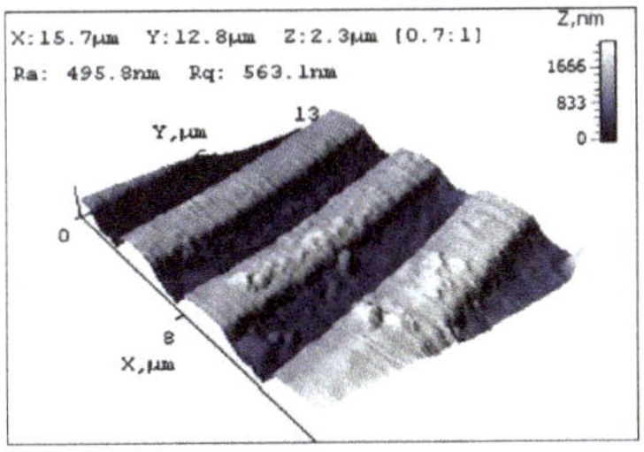

(i)

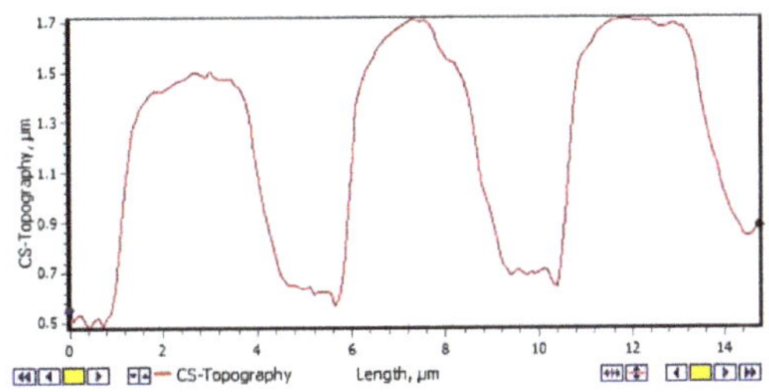

(j)

Figure 9. AFM topography images and surface profiles of embossed microstructures in thermoplastics with topography: (**a**,**b**) PETG; (**c**,**d**) PP; (**e**,**f**) PVC; (**g**,**h**) SAN; and (**i**,**j**) a nickel master grating.

Thus, results of SEM and AFM measurements imply that the most qualitative microstructures were obtained from PP and PETG thermoplastics. This might influence not only on the thermal imprint procedure, but also on the thermal properties of the plastics themselves. Thus, PVC and SAN are more brittle materials compared to PP and PETG, and the results of the imprinted microstructures were less accurate.

3.3.2. Optical Properties of Imprinted Microstructures

The quality of microstructures is defined not only by geometrical parameters of the formed gratings, but also by optical properties. In this paper, diffraction efficiency measurements in different peaks were performed using a He-Ne laser diffractometer. Due to the non-optical nature, the PP material was not examined. So, only diffraction efficiencies of PETG, PVC and SAN microstructures were measured and evaluated.

Measurements of diffraction efficiency were performed for 12 samples of each thermoplastic, fabricated at different embossing conditions (load, embossing time and temperature) (Table 5). Thus, the best relative diffraction efficiency RDE = 34.62% was observed in the PETG microstructure, fabricated at the following embossing conditions: load of 2000 N for 10 s at 125 °C (Table 5). For the SAN microstructure, the best values of RDE = 29.04% were obtained when embossing parameters were 5000 N for 10 s at 130 °C. The PVC microstructure showed best results of RDE = 22.44% when a load of 5000 N was applied for 10 s at 80 °C during a thermal imprint procedure. Applying theoretical calculations and experimental maximum diffraction efficiency of plastics, the grating depth of the plastics were obtained: SAN—0.62 µm, PETG—0.57 µm and PVC—0.72 µm. The relative error between the measured spectral diffraction efficiencies using the diffractometer is 1.24% in the case of SAN, 1.13% in the case of PETG and for PVC the relative error is 1.57%. The load and time are constant values during the hot imprint process.

Table 5. Diffraction efficiency measurement results.

SAN				PETG				PVC			
Load (N)	Time (Sec)	Temp. (°C)	RDE (%)	Load (N)	Time (Sec)	Temp. (°C)	RDE (%)	Load (N)	Time (Sec)	Temp. (°C)	RDE (%)
5000	10	100	27.07	4000	10	100	24.31	5000	10	100	5.39
5000	10	120	26.49	5000	5	100	31.43	5000	5	100	12.63
5000	15	120	24.94	5000	10	100	30.50	4000	10	100	7.28
5000	10	140	22.52	4000	5	100	23.49	4000	5	100	11.59
4000	10	140	23.28	**2000**	**10**	**125**	**34.62**	3000	10	100	5.43
4000	5	140	17.85	2000	10	100	28.81	3000	5	100	11.66
3000	10	140	28.12	2000	5	100	32.31	4000	10	125	10.44
3000	5	140	26.85	3000	10	100	30.67	4000	5	125	7.38
3000	10	130	28.23	2000	10	90	24.85	5000	5	90	6.03
5000	2	130	27.76	2000	5	90	29.86	**5000**	**10**	**80**	**22.44**
5000	**10**	**130**	**29.04**	2000	10	80	23.54	2000	10	125	6.74
4000	10	130	25.25					2000	5	125	6.65

To prove the relevance of the experimental results, theoretical diffraction efficiencies of microstructures were calculated. The following refractive indexes of plastics were used for the calculations: SAN—1.572 [16], PETG—1.57 [17] and PVC—1.531 [18]. The direct influence of the grating depth on diffraction efficiency was determined. Thus, results implied that the grating depth can be determined from the diffraction efficiencies using the theoretical calculations (Figure 10).

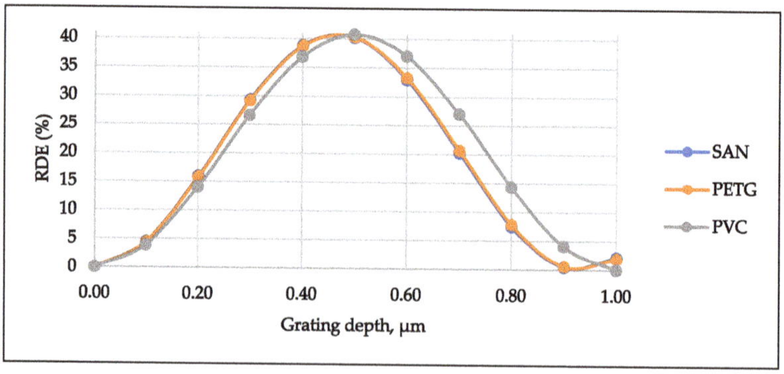

Figure 10. Dependence of the theoretical diffraction efficiency of plastics on the grating depth.

It may be concluded that the diffraction efficiency of all microstructures (PETG, PVC and SAN) was higher when the temperature exceeded the plastic glass transition temperature, i.e., exceeding the glass transition temperature impairs the properties of thermoplastics.

4. Discussion

Microfluidics requires high-aspect-ratio microchannels to be designed in order to avoid uneven flow of the liquid, to achieve a larger active surface area, to create a massive MEMS device with parallelization functionality or to achieve higher system throughput [19,20]. Using the thermal imprint procedure, there are some important factors which should be evaluated in advance, and final element modeling is a great way to do it. The advantages of the finite element model give the opportunity to change the geometry, i.e., in this case, it is much easier to find the parameters required for forming qualitatively replicated structures by the hot imprint procedure. Thus, the results confirmed that is better to form the microstructure when the temperature is near the glass transition temperature and when the plastic can flow slowly into the mold. Time, here, also plays an important role leading to better surface morphology and optical properties.

5. Conclusions

The behavior of thermoplastics can be simulated using multilinear isotropic hardening and applying the created finite element model. Using the results of the tensile experiment, successfully interpolated stress and strain distributions at different temperatures of PP, PETG, PVC and SAN thermoplastics were obtained.

Using the numerical simulations, optimal working parameters, stress and strain distribution were defined and applied in the thermal imprint process for formation of microstructures. AFM results showed that the PETG thermoplastic microstructure had the best surface properties compared to other PP, PVC and SAN microstructures. Its optical properties were proved with diffraction efficiency measurements, which showed the best quality of the PETG microstructure under the following conditions: grating formation under load of 2000 N for 10 s at 125 °C. Thus, the best diffraction efficiency measurement results for the PVC microstructure were observed when it was formed at 5000 N for 10 s at 80 °C. For the SAN microstructure, the best results were obtained when it was formed under a load of 5000 N for 10 s at 130 °C.

The created finite element model is a useful tool for fabrication of high-aspect-ratio microstructures with great surface morphological and optical properties. The model may easily help to control the thermal imprint process to ensure the quality of microstructures when slow formation occurs.

Author Contributions: Conceptualization, J.C.; methodology, J.C. and G.J.; investigation, J.C. and P.G.; calculation and analysis, J.C. and P.G.; writing—original draft preparation, J.C., S.U. and A.P.; writing—review and editing, S.U. and A.P.; visualization, S.U.; validation, G.J. All authors have read and agreed to the published version of the manuscript.

Funding: This research was funded by grant No. 0.1.2.2-CPVA-K-703-03-0015 "Development of new technology for the formation of microstructures in functional materials" from the European Regional Development Fund.

Conflicts of Interest: The authors declare no conflict of interest.

References

1. Ivanov, A.; Cheng, K. Non-traditional and hybrid processes for micro and nano manufacturing. *Int. J. Adv. Manuf. Technol.* **2019**, *105*, 4481–4482. [CrossRef]
2. Daniel, M.; Stefan, H.; Günter, R.; Felix, S.; Roland, Z. Microfluidic lab-on-a-chip platforms: Requirements, characteristics and applications. *J. Chem. Soc. Rev.* **2010**, *39*, 1153–1182.
3. Fabrication of Nanofluidic Biochips with Nanochannels for Applications in DNA Analysis. Available online: https://www.webofscience.com/wos/woscc/full-record/WOS:000308874900002 (accessed on 18 August 2022).
4. Label-Free Optical Single-Molecule Micro- and Nanosensors. Available online: https://www.webofscience.com/wos/woscc/full-record/WOS:000453926000013 (accessed on 24 August 2022).
5. Choe, S.-W.; Kim, B.; Kim, M. Progress of Microfluidic Continuous Separation Techniques for Micro-/Nanoscale Bioparticles. *Biosensors* **2021**, *11*, 464. [CrossRef] [PubMed]
6. Lee, C.-Y.; Chang, C.-L.; Wang, Y.-N.; Fu, L.-M. Microfluidic Mixing: A Review. *Int. J. Mol. Sci.* **2011**, *12*, 3263–3287. [CrossRef] [PubMed]
7. Heckele, M.; Schomburg, W.K. Review on micro molding of thermoplastic polymers. *J. Micromech. Microeng.* **2004**, *14*, R1–R14. [CrossRef]
8. Wilson, S.A.; Jourdain, R.P.; Zhang, Q.; Dorey, R.A.; Bowen, C.R.; Willander, M.; Wahab, Q.U.; Al-hilli, S.M.; Nur, O.; Quandt, E.; et al. New materials for micro-scale sensors and actuators: An engineering review. *Mater. Sci. Eng. R Rep.* **2007**, *56*, 1–129. [CrossRef]
9. Liu, J.; Jin, X.; Sun, T.; Xu, Z.; Liu, C.; Wang, J.; Chen, L.; Wang, L. Hot embossing of polymer nanochannels using PMMA moulds. *Microsyst. Technol.* **2013**, *19*, 629–634. [CrossRef]

10. Han, K.S.; Lee, H.; Kim, D.; Lee, H. Fabrication of anti-reflection structure on protective layer of solar cells by hot-embossing method. *Sol. Energy Mater. Sol. Cells* **2009**, *93*, 1214–1217. [CrossRef]
11. Wen, G.; An, L. Pressure-dependent glass-transition temperatures of poly(methyl methacrylate)/poly(styrene-*co*-acrylonitrile) blends. *J. Appl. Polym. Sci.* **2003**, *90*, 959–962. [CrossRef]
12. Wang, W.; Zeng, Y. (Eds.) *Polypropylene—Polymerization and Characterization of Mechanical and Thermal Properties*; IntechOpen: London, UK, 2020. [CrossRef]
13. Sepahi, M.T.; Abusalma, H.; Jovanovic, V.; Eisazadeh, H. Mechanical Properties of 3D-Printed Parts Made of Polyethylene Terephthalate Glycol. *J. Mater. Eng. Perform.* **2021**, *30*, 6851–6861. [CrossRef]
14. Thermoplastic Materials: Properties, Manufacturing Methods, and Applications. Available online: https://ebookcentral.proquest.com/lib/ktu-ebooks/detail.action?pq-origsite=primo&docID=1447177 (accessed on 17 August 2022).
15. Pouzada, A.S.; Ferreira, E.; Pontes, A. Friction properties of moulding thermoplastics. *Polym. Test.* **2006**, *25*, 1017–1023. [CrossRef]
16. Dispersion Properties of Optical Polymers. Available online: http://przyrbwn.icm.edu.pl/APP/PDF/116/a116z442.pdf (accessed on 12 August 2022).
17. El-Farahaty, K.; Sadik, A.; Hezma, A. Study of Optical and Structure Properties of Polyester (PET) and Copolyester (PETG) Fibers by Interferometry. *Int. J. Polym. Mater.* **2007**, *56*, 715–728. [CrossRef]
18. Mortazavi, S.H.; Ghoranneviss, M.; Faryadras, S. Effect of Low Pressure Nitrogen DC Plasma on Optical Properties of Biaxial-Oriented Polypropylene (BOPP), Poly Methyl Methacrylate (PMMA) and Poly Vinyl Chloride (PVC) Films. *J. Fusion Energy* **2012**, *31*, 211–215. [CrossRef]
19. Nasseri, B.; Akar, S.; Naseri, E. Chapter 3—Microchannels for microfluidic systems. In *Biomedical Applications of Microfluidic Devices*; Hamblin, M.R., Karimi, M., Eds.; Academic Press: Cambridge, MA, USA, 2021; pp. 37–75.
20. Yi, H.; Zhu, C.; Fu, T.; Ma, Y. Interfacial evolution and dynamics of liquid bridge during droplet coalescence in rectangular microchannels: Effect of aspect ratio. *J. Taiwan Inst. Chem. Eng.* **2021**, *123*, 59–67. [CrossRef]

micromachines

Communication

Fabrication of Transparent and Flexible Digital Microfluidics Devices

Jianchen Cai [1], Jiaxi Jiang [2], Jinyun Jiang [1], Yin Tao [1], Xiang Gao [1], Meiya Ding [1] and Yiqiang Fan [2,*]

1 College of Mechanical Engineering, Quzhou University, Quzhou 324000, China; cai198666@126.com (J.C.); jiangjinyun2@163.com (J.J.); yintaoyy120@gmail.com (Y.T.); a17535873277@163.com (X.G.); ding1600615038@163.com (M.D.)
2 College of Mechanical and Electrical Engineering, Beijing University of Chemical Technology, Beijing 100029, China; xsjiangjiaxi@163.com
* Correspondence: yiqiang.fan50@gmail.com; Tel.: +86-1851-3899-9080

Abstract: This study proposed a fabrication method for thin, film-based, transparent, and flexible digital microfluidic devices. A series of characterizations were also conducted with the fabricated digital microfluidic devices. For the device fabrication, the electrodes were patterned by laser ablation of 220 nm-thick indium tin oxide (ITO) layer on a 175 μm-thick polyethylene terephthalate (PET) substrate. The electrodes were insulated with a layer of 12 μm-thick polyethylene (PE) film as the dielectric layer, and finally, a surface treatment was conducted on PE film in order to enhance the hydrophobicity. The whole digital microfluidic device has a total thickness of less than 200 μm and is nearly transparent in the visible range. The droplet manipulation with the proposed digital microfluidic device was also achieved. In addition, a series of characterization studies were conducted as follows: the contact angles under different driving voltages, the leakage current density across the patterned electrodes, and the minimum driving voltage with different control algorithms and droplet volume were measured and discussed. The UV–VIS spectrum of the proposed digital microfluidic devices was also provided in order to verify the transparency of the fabricated device. Compared with conventional methods for the fabrication of digital microfluidic devices, which usually have opaque metal/carbon electrodes, the proposed transparent and flexible digital microfluidics could have significant advantages for the observation of the droplets on the digital microfluidic device, especially for colorimetric analysis using the digital microfluidic approach.

Keywords: digital microfluidics; lab-on-a-chip; ITO; PET

Citation: Cai, J.; Jiang, J.; Jiang, J.; Tao, Y.; Gao, X.; Ding, M.; Fan, Y. Fabrication of Transparent and Flexible Digital Microfluidics Devices. *Micromachines* **2022**, *13*, 498. https://doi.org/10.3390/mi13040498

Academic Editor: Pingan Zhu

Received: 4 March 2022
Accepted: 22 March 2022
Published: 23 March 2022

1. Introduction

Compared with the conventional microfluidic devices that continuously handle the fluid flow inside of the microchannel on polymer- or silicon/glass-based microfluidic chips, the digital microfluidic chips (DMF) can precisely manipulate the discrete fluid flow (i.e., droplet) without the requirement of physical pumps, valves, or complex microchannel structures. Each droplet in the digital microfluidic approach is an isolated reaction chamber containing reagents. The precise handling of the droplet down to picoliter can be achieved with digital microfluidic devices [1]. Digital microfluidics has been widely used in various applications, such as cell manipulation [2], virus detection [3], nucleic acid amplification [4], and ion detection [5]. Several commercial digital microfluidic products were also put on the optical and biomedical market in the past few years [6].

Various droplet driving methods have been used in digital microfluidics, including magnetic [7,8], acoustic [9], gravitational [10], and electrowetting-on-dielectric (EWOD) [11]. Currently, the EWOD is the most widely used approach for digital microfluidics with the advantage of precise control of the droplet movement. The electrowetting induces the interfacial tension gradient inside of the droplet in order to trigger the droplet movement.

Multiple functions can be achieved with electrowetting control, such as droplet dispensing, translocation, merging, and splitting.

The digital microfluidic devices with single-plate configuration usually consist of the substrate, electrodes, insulation dielectric layer, and the hydrophobic layer on the top surface of the devices. The substrate of the digital microfluidic device is of decisive significance for digital microfluidic devices. The substrate is usually made of glass or silicon material, with the electrodes (e.g., chromium, chromium, gold) fabricated with photolithography method [12,13], the dielectric layer is commonly fabricated with chemical vapor deposition of parlance or silicon nitride [14]. The glass/silicon-based digital microfluidics have high electrode patterns precision but are costly on materials and processing instruments. On the other hand, various low-cost approaches for digital microfluidics have also been invented, as follows: electrodes were fabricated on PCB (printed circuit board) [15]; screen-printing was used for the fabrication of carbon electrodes on thin polymer films or paper as substrate [16]; xurography or even hand-painting were also used for the low-cost digital microfluidic approach [17].

In this study, a thin, transparent, and flexible fabrication approach is proposed for digital microfluidic devices. The commercially available ITO (indium tin oxide) coated PET (polyethylene terephthalate) film was used as the substrate. Additionally, the Nd: YLF laser was used for the patterning of the conductive ITO layer on PET substrate in order to form electrodes. Finally, a thin PE (polyethylene) film was used as the dielectric layer with hydrophobic surface treatment by Rain-X water repellent.

Compared with the previous studies on digital microfluidics using the photolithography method for patterning electrodes on silicon/glass substrate, the fabricated digital microfluidic device has a total thickness of less than 200 μm and is flexible and transparent in the visible range. On the aspect of fabrication and cost, the proposed method is much more rapid, with less requirement on the highly sophisticated instrument, also with a lower cost of materials. The proposed fabrication approach for digital microfluidic devices provides a new alternative to the conventional fabrication method for digital microfluidic devices, with the advantage of transparency in the visible range, and may have significant importance for observing the colorimetric change in the droplets in some applications.

2. Materials and Methods

2.1. Materials and Instruments

The ITO-coated PET film was obtained from MSE supplies, Arizona, AZ, USA, the PET film had a thickness of 175 μm and was covered with an ITO layer of around 220 nm in thickness. The measured resistance of the ITO layer ranged from 5.6 to 6.1 Ohm/sq. The polyethylene film (Glad cling wrap, W300N, Clorox China Limited, Guangzhou, China) was used as a dielectric (insulation) layer, with a thickness of around 12 μm. Rain-X water repellent, which was used for surface treatment, was obtained from ITW GlOBAL Brands, San Luis Obispo, CA, USA. Sodium chloride was sourced from Shanghai Aladdin Biochemical Technology Co., Shanghai, China. All materials and chemicals were used as received.

Several fabrication and testing instruments were used in this study. Nd: YLF laser (LSF20D, HGTECH, Wuhan, China), with a wavelength of 1064 nm, was used for the patterning of the ITO layer on PET substrate. The laser worked at the multi-plus overlapped mode and the pulse duration was 10 ps with the repetition rate adjustable from 0 to 200 kHz. The oscilloscope used in this study was Tektronix TDS1012B (Tektronix, Inc., Beaverton, OR, USA), and the function generator (UNI-T UTG9005C) was sourced from Uni-Trend Technology Co., Ltd., Dongguan, China. ATA-2161 high-voltage amplifier is capable of amplifying the AC/DC signal with gain adjustable from 0 to 240, up to 1600 Vpp. To control the voltage supply to each of the electrodes in the proposed digital microfluidic device, a control circuit with STC 8051 microcontroller (STCmicro Technology Co., Ltd., Beijing, China) was also designed and assembled. Olympus ols5000 laser scanning confocal microscope (Olympus Corporation, Tokyo, Japan) was used to observe and measure the laser fabricated electrodes. The optical images of the system setup and fabricated devices were taken with a Nikon D3000 Digital SLR Camera (Nikon Corporation, Tokyo, Japan).

2.2. Fabrication

The fabrication process of the proposed digital microfluidic device is shown in Figure 1a, the PET substrate (175 μm in thickness) covered with ITO layer (~220 nm in thickness) was selectively laser-ablated for the fabrication of electrodes array. It is worth pointing out that, other than the laser ablation that is usually strong enough to directly evaporate the materials, the role of the Nd: YLF laser in this study was to induce the thermal stress between the ITO layer and PET substrate, which finally caused the ITO film fracture and ejection from the PET substrate. The following section will discuss more details of the laser ablation process.

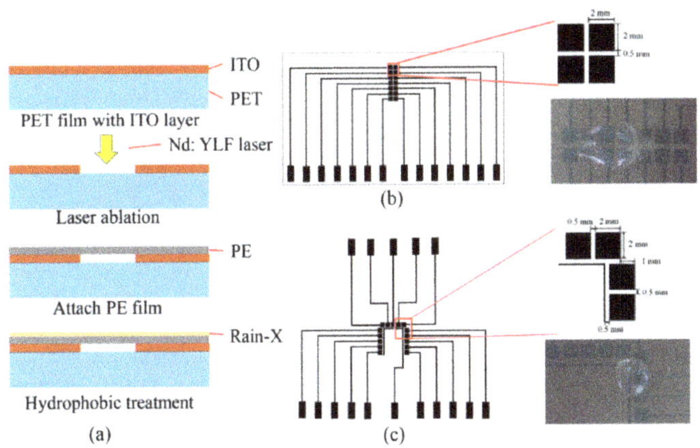

Figure 1. The design and fabrication process of the proposed digital microfluidic devices. (**a**): The fabrication process of the digital microfluidic device, a pulsed Nd: YAG laser was used to pattern the ITO layer on PET substrate to form electrodes array, then the ITO layer was insulated with a layer of PE film, finally, Rain-X water repellent was applied on the surface of PE film to enhance the hydrophobicity. (**b**): The digital microfluidic device for droplets merging. (**c**): The digital microfluidic device for droplets moving straight and turns. Video footages for the two proposed digital microfluidic devices are provided in the Supplementary Files.

After laser ablation to fabricate electrodes on PET substrate, a layer of 12 μm-thick PE film (i.e., cling wrap) was used to insulate the electrodes. The PE film was cut in the same size as the PET substrate and gently attached by hand. The "bonding" between the PE film and PET substrate was achieved with the help of electrostatic charge as follows: by unrolling the PE film, the PE film becomes charged by losing or gaining electrons, when in touch with another insulator (i.e., PET substrate), the electrostatic charge induces an opposite charge and bonds the two layers.

Two types of devices were designed and fabricated for the demonstration in this study, as shown in Figure 1b,c. Figure 1b shows the design and fabricated digital microfluidic device for single droplet manipulation and two droplets merging. Figure 1c shows another electrode configuration that enables droplets to move straight and turn. The electrodes in Figure 1b,c have the same dimension of 2 mm by 2 mm with 0.5 mm spacing. The Supplementary Files also provide video footage for the droplet movement in Figure 1b,c.

The ITO-based electrodes on PET substrate were defined by laser ablation. The fabrication process of the electrodes is shown in Figure 2a. Compared with other laser ablation methods (e.g., CO_2 laser) that directly meltdown and vaporize the material on the radiated spot, the Nd: YLF laser only induces the tensile stress at the laser radiated spot. When the tensile stress strength is exceeded at the laser radiated spot, the ITO film will fracture and eject from the PET substrate at the laser radiated spot [18], and finally forms the desired ITO patterns on the PET substrate.

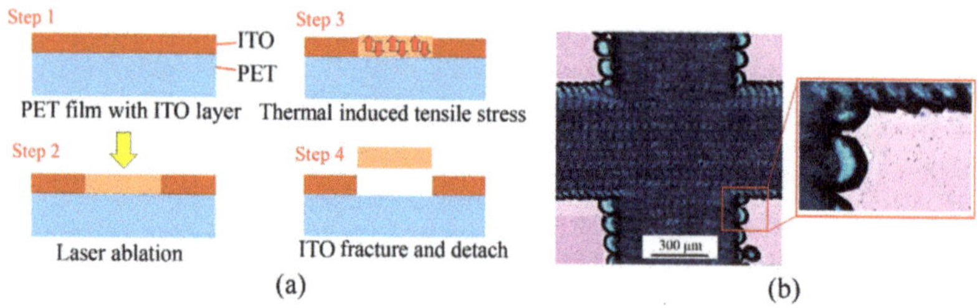

Figure 2. Laser ablation process on the ITO-covered PET substrate. (**a**): Laser fabrication process for patterning the ITO electrodes on PET substrate. (**b**): Optical microscope image of the laser-ablated areas on the PET substrate.

The optical image of the patterned ITO electrodes on PET substrate is shown in Figure 2b, the relatively dark area was laser scanned and the covered ITO material has been removed, the bright area is the unaffected part that the ITO layer, and PET substrate was still firmly attached. In order to effectively remove the ITO layer without damaging the PET substrate, an overlap laser scan method was used in this study. During the laser scanning process, an overlap rate of 90% was used between the adjacent pulses (two adjacent laser radiated spots had an overlap of 90%).

During laser ablation, the PET substrate unavoidably received the thermal energy, which may cause the meltdown and solidification of the PET material along the laser-scanned route, the visible trace can be found along the laser-scanned route, as shown in Figure 2b. To minimize the thermal damage to the PET substrate during laser ablation, the pulse energy was set at 1 J/cm^2 in this study, after several trials. The repeatability of the proposed patterning method was relatively reliable, the deviation of 8 repeated experiments was less than 10%.

The system setup for the transparent and flexible digital microfluidic device is shown in Figure 3. As shown in Figure 3, a function generator was used to generate a sinusoidal signal with a frequency of 1 kHz, then the generated signal was further amplified with a high-voltage amplifier. The final output voltage was adjustable from 0 to 500 V_{rms} in this study.

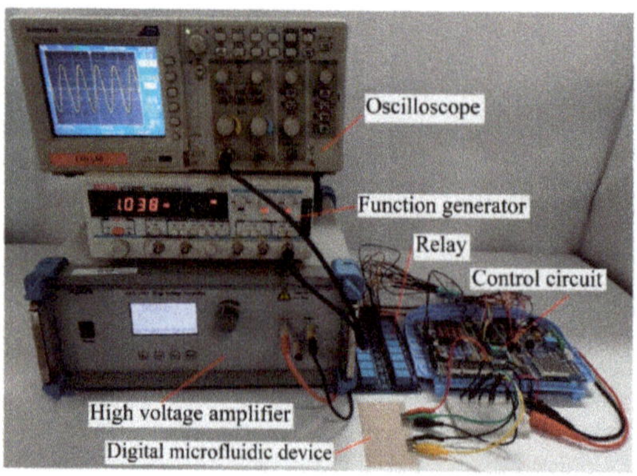

Figure 3. Testing system setup for the digital microfluidic device. The function generator provides the AC signal and is amplified with a high-voltage amplifier. A control circuit with microcontroller was used to control the AC voltage supply to each electrode on the digital microfluidic device.

To independently control the voltage supply to each electrode, a control circuit with an STC 80C51 microcontroller was used to control a series of relays (blue blocks in Figure 3) that can independently control the voltage supply to each of the ITO electrodes. The control circuit is also able to adjust the gain of the high-voltage amplifier that finally controls the voltage output. An oscilloscope was used to monitor the signal output from the function generator.

3. Result and Discussion

3.1. Surface Properties

The fabricated digital microfluidic device is shown in Figure 4, the device has a total thickness of less than 200 μm. The device is flexible (bendable) and transparent in the visual range. The insert shows the enlarged image of the laser-ablated ITO electrodes on the PET substrate.

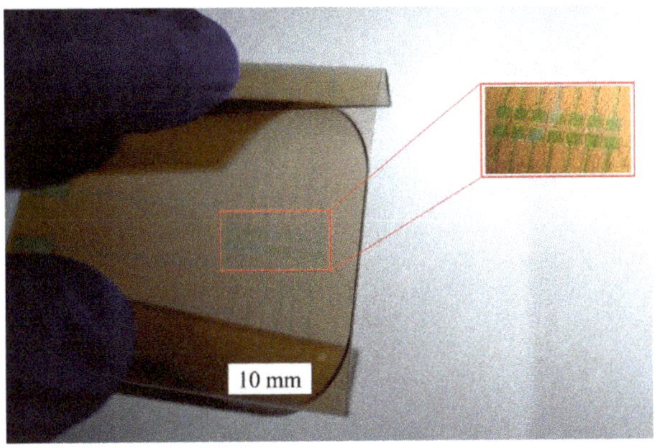

Figure 4. The fabricated transparent and flexible digital microfluidic device.

For easier manipulation of the droplets, it is necessary to enhance the hydrophobicity on the surface of the PE film, as illustrated in Figure 1a, a layer of commercial water repellent Rain-X was sprayed on the surface of the PE film and was allowed to air-dry at room temperature in order to increase the water contact angle from 78.3° to 103.2° (as shown in Figure 5). The device fabrication is completed after surface treatment.

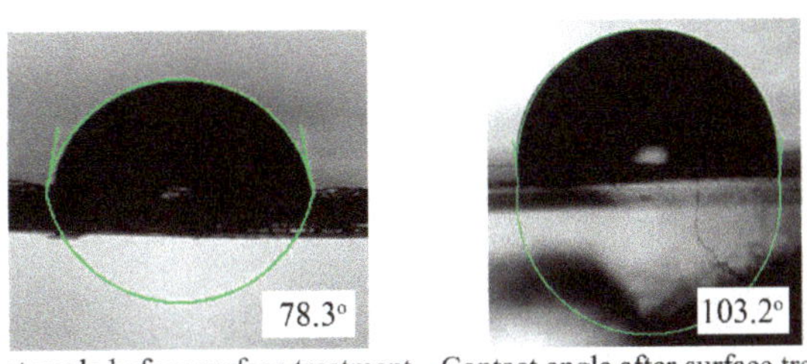

Figure 5. Contact angle measurement on the PE film before and after Rain-X surface treatment.

A laser scanning confocal microscope was used to explore the profile of the laser fabricated ITO electrodes on the PET substrate. As shown in Figure 6, the ITO electrodes are about 500 nm higher than the PET substrate (the original ITO layer had a thickness of 220 nm), which indicates that some of the PET material may have melted and been vaporized during the laser scan process. Another interesting finding is the formation of the bugles at the edge of the ITO patterns, during the laser ablation process, the material on the laser radiated spot meltdown and vaporize, and some of the PET material was ejected and re-solidified on the edge of the scan route. Such a bulge forming phenomenon after laser ablation is also commonly reported in the CO_2 laser ablation on thermoplastics [19]. In Figure 6, the inset on the left is the image captured by an optical microscope at the gap between the electrodes, the insert on the right illustrates the formation principle of the bulges.

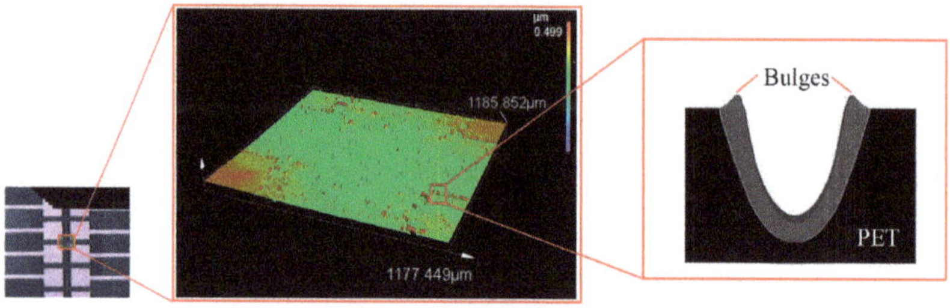

Figure 6. Laser confocal microscope measurement of the laser ablated surface.

The UV–VIS spectrums of the ITO-covered PET substrate and the PET substrate after laser ablation (covered ITO layer has been removed) are shown in Figure 7 (measured with UV-2600i UV–VIS Spectrophotometer, Shimadzu, Kyoto, Japan). As shown in Figure 7, the absorption of the light in the visible range (380–750 nm) is relatively low, indicating the relatively good transparency of the proposed digital microfluidic device.

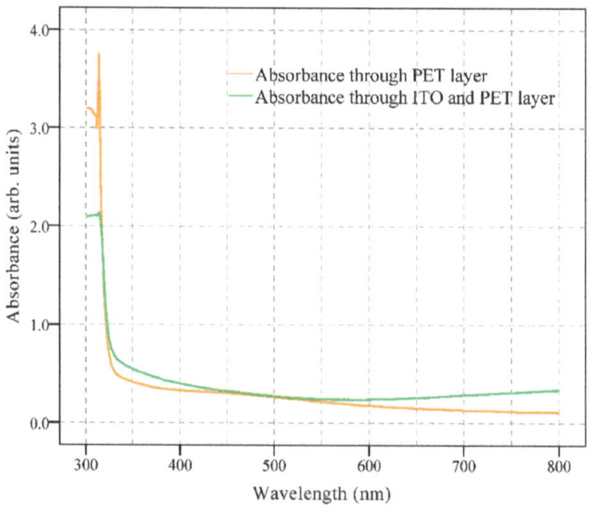

Figure 7. UV–VIS measurement of the PET substrate after laser ablation and ITO-covered PET substrate.

3.2. Leakage Current and Droplet Control

The leakage current density across the proposed digital microfluidic device was measured with the AC voltage supply, ranging from 250 V_{rms} to 450 V_{rms}. As shown in the inset of Figure 8, two electrodes were placed on the Rain-X-covered PE film and the bottom of the PET substrate and the current was measured five times with the voltage increased from 250 V_{rms} to 450 V_{rms} with an increment of 25 V. The error bar in Figure 8 is the statistical representation of the measured leakage current density variability after five measurements at each data point. The leakage current density increased with a higher voltage supply. Generally, the leakage current is ignorable within the operation range of the proposed digital microfluidic device. The low leakage current also indicates the thin PE film is acceptable as the dielectric layer.

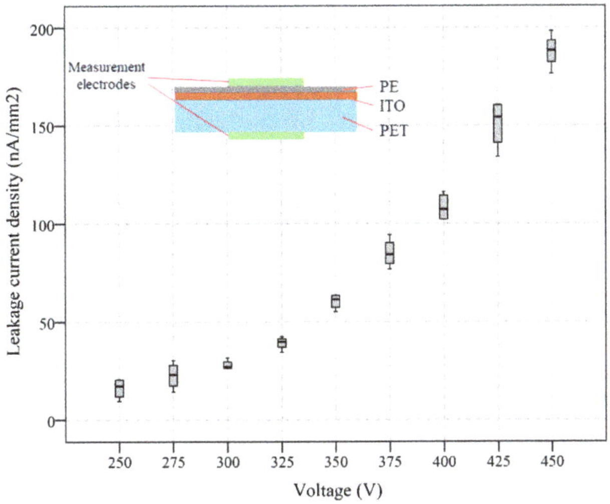

Figure 8. Leakage current density across the fabricated digital microfluidic device.

The measured contact angles under different AC voltage (with a frequency of 1 kHz, droplet volume of 10 μL) with three different solutions are shown in Figure 9a. The DI water, 0.1 M, and 1 M sodium chloride were used for the contact angle measurement. The 0 V shows the contact angles without the voltage supply, when the voltage was supplied and increased, the contact angles of all three of the solutions decreased almost linearly and gradually reached contact angle saturation.

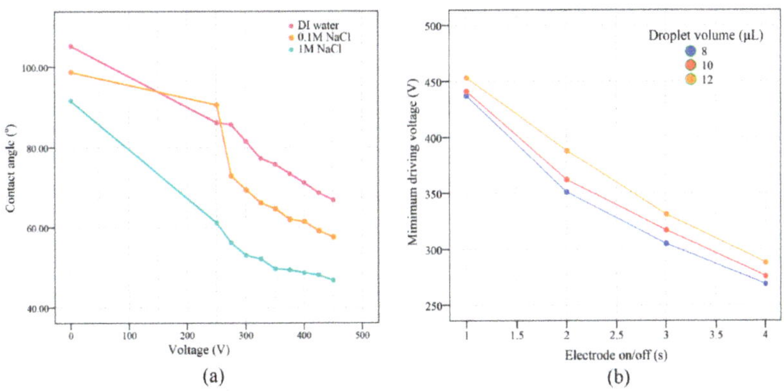

Figure 9. (**a**): Measured contact angle under different voltages (the frequency of the AC voltage supply is 1 kHz, with droplet volume of 10 μL). (**b**): Minimum driving voltages with the different electrode on/off time in the proposed digital microfluidic device.

The minimum droplet driving voltages with different volumes of DI water droplets under different electrode on/off time is shown in Figure 9b. The minimum driving voltage depends on many factors, including the volume of the droplet, the thickness of the dielectric layer, and the gap between the electrodes etc. In this study, we explored the minimum driving voltages that were influenced by the electrode on/off time. As previously described, the voltage supply to each of the electrodes was independently controlled by a series of relays. We found in our experiment that the electrode on/off time also significantly influenced the minimum driving voltage, as shown in Figure 9b, and a slower electrode on/off action (droplet moving slowly) required a lower driving voltage while the faster electrode on/off action (droplet moving fast) required higher driving voltage.

4. Conclusions

This study proposed a fabrication method for a thin, transparent, and flexible digital microfluidic device. The ITO layer on PET substrate was patterned with laser ablation for the fabrication of electrodes, a thin layer of PE film was used as the dielectric layer with the enhanced surface hydrophobicity with a commercial water repellent. The whole device has a thickness lower than 200 μm, which is easily bendable and also transparent in the visible range. In order to demonstrate the proposed fabrication technique, several digital microfluidic devices were fabricated with various droplet manipulation functions achieved. The leaking current density, the contact angle under different voltage supply, and the minimum driving voltage were also discussed in this study.

The limitation of the proposed study comes from two aspects. Compared with the conventional digital microfluidic devices that are fabricated on quartz or silicon wafer, the minimum achievable microstructures on ITO film are limited, thus, the gap between the electrodes is relatively wide and requires a higher driving voltage. In addition, compared with the conventional digital microfluidic devices, the chemical resistance of the ITO film also has limitations on some organic solvents.

The proposed transparent digital microfluidics suggests an alternative to conventional digital microfluidic devices on glass or silicon wafers. The transparency of the proposed digital microfluidic device could be of significant importance for observing the droplets inside of the device.

Supplementary Materials: The following supporting information can be downloaded at: https://www.mdpi.com/article/10.3390/mi13040498/s1, Video S1: Droplet manipulation with digital microfluidic device; Video S2: Droplets fusion on the proposed digital microfluidic device.

Author Contributions: Conceptualization, Y.F.; investigation, J.C. and J.J. (Jiaxi Jiang); software, Y.T.; validation, M.D.; original draft preparation, Y.F. and X.G.; visualization, J.J. (Jinyun Jiang). All authors have read and agreed to the published version of the manuscript.

Funding: This research received no external funding.

Data Availability Statement: The data that support the findings of this study are available upon reasonable request from the authors.

Conflicts of Interest: The authors declare no conflict of interest.

References

1. Ooi, C.H.; Vadivelu, R.; Jin, J.; Sreejith, K.R.; Singha, P.; Nguyen, N.-K.; Nguyen, N.-T. Liquid marble-based digital microfluidics–fundamentals and applications. *Lab Chip* **2021**, *21*, 1199–1216. [CrossRef] [PubMed]
2. Pang, L.; Ding, J.; Fan, S.-K. Digital microfluidics for single cell manipulation and analysis. In *Handbook of Single-Cell Technologies*; Springer: Berlin/Heidelberg, Germany, 2021; pp. 185–205.
3. Sun, Z.; Lin, K.-F.; Zhao, Z.-H.; Wang, Y.; Hong, X.-X.; Guo, J.-G.; Ruan, Q.-Y.; Lu, L.-Y.; Li, X.; Zhang, R. An automated nucleic acid detection platform using digital microfluidics with an optimized Cas12a system. *Sci. China Chem.* **2022**, *65*, 630–640. [CrossRef] [PubMed]
4. Coelho, B.; Veigas, B.; Fortunato, E.; Martins, R.; Águas, H.; Igreja, R.; Baptista, P.V. Digital microfluidics for nucleic acid amplification. *Sensors* **2017**, *17*, 1495. [CrossRef]
5. Huang, S.; Connolly, J.; Khlystov, A.; Fair, R.B. Digital microfluidics for the detection of selected inorganic ions in aerosols. *Sensors* **2020**, *20*, 1281. [CrossRef] [PubMed]
6. Li, J. Current commercialization status of electrowetting-on-dielectric (EWOD) digital microfluidics. *Lab Chip* **2020**, *20*, 1705–1712. [CrossRef] [PubMed]

7. Hassan, M.R.; Zhang, J.; Wang, C. Digital microfluidics: Magnetic transportation and coalescence of sessile droplets on hydrophobic surfaces. *Langmuir* **2021**, *37*, 5823–5837. [CrossRef] [PubMed]
8. Guo, J.; Lin, L.; Zhao, K.; Song, Y.; Huang, M.; Zhu, Z.; Zhou, L.; Yang, C. Auto-affitech: An automated ligand binding affinity evaluation platform using digital microfluidics with a bidirectional magnetic separation method. *Lab Chip* **2020**, *20*, 1577–1585. [CrossRef] [PubMed]
9. Agostini, M.; Cecchini, M. Ultra-high-frequency (UHF) surface-acoustic-wave (SAW) microfluidics and biosensors. *Nanotechnology* **2021**, *32*, 312001. [CrossRef] [PubMed]
10. Gilet, T.; Terwagne, D.; Vandewalle, N. Digital microfluidics on a wire. *Appl. Phys. Lett.* **2009**, *95*, 014106. [CrossRef]
11. Grant, N.; Geiss, B.; Field, S.; Demann, A.; Chen, T.W. Design of a hand-held and battery-operated digital microfluidic device using EWOD for lab-on-a-chip applications. *Micromachines* **2021**, *12*, 1065. [CrossRef] [PubMed]
12. Dimov, N.; McDonnell, M.B.; Munro, I.; McCluskey, D.K.; Johnston, I.D.; Tan, C.K.; Coudron, L. Electrowetting-based digital microfluidics platform for automated enzyme-linked immunosorbent assay. *JoVE* **2020**, e60489. [CrossRef] [PubMed]
13. Geng, H.; Cho, S.K. Antifouling digital microfluidics using lubricant infused porous film. *Lab Chip* **2019**, *19*, 2275–2283. [CrossRef] [PubMed]
14. Coelho, B.J.; Veigas, B.; Águas, H.; Fortunato, E.; Martins, R.; Baptista, P.V.; Igreja, R. A digital microfluidics platform for loop-mediated isothermal amplification detection. *Sensors* **2017**, *17*, 2616. [CrossRef] [PubMed]
15. Chen, S.; He, Z.; Choi, S.; Novosselov, I.V. Characterization of Inkjet-Printed Digital Microfluidics Devices. *Sensors* **2021**, *21*, 3064. [CrossRef] [PubMed]
16. Fan, Y.; Kong, X.; Chai, D.; Wei, B.; Zhang, Y. Low-cost and flexible film-based digital microfluidic devices. *Micro Nano Lett.* **2020**, *15*, 165–167. [CrossRef]
17. Soum, V.; Kim, Y.; Park, S.; Chuong, M.; Ryu, S.R.; Lee, S.H.; Tanev, G.; Madsen, J.; Kwon, O.-S.; Shin, K. Affordable fabrication of conductive electrodes and dielectric films for a paper-based digital microfluidic chip. *Micromachines* **2019**, *10*, 109. [CrossRef] [PubMed]
18. Karnakis, D.; Kearsley, A.; Knowles, M. Ultrafast laser patterning of OLEDs on flexible substrate for solid-state lighting. *J. Laser Micro/Nanoeng.* **2009**, *4*, 218–223. [CrossRef]
19. Chung, C.-K.; Lin, Y.-C.; Huang, G. Bulge formation and improvement of the polymer in CO_2 laser micromachining. *J. Micromech. Microeng.* **2005**, *15*, 1878. [CrossRef]

micromachines

Article

Picoliter Droplet Generation and Dense Bead-in-Droplet Encapsulation via Microfluidic Devices Fabricated via 3D Printed Molds

Tochukwu D. Anyaduba [1,2], Jonas A. Otoo [1] and Travis S. Schlappi [1,*]

1 Keck Graduate Institute, Riggs School of Applied Life Sciences, Claremont, CA 91711, USA
2 Abbott Rapid Diagnostics, 4545 Towne Center Ct, La Jolla, San Diego, CA 92121, USA
* Correspondence: travis_schlappi@kgi.edu

Abstract: Picoliter-scale droplets have many applications in chemistry and biology, such as biomolecule synthesis, drug discovery, nucleic acid quantification, and single cell analysis. However, due to the complicated processes used to fabricate microfluidic channels, most picoliter (pL) droplet generation methods are limited to research in laboratories with cleanroom facilities and complex instrumentation. The purpose of this work is to investigate a method that uses 3D printing to fabricate microfluidic devices that can generate droplets with sizes <100 pL and encapsulate single dense beads mechanistically. Our device generated monodisperse droplets as small as ~48 pL and we demonstrated the usefulness of this droplet generation technique in biomolecule analysis by detecting Lactobacillus acidophillus 16s rRNA via digital loop-mediated isothermal amplification (dLAMP). We also designed a mixer that can be integrated into a syringe to overcome dense bead sedimentation and found that the bead-in-droplet (BiD) emulsions created from our device had <2% of the droplets populated with more than 1 bead. This study will enable researchers to create devices that generate pL-scale droplets and encapsulate dense beads with inexpensive and simple instrumentation (3D printer and syringe pump). The rapid prototyping and integration ability of this module with other components or processes can accelerate the development of point-of-care microfluidic devices that use droplet-bead emulsions to analyze biological or chemical samples with high throughput and precision.

Keywords: microfluidics; picoliter droplets; rapid prototyping; bead encapsulation; 3D printing

Citation: Anyaduba, T.D.; Otoo, J.A.; Schlappi, T.S. Picoliter Droplet Generation and Dense Bead-in-Droplet Encapsulation via Microfluidic Devices Fabricated via 3D Printed Molds. *Micromachines* **2022**, *13*, 1946. https://doi.org/10.3390/mi13111946

Academic Editor: Pingan Zhu

Received: 21 October 2022
Accepted: 5 November 2022
Published: 10 November 2022

1. Introduction

Droplet microfluidics uses devices with channels dimensions tens or hundreds of microns wide to generate and manipulate discrete μL or less volumes. Dividing a sample of interest into fL to μL scale volumes reduces reagent usage, increases the sensitivity of chemical analyses, and provides enhanced control over reagent delivery, mixing, and chemical interactions [1]. There are many applications of droplet microfluidics in chemistry, biology, and biomedical engineering, such as therapeutic agent delivery, biomedical imaging, biomolecule synthesis, diagnostic chips, drug discovery, cell culture, biochemical characterization, and single cell analysis [2]. The implementation of droplet microfluidics in these applications are accomplished through lab-on-a-chip devices. These lab-on-a-chip devices may require droplet manipulation processes such as mixing, fission and/or fusion, sorting, and transportation of droplets [3,4], which can be accomplished via electrowetting, magnetic actuation, dielectrophoresis, surface acoustic waves, optical methods, or thermal methods [3–7]. However, due to the complicated processes used to fabricate channels that are tens or hundreds of microns wide, most droplet microfluidic methods are limited to research in laboratories with cleanroom facilities and complex instrumentation (e.g., photolithography with silicon wafers [8–10] or wet etching [11–13]). The few droplet generation technologies commercially available for diagnostic use are expensive ($89 k–$100 k for an instrument and $24–$240 per disposable cartridge) and not integrated with other assay steps such as chemical reaction incubation and droplet analysis [14].

To make the droplet generation process simpler, less time-intensive, and less expensive, many innovative methods have been created. Some researchers have used glass capillaries to generate pL-scale droplets. For example, Li et al. bonded microscope glass slides to pulled glass capillaries to generate monodisperse multiple emulsions [15], Gu et al. created and manipulated pL droplets for single cell assays with a 75 μm fused-silica capillary [16], and Li et al. used an asymmetrical beveled capillary to generate pL to nL droplets and execute a digital PCR assay [17]. While the instrumentation costs for these devices are lower than for photolithography, devices made from glass capillaries are difficult to integrate into other upstream or downstream modules and not amenable to rapid prototyping due to the intricate procedures for fabricating capillaries <1 mm in diameter. Other groups rely on micromachining to generate droplets, such as direct milling of polycarbonate [18,19] or micromachining in PMMA [20]. These methods have demonstrated consistent and controllable droplet generation; however, the droplet sizes are large (>1 nL) or when ply-sized droplets are achieved, a centrifuge is needed to create the droplets in a reaction tube, which precludes its ability to be integrated into other microfluidic modules [20].

3D printing is now commonly used to create molds for PDMS devices, which eliminates the need for cleanroom facilities, photolithography, or etching and enables rapid prototyping and fabrication [21–24]. Researchers have also used 3D printers to build monolithic devices out of resin for droplet generation, albeit with larger channel dimensions and therefore larger droplets (>1 nL) [25–31]. Picoliter-scale droplets are important for several applications, such as increasing the precision, sensitivity and dynamic range of digital PCR [32], or preventing cross contamination and target dilution in single cell analysis [16,33]. The small channel sizes required for pL-scale droplets are typically fabricated with complex processes insides a cleanroom, usually photolithography [8–10], and have not been made with 3D printed molds or 3D printed monolithic devices. The methods described above have significantly advanced droplet generation for the picoliter scale via photolithography or glass capillaries, and the nanoliter scale via 3D printing; however, there remains a need for <100 pL droplet generation from a rapid prototyping method (e.g., 3D printed molds) that can be easily integrated into other sample preparation, analysis, and detection modules.

An important area of investigation in droplet microfluidics are methods that encapsulate a single bead in a droplet (BiD). These BiD platforms have enabled exciting advancements in biomedical research and diagnostics, including genome sequencing [34], enzyme evolution and screening [35,36], detecting rare genetic mutations [37,38] single cell analysis [39], and molecular diagnostics [40]. While these devices have high throughput and multiplexing capabilities, they are limited to laboratories with sophisticated instrumentation for photolithography and bead encapsulation. Additionally, they have shown Poisson or better distributions of BiDs for particles with a similar density to water, such as gel particles [41], polystyrene beads [42–44], agarose beads [39], or biological cells [39,44,45]. Particles with a higher density than water sediment to the bottom before being encapsulated in droplets and cause the first fraction of droplets to have more than 1 bead per droplet and the remaining fraction to not have any beads. To use beads of varying densities in BiD platforms, this sedimentation effect must be overcome.

The purpose of this work is to overcome current limitations of droplet microfluidic devices by creating a droplet generation device with the following features: (i) a simple and inexpensive fabrication process that is amenable to rapid prototyping and integration with other modules, (ii) droplet volumes <100 pL, and (iii) the ability to encapsulate dense beads in aqueous droplets with a Poisson-like distribution. We found that using 3D printing to create a mold instead of photolithography or etching is a suitable fabrication method to accomplish this purpose. Our device generated monodisperse droplets as small as ~48 pL and we demonstrated the usefulness of this droplet generation technique in biomolecule detection by quantifying nucleic acids via digital loop-mediated isothermal amplification (dLAMP). We also designed a mixer that can be integrated into a syringe to overcome dense bead sedimentation and found that the BiD emulsions created from our device had less than 2% of the droplets populated with more than 1 bead when the average input concentration was 0.15 beads/droplet, in line with Poisson statistical projections. This study will enable researchers to create devices that generate pL-scale droplets and encapsulate dense beads with inexpensive and simple instrumentation (3D printer and syringe pump). The rapid prototyping and integration ability of this method can accelerate the development of point-of-care microfluidic devices that generate droplet-bead emulsions and analyze samples with high throughput and precision.

2. Materials and Methods

2.1. Device Fabrication

3D models of the master molds were designed using SolidWorks CAD software (Dassault Systems, Velizi-villacoublay, France) to have flow channel dimensions of 100 μm × 100 μm and inlet/outlet ports of 750 μm (Figure 1A). Stereolithography (SLA) files were prepared for 3D printing by orienting them at a 45° angle and avoiding cups and overhangs in Form Labs' Preform software. The models were then printed using the Form3 SLA 3D printer (Form Labs) in Clear resin (FLGPCL04) at a layer thickness of 25 μm. The printed master molds were thoroughly cleaned with isopropyl alcohol to remove excess resin, then UV-cured for 30 min.

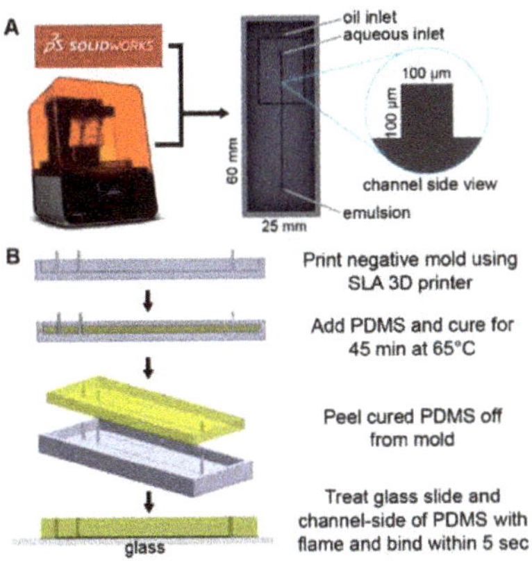

Figure 1. Microfluidic device design and fabrication. (**A**) A solid master mold was designed with Solidworks CAD software and printed with FormLabs Form3 SLA printer. (**B**) PDMS device fabrication process.

To make polydimethylsiloxane (PDMS), SYLGARDTM 184 Silicone Elastomer Base and SYLGARDTM 184 Silicone Elastomer Curing Agent (Dow Corning, Midland, MI, USA) are combined at 10:1 *w/w* ratio to make up ~3 gm needed to fill each mold. Prior to pouring the mixture into the mold, it is degassed in a Cole Parmer Diblock oven at room temperature until no bubbles can be seen in the PDMS mixture. After filling the molds with the degassed PDMS, the degassing process is repeated to ensure complete filling of the corners of the channels before curing at 65 °C for 45 min. Once cured, the PDMS is gently peeled from the master mold and bonded onto glass microscope slides (Amscope BS-72P 100S-22) after surface activation using flame treatment as an alternative to oxygen plasma bonding [46] (Figure 1B). The device is then placed in an 85 °C oven overnight to allow the PDMS to harden. Next, the devices are examined for binding strength of the PDMS by gently prying at them. They are also checked for channel dimensions under a microscope. A ± 10% tolerance is allowed for the channel widths measured from micrographs prior to the attachment of the flow tubing (Scientific Commodities, Lake Havasu City, AZ, USA,

BB31695 PE/3). The tubing is attached to the chip by plumbing them into the inlet and outlet ports, making sure to leave a clearance space between the tubing nozzle and the slide surface. The tubing is further held in place using cold weld steel-reinforced epoxy (JB Weld, Marietta, GA, USA).

2.2. Droplet Generation

Droplets were generated using the designed flow-focusing PDMS microfluidic devices described above. The oil phase consisted of mineral oil (Sigma Aldrich M3516-1L), 0.1 wt% Triton X-100 (Fisher Scientific, Waltham, MA, USA), and 3 wt% ABIL EM 90 (Evonik, Essen, Germany), and was pumped at various volumetric flow rates (20, 25, 50, 75, 100 μL/min). The aqueous phase (DI water) was maintained at a volumetric flow rate of 1 μL/min. The oil and aqueous phases were pumped to an intersection in the device by syringe pumps (KD Scientific, Holliston, MA, USA), at which point droplets were generated and subsequently collected from the outlet in Eppendorf tubes. A fraction of the droplets were imaged using confocal imaging (Leica SP5, Wetzlar, Germany) and the respective planar areas of the droplets were deduced using ImageJ software after segmentation processing. The spherical diameter of each droplet is calculated from the deduced area.

2.3. Droplet Digital Loop-Mediated Isothermal Amplification for DNA Quantification

Lactobacillius acidophilus (L. acid.) obtained from MicroKwik vials (Carolina Biological Supply, Burlington, NC, USA) was cultured in de Man, Rogosa and Sharpe (MRS) agar formulated in-house using Millipore-Sigma formulation (CCW4691). The QuickExtractTM one-step DNA extraction kit (Lucigen, Middleton, WI, USA) was used to extract DNA from the colonies. Extracted genomic DNA was quantified via absorbance measurements from a Nanodrop One instrument (ThermoFisher Scientific, Waltham, MA, USA) and diluted in nuclease-free water to concentrations ranging from 0 to 9.5 × 106 copies/mL.

LAMP master mix was prepared with final concentrations of 1× isothermal amplification buffer (New England Biolabs, NEB), 8 mM of MgSO4 (NEB), 1.4 mM dNTPs (NEB), 320 U/ mL Bst 2.0 WarmStart polymerase (NEB), primer mix, and 1× SybrGreen (Life Technologies). The primer mix was designed in-house to target the *L. acidophilus* 16S rRNA gene and consisted of 1.6 μM each of forward inner primer (CTGCACTCAA-GAAAAACAGTTTCCGAGTCTGATGTGAAAGCCCTC) and backward inner primer (AA-GAGGAGAGTGGAACTCCATGTGAGACCAGAGAGCCGCCTT), 0.2 μM each of forward outer primer (TAAAGCGAGCGCAGGC) and backward outer primer (CCTCAGCGTCAGTTGC), 0.4 μM each of forward loop primer (GCAGTTCCTCGGTTAAGCC) and backward loop primer (ATGCGTAGATATATGGAAGAACACC) (Integrated DNA Technologies, Clarville, IA, USA). *L. acid* DNA dilutions were added to LAMP master mix to yield final concentrations of 0, 1.0·× 107, 2.5·× 107, 5.0 × 107, 4.0·× 108 DNA copies/mL (quantified by Nanoquant absorbance measurements). Four replicates of each dilution (10 uL/well) were amplified at 68 °C for 60 min using a LightCycler®96 Instrument (Roche, Basel, Switzerland) as positive controls.

The LAMP mix + *L. acid* DNA samples were infused into a droplet generation device as described in "Droplet Generation", with oil flow rate 75 μL/min and aqueous flow rate 1 μL/min. Droplets from the microfluidic devices were collected in amber SepCap vials (Thermoscientific, Waltham, MA, USA C4015-99) and incubated at 68°C for 60 min using a Multi-Therm shaker (Benchmark Scientific, Sayreville, NJ, USA). After incubation, the droplets were imaged using a Leica SP5 confocal microscope, and images were analyzed with Image J to determine the relative fluorescence intensity (RFI) of each droplet. A threshold was determined by computing $\mu_{NTC} + 3{\cdot}\sigma_{NTC}$, where μ_{NTC} is the mean and σ_{NTC} is the standard deviation of the RFI of the 0 cop/mL sample droplets. Droplets with RFI greater than the threshold were classified as positive while the droplets less than or equal to the threshold are classified as negative. One can then use Poisson statistics with the number of positive and negative droplets to calculate a concentration for each sample [47].

2.4. Bead Mixer

A blind hole with a diameter of about 9 mm was drilled into the side of a 3 mL plastic syringe (CareTouch, Westminster, CO, USA) at the 0.5 mL mark. A small DC motor with a plastic impeller which was originally designed for a bead-beating sample preparation device (Claremont Bio 01.340.48 OmniLyse®Kit) was retrieved and carefully positioned into the syringe through the blind hole. The motor with the impeller was affixed to the syringe with cold weld steel-reinforced epoxy (JB Weld, Marietta, GA, USA) such that the blind hole was completely sealed and airtight. The epoxy was allowed to set for 48 to 72 h. The impeller mixer was powered by a 1.5 V DC power supply (SI, Figure S1).

2.5. Bead-in-Droplet Emulsions

Hard shell Polymethyl Methacrylate (PMMA) beads (PolyAn Microshperes Po-105 00 020 and Alpha Nanotech colloidal PMMA) of 20 µm in diameter were used in the bead encapsulation experiment. A mixture of the beads and 0.1 $\%v/v$ Tween 20 in nuclease-free water at working concentrations of 0.15, 0.2 and 0.3 beads/droplet (λ) were used as the dispersed phase for the experiments. A mixture of mineral oil (Sigma Aldrich-M3516-1L), 0.1 wt% Triton X-100 (Fisher Scientific, Waltham, MA, USA) and 3 wt% ABIL EM 90 (Evonik, Essen, Germany) was used as the continuous phase. The dispersed phase (bead suspension) was aspirated into a modified syringe and loaded onto a syringe pump (KD Scientific, Holliston, MA, USA, KDS100). A 1.5 V DC power supply was connected to the mixer to keep the beads solution homogenous. The continuous phase was put into a 10 mL plastic syringe (CareTouch, Westminster, CO, USA) and loaded onto a syringe pump. The continuous and dispersed phases were introduced into the droplet generation device using syringe pumps at flow rates of 30 µL/min and 1–7 µL/min, respectively. A period of about 5 min was allowed for the cartridge to be primed and for the droplet generation to be stabilized. The droplets were collected from the cartridge into 1 mL amber SepCap vials (Thermoscientific, Waltham, MA, USA, C4015-99). The excess oil from the continuous phase was poured off and the droplets were put onto a microscope slide and mounted onto a microscope (Omax microscope 3152102) for imaging. Micrographs of the droplets were taken using the Amscope microscope camera md35 and Amsocpe software version 4.

2.6. Image Analysis

The images were opened in Image J. The scale was set according to the scale bar on the images and the unit was set to µm. The images were converted to 8-bit gray scale images and speckles and noise were filtered from the images. The threshold of the images was adjusted to convert them to binary images. The images were converted to mask to invert the black to white, making the droplets appear white. The droplets were then analyzed to calculate the area of each droplet. The diameter and volume of each droplet were calculated from the area of the droplets. The droplets containing beads were manually counted and the number of beads in each droplet was recorded. The data were compiled in Excel (Microsoft Office) and parsed into Python 3.0 for further analysis and visual presentation.

3. Results and Discussion

3.1. Picoliter-Scale Droplet Generation

The physics of droplet generation via flow focusing has been well documented with theory and experiments showing an inverse logarithmic relationship between Capillary number ($Ca = \mu_{ave}(2Q_o + Q_w)/\sigma hw$) and non-dimensionalized droplet diameter, D_d/D_h, where μ_{ave} is the average viscosity of the two fluids, Q_o is the oil flow rate, Q_w is the water flow rate, σ is the surface tension, h is the channel height, w is the channel width, D_d is the diameter of the droplet, and D_h is the hydraulic diameter of the channel, $2hw/(h+w)$ [48,49]. These flow focusing studies demonstrate that <100 pL droplets can theoretically be generated with $Ca > 0.001$ (faster flow rates (Q_o,Q_w) relative to channel dimensions (h,w)) and 144 µm > D_h > 39 µm, or with $Ca < 0.001$ (slower flow rates (Q_o,Q_w) relative to channel dimensions (h,w)) and 14 µm < D_h < 39 µm [48] (SI, Section S2). Experimentally, the authors test devices with maximum channel heights of 27 μm [48] or widths of 71 μm [49]. In these studies and others [8–13], pL droplets are generated by using small channel widths (<100 μm) facilitated by photolithographic processes in cleanrooms. As our objective was to develop a device that generates pL droplets without complex fabrication processes, we were limited to the channel widths 100 μm or greater that an SLA 3D printer is capable of printing in a mold. Therefore, our device design would need to be in the $Ca > 0.001$ regime with faster flow rates relative to channel dimensions.

With the limits on our device's physical features established, we 3D printed a mold and made a PDMS cast of 100 μm channel width and 100 μm channel height without a cleanroom, photolithography processes, or complex instrumentation (Figure 1). We chose oil and water flow rates such that the droplet generation device would have $Ca \gg 0.001$, with Q_o = 25 to 100 μL/min and Q_w = 1 μL/min (SI, Section S2), which resulted in droplets of diameters 45 to 112 μm (48 to 736 pL) (Figure 2). The droplets generated from this device are monodisperse (Figure 2B, coefficient of variation (CV) from 2–12%), which is in the range of droplets generated from other devices [50,51]. As expected, there is an inverse power relationship between droplet volume and oil flow rate [49], showing that devices fabricated with 3D printed molds give similar consistency and expected performance at the picoliter scale as devices made with photolithography in a cleanroom. Because this device is made from a 3D printed mold, researchers can iterate prototypes rapidly without undergoing the time and resource-consuming processes of photolithography; additionally, the droplet generation module can be part of a larger 3D printed mold that includes modules for executing other upstream or downstream assay processes.

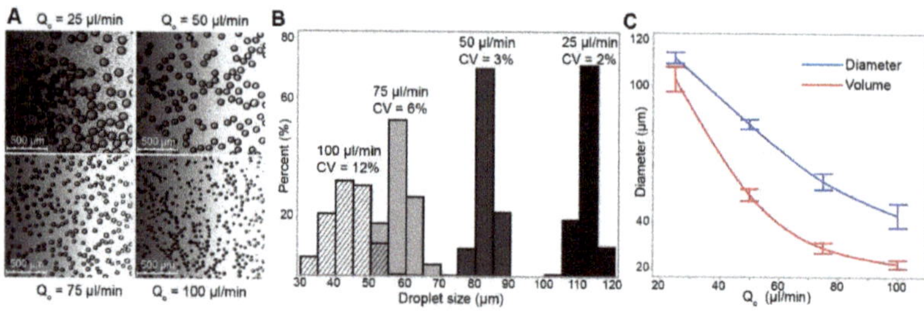

Figure 2. Picoliter-scale droplet generation. (**A**) Micrograph of the droplets retrieved from microfluidic cartridge outlet. (**B**) Droplet diameter distribution and CV at each flow condition. (**C**) The droplet diameter changes with volumetric flow rate of the oil phase. The volumetric flow rate of the aqueous phase was kept constant at 1 μL/min.

3.2. Droplet Digital Loop-Mediated Isothermal Amplification

To explore the utility of this droplet generation device in molecular diagnostic applications, droplet digital loop-mediated isothermal amplification (ddLAMP) was performed to detect and quantify a DNA target. Digital LAMP is an emerging nucleic acid (NA) amplification method that can quantify the NA concentration of a sample with high accuracy and precision, even in the midst of temperature, reaction time, or imaging variance [52]. NA quantification via dLAMP is useful in several applications, such as viral load measurements for HIV [53], hepatitis C virus genotyping [54], and rapid antibiotic susceptibility testing [55]. Current dLAMP methods partition the sample into pL to nL droplets with microfluidic devices made using photolithography [56,57], wet etching [52–55], or fused-silica capillaries [58]. Our droplet generation device made from a 3D printed mold could make dLAMP more accessible by eliminating the need for complex facilities or instruments and enabling integration with other amplification or detection modules.

We tested the feasibility of encapsulating LAMP reagents with target DNA and primers into droplets with our device (Materials and Methods). After generation, the droplets were incubated at 68 °C for 60 min for amplification of DNA via LAMP and SybrGreen fluorescence was measured to indicate the presence or absence of amplification product within each droplet (Figure 3A). Five DNA dilutions were tested, and the positive droplet percentage was plotted against the prediction from Poisson statistics (Figure 3B), assuming a 10% LAMP efficiency and 300 pL droplet volume (SI, Section S3).

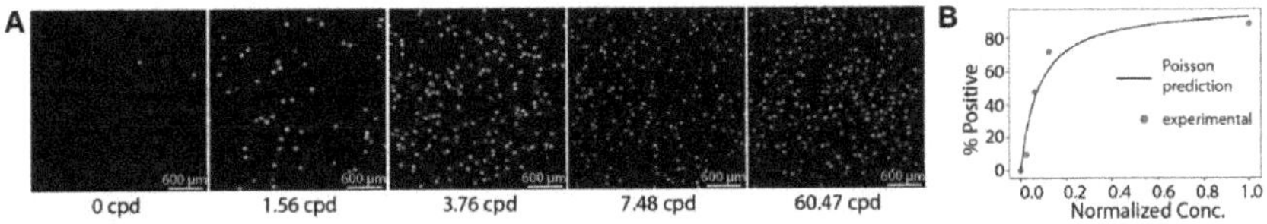

Figure 3. Droplet digital LAMP. (**A**) Post-amplification micrographs of droplets. (**B**) Agreement between Poisson predicted positive droplet percentage and experimental data.

3.3. *Dense Bead-in-Droplet Emulsions*

Interest in using microparticles as delivery systems in various technologies has been widely researched, especially in combination with microdroplets for biological applications [59–62]. This is due to the high surface-to-volume ratio and the ease of immobilizing biorecognition molecules on them, as well as the potential for compartmentalized single-molecule assays [63–65]. Single particle encapsulation in droplets, however, faces two major challenges: sedimentation due to particle density [62], and mechanistic single particle encapsulation [41,66].

Particle density poses a challenge when loading microparticles into encapsulation devices because the higher density particles (>1 gm/mL) sediment in the syringe and delivery tubing, causing nonhomogeneous distribution of microparticles in droplets (SI, Figures S1A and S2). This can be solved by the dissipation of the bead density by suspending them in denser fluids such as glycerol [62]; however, such fluids may not be compatible with the intended bio-application. For example, glycerol at 50% v/v inhibits NA amplification, thereby defeating the purpose of using microbeads for NA applications (SI, Figure S4). To circumvent this challenge, researchers used gel beads with similar density to water, which ensured a binary distribution of beads in the droplets [65,67–69]. However, this method is time-consuming, requiring a particle velocity of ~50 µm/h [41] to achieve single-particle encapsulation; furthermore, some multiplexed nucleic acid detection methods are not compatible with beads made in gel form [70–72].

Price et al. presented a potentially simple solution by exploiting the sedimentation potential of the beads using a hopper system [62]. They, however, showed that it took 0.8 h (17 µm Tetangel resin beads) and 3.8 h (2.8 µm magnetic beads) for bead introduction before achieving single bead encapsulation. Kim et al. successfully developed a pneumatic system which is capable of trapping and releasing beads, thus creating a deterministic encapsulation of a defined number of beads per droplet [62]. This system is not simple to develop or operate, thus, unfit for low-cost point-of-care devices that can integrate with other modules.

Our goal was to present a simple, easy-to-fabricate method to encapsulate single dense beads in droplets that can be used for further downstream analysis. It is important to encapsulate single beads as opposed to multiple beads to avoid cross-contamination or confusion of which target molecule or bead is in the droplet. The idea is to vertically orient the syringe pump while keeping the beads suspended by mechanical agitation (which prevents loss of beads due to sedimentation in the flow tubing and in the syringe) (Figure S1B), then pump the contents directly into the droplet generation cartridge (Figure 1B). Using this principle, we set up bead encapsulation with the droplet generation device such that $\lambda \approx 0.15$, 0.2 and 0.3 beads/droplet, where λ represents the average number of beads per droplet input into the device (Figure 4). We observed that our dense bead encapsulation method agreed well with Poisson predictions (Figure 4B). Importantly, the droplet generation device resulted in <2% of droplets containing more than 1 bead at $\lambda \approx 0.15$, <4% of droplets containing more than 1 bead at $\lambda \approx 0.2$, and <6% of droplets containing more than 1 bead at $\lambda \approx 0.3$.

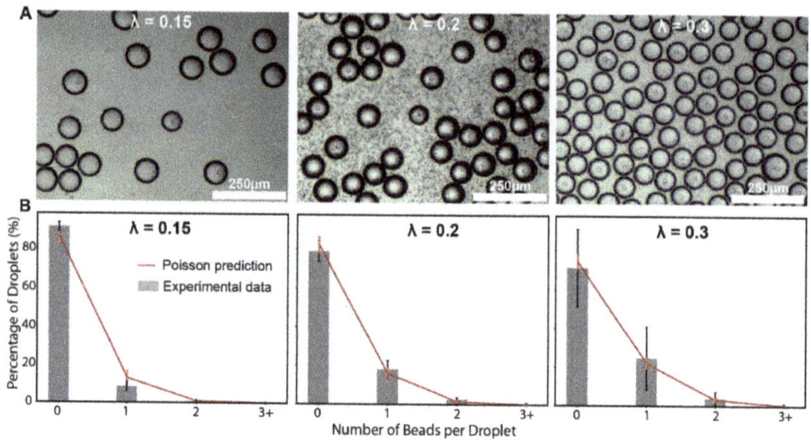

Figure 4. Dense bead in droplet (BiD) encapsulation at varying bead concentrations (λ = 0.15, 0.2, 0.3). (**A**) Microscope images of BiD. (**B**) Poisson-predicted bead in droplets distribution in comparison to the observed experimental data.

4. Conclusions

Using design principles from droplet microfluidic device literature, we designed and developed a microfluidic device fabricated without complex equipment or cleanroom facilities that can generate sub-100 pL droplets and encapsulate dense beads with a Poisson-like distribution. Because the device is made from a 3D printed mold, researchers can iterate prototypes rapidly without undergoing the time and resource-consuming processes of photolithography; additionally, the droplet generation module can be part of a larger 3D printed mold that includes modules for executing other upstream or downstream assay processes, such as sample preparation, NA amplification, or single cell analysis.

While simple instrumentation was used to fabricate the microfluidic device, we still needed a syringe pump for operation of the device to generate consistent and controlled droplet sizes. Further improvements need to be made to our design to make it more amenable to point-of-care settings, such as a pumping lid [73], or other equipment-free pumping mechanisms [74]. Another limitation is that due to the 3D printer's minimum channel dimension (~100 μm), the lowest droplet diameter achieved was 45 μm (48 pL). Lower sizes could be possible in the future with the next generation of 3D printers that print channels down to 15 μm [75].

Other microfluidic devices have encapsulated beads in a non-random distribution and thus have a much higher percentage of droplets with a single bead [41,43], though the beads in those studies have a similar density to water. While the phenomenon for the non-random distribution is unexplained, similar designs could potentially be used with the dense bead mixing method studied here for higher percentages of droplets with single beads. In its current form, this device enables research and innovation into assays or methods that need to use beads with a density greater than water and thus overcome the sedimentation effect, such as PMMA or magnetic beads. Because it can easily be printed and combined with others as part of a larger device, microfluidic sorting mechanisms can also be used to concentrate the beads downstream if desired [76].

Future research directions from this work can include: eliminating the need for a syringe pump for <100 pL droplet generation, adapting the device to other biological assay applications beyond digital LAMP, beating Poisson encapsulation statistics for dense beads to reduce the waste of empty droplets, or adapting the BiD method for tagging multiple biomarkers. Due to the simple instrumentation used, this work enables rapid prototyping for a variety of biological applications of droplet microfluidic devices and dense bead encapsulation.

Supplementary Materials: The following supporting information can be downloaded at: https://www.mdpi.com/article/10.3390/mi13111946/s1, Section S1: Syringe with mixer to overcome sedimentation effect of dense beads; Section S2: Capillary number calculations for picoliter-scale droplet generation design; Section S3: Poisson prediction of positive droplet percentage; Section S4: Effect of Glycerol on LAMP Amplification; Section S5: Pitalls of 3D Printing Fabrication of Microfluidic Cartridges. Figure S1: (A) Tube connecting syringe containing bead suspension to the droplet generation cartridge; red arrow shows region of bead sedimentation. (B) Syringe design for mechanical resuspension and homogenization of dense particles for vertical delivery. The DC motor is powered using a 3V battery. Figure S2: Without the syringe mixer in Figure S1, bead sedimentation happens in the syringe and tubing, leading to the encapsulation of multiple beads per droplet. Figure S3: Denser fluids, such as glycerol, may improve bead buoyancy but it inhibits LAMP amplification (blue trace vs. red trace). Bead Density = 1.18 g/ cm^3, Glycerol Density = 1.26 g/cm^3. Figure S4: Microcapillary lines imprinted by 3D printed mold. This is often due to printer-head misalignment that often occurred during prolonged prints. Figure S5: Micrograph showing curved vertices imprinted from 3D-printed mold. Figure S6: Irregularities in chamber dimensions due to myriad factors, including incompletely cured PDMS and build-up PDMS deposit due to mold reuse. Note that the displayed images contain channels designed to have widths of 50 and 100 μm. Figure S7: Frosted PDMS molded on improperly cleaned 3D printed mold. Figure S8: Image of final fabricated PDMS device and 3D printed mold.

Author Contributions: Conceptualization, T.D.A. and T.S.S.; methodology, T.D.A.; software, T.D.A.; validation, T.D.A. and J.A.O.; formal analysis, T.D.A., J.A.O. and T.S.S.; investigation, T.D.A. and J.A.O.; resources, T.S.S.; data curation, T.D.A. and J.A.O.; writing, T.S.S., T.D.A. and J.A.O.; visualization, T.D.A., T.S.S. and J.A.O.; supervision, T.S.S.; project administration, T.S.S. and T.D.A.; funding acquisition, T.S.S. and T.D.A. All authors have read and agreed to the published version of the manuscript.

Funding: Research reported in this publication was supported by the National Institute Of Biomedical Imaging And Bioengineering of the National Institutes of Health under Award Number K01EB027718. The content is solely the responsibility of the authors and does not necessarily represent the official views of the National Institutes of Health.

Data Availability Statement: The data presented in this study are available in this article and Supplementary Materials.

Conflicts of Interest: The authors declare no conflict of interest.

References

1. Teh, S.Y.; Lin, R.; Hung, L.H.; Lee, A.P. Droplet microfluidics. *Lab Chip.* **2008**, *8*, 198–220. [CrossRef] [PubMed]
2. Sohrabi, S.; Kassir, N.; Keshavarz Moraveji, M. Droplet microfluidics: Fundamentals and its advanced applications. *RSC Adv.* **2020**, *10*, 27560–27574. [CrossRef]
3. Zhu, G.P.; Wang, Q.Y.; Ma, Z.K.; Wu, S.H.; Guo, Y.P. Droplet Manipulation under a Magnetic Field: A Review. *Biosensors* **2022**, *12*, 156. [CrossRef] [PubMed]
4. Yang, C.G.; Xu, Z.R.; Wang, J.H. Manipulation of droplets in microfluidic systems. *TrAC—Trends Anal. Chem.* **2010**, *29*, 141–157. [CrossRef]
5. Park, S.Y.; Kalim, S.; Callahan, C.; Teitell, M.A.; Chiou, E.P.Y. A light-induced dielectrophoretic droplet manipulation platform. *Lab Chip.* **2009**, *9*, 3228–3235. [CrossRef]
6. Hartmann, J.; Schür, M.T.; Hardt, S. Manipulation and control of droplets on surfaces in a homogeneous electric field. *Nat. Commun.* **2022**, *13*, 289. [CrossRef]
7. Zaman, M.A.; Padhy, P.; Ren, W.; Wu, M.; Hesselink, L. Microparticle transport along a planar electrode array using moving dielectrophoresis. *J. Appl. Phys.* **2021**, *130*, 034902. [CrossRef]
8. Mazutis, L.; Araghi, A.F.; Miller, O.J.; Baret, J.-C.; Frenz, L.; Janoshazi, A.; Taly, V.; Miller, B.J.; Hutchison, J.B.; Link, D.; et al. Droplet-based microfluidic systems for high-throughput single DNA molecule isothermal amplification and analysis. *Anal. Chem.* **2009**, *81*, 4813–4821. [CrossRef]
9. Zhang, K.; Kang, D.K.; Ali, M.M.; Liu, L.; Labanieh, L.; Lu, M.; Riazifar, H.; Nguyen, T.N.; Zell, J.A.; Digman, M.A.; et al. Digital quantification of miRNA directly in plasma using integrated comprehensive droplet digital detection. *Lab Chip.* **2015**, *15*, 4217–4226. [CrossRef]
10. Kintses, B.; Hein, C.; Mohamed, M.F.; Fischlechner, M.; Courtois, F.; Lainé, C.; Hollfelder, F. Picoliter cell lysate assays in microfluidic droplet compartments for directed enzyme evolution. *Chem. Biol.* **2012**, *19*, 1001–1009. [CrossRef]
11. Perroud, T.D.; Meagher, R.J.; Kanouff, M.P.; Renzi, R.F.; Wu, M.; Singh, A.K.; Patel, K.D. Isotropically etched radial micropore for cell concentration, immobilization, and picodroplet generation. *Lab Chip.* **2009**, *9*, 507–515. [CrossRef] [PubMed]

12. Beer, N.R.; Hindson, B.J.; Wheeler, E.K.; Hall, S.B.; Rose, K.A.; Kennedy, A.I.M.; Colston, B.W. Reaction in Picoliter Droplets. *Anal. Chem.* **2007**, *79*, 8471–8475. [CrossRef] [PubMed]
13. Zhu, Y.; Zhang, Y.X.; Cai, L.F.; Fang, Q. Sequential operation droplet array: An automated microfluidic platform for picoliter-scale liquid handling, analysis, and screening. *Anal. Chem.* **2013**, *85*, 6723–6731. [CrossRef]
14. Baker, M. Digital PCR hits its stride. *Nat. Methods* **2012**, *9*, 541–544. [CrossRef]
15. Li, E.Q.; Zhang, J.M.; Thoroddsen, S.T. Simple and inexpensive microfluidic devices for the generation of monodisperse multiple emulsions. *J. Micromech. Microeng.* **2014**, *24*, 015019. [CrossRef]
16. Gu, S.Q.; Zhang, Y.X.; Zhu, Y.; Du, W.B.; Yao, B.; Fang, Q. Multifunctional picoliter droplet manipulation platform and its application in single cell analysis. *Anal. Chem.* **2011**, *83*, 7570–7576. [CrossRef]
17. Li, H.T.; Wang, H.F.; Wang, Y.; Pan, J.Z.; Fang, Q. A minimalist approach for generating picoliter to nanoliter droplets based on an asymmetrical beveled capillary and its application in digital PCR assay. *Talanta* **2020**, *217*, 120997. [CrossRef] [PubMed]
18. Postek, W.; Kaminski, T.; Garstecki, P. A passive microfluidic system based on step emulsification allows to generate libraries of nanoliter-sized droplets from microliter droplets of varying and known concentration of sample. *Lab Chip.* **2017**, *17*, 1323–1331. [CrossRef]
19. Churski, K.; Nowacki, M.; Korczyk, P.M.; Garstecki, P. Simple modular systems for generation of droplets on demand. *Lab Chip.* **2013**, *13*, 3689. [CrossRef]
20. Schulz, M.; Probst, S.; Calabrese, S.; Homann, A.R.; Borst, N.; Weiss, M.; Von Stetten, F.; Zengerle, R.; Paust, N. Versatile tool for droplet generation in standard reaction tubes by centrifugal step emulsification. *Molecules* **2020**, *25*, 1914. [CrossRef]
21. Mohamed, M.G.A.; Kumar, H.; Wang, Z.; Martin, N.; Mills, B.; Kim, K. Rapid and inexpensive fabrication of multi-depth microfluidic device using high-resolution LCD stereolithographic 3D printing. *J. Manuf. Mater. Process.* **2019**, *3*, 26. [CrossRef]
22. Hwang, Y.; Paydar, O.H.; Candler, R.N. 3D printed molds for non-planar PDMS microfluidic channels. *Sens. Actuators A Phys.* **2015**, *226*, 137–142. [CrossRef]
23. Saggiomo, V.; Velders, A.H. Simple 3D Printed Scaffold-Removal Method for the Fabrication of Intricate Microfluidic Devices. *Adv. Sci.* **2015**, *2*, 1500125. [CrossRef] [PubMed]
24. Li, Z.; Yang, J.; Li, K.; Zhu, L.; Tang, W. Fabrication of PDMS microfluidic devices with 3D wax jetting. *RSC Adv.* **2017**, *7*, 3313–3320. [CrossRef]
25. Zhang, J.M.; Li, E.Q.; Aguirre-Pablo, A.A.; Thoroddsen, S.T. A simple and low-cost fully 3D-printed non-planar emulsion generator. *RSC Adv.* **2016**, *6*, 2793–2799. [CrossRef]
26. Zhang, J.M.; Aguirre-Pablo, A.A.; Li, E.Q.; Buttner, U.; Thoroddsen, S.T. Droplet generation in cross-flow for cost-effective 3D-printed "plug-and-play" microfluidic devices. *RSC Adv.* **2016**, *6*, 81120–81129. [CrossRef]
27. Bhargava, K.C.; Thompson, B.; Malmstadt, N. Discrete elements for 3D microfluidics. *Proc. Natl. Acad. Sci. USA* **2014**, *111*, 15013–15018. [CrossRef]
28. Shallan, A.I.; Smejkal, P.; Corban, M.; Guijt, R.M.; Breadmore, M.C. Cost-effective three-dimensional printing of visibly transparent microchips within minutes. *Anal. Chem.* **2014**, *86*, 3124–3130. [CrossRef]
29. Donvito, L.; Galluccio, L.; Lombardo, A.; Morabito, G.; Nicolosi, A.; Reno, M. Experimental validation of a simple, low-cost, T-junction droplet generator fabricated through 3D printing. *J. Micromech. Microeng.* **2015**, *25*, 035013. [CrossRef]
30. Jiao, Z.; Zhao, L.; Tang, C.; Shi, H.; Wang, F.; Hu, B. Droplet-based PCR in a 3D-printed microfluidic chip for miRNA-21 detection. *Anal. Methods.* **2019**, *11*, 3386–3393. [CrossRef]
31. Ji, Q.; Zhang, J.M.; Liu, Y.; Li, X.; Lv, P.; Jin, D.; Duan, H. A Modular Microfluidic Device via Multimaterial 3D Printing for Emulsion Generation. *Sci. Rep.* **2018**, *8*, 4791. [CrossRef] [PubMed]
32. Heyries, K.A.; Tropini, C.; Vaninsberghe, M.; Doolin, C.; Petriv, O.I.; Singhal, A.; Leung, K.; Hughesman, C.B.; Hansen, C.L. Megapixel digital PCR. *Nat. Methods* **2011**, *8*, 649–651. [CrossRef] [PubMed]
33. Chiu, D.T.; Lorenz, R.M.; Jeffries, G.D.M. Droplets for ultrasmall-volume analysis. *Anal. Chem.* **2009**, *81*, 5111–5118. [CrossRef] [PubMed]
34. Margulies, M.; Egholm, M.; Altman, W.E.; Attiya, S.; Bader, J.S.; Bemben, L.A.; Berka, J.; Braverman, M.S.; Chen, Y.-J.; Chen, Z.; et al. Genome sequencing in microfabricated high-density picolitre reactors. *Nature* **2005**, *437*, 376–380. [CrossRef]
35. Tawfik, D.S.; Griffiths, A.D. Man-made cell-like compartments for molecular evolution. *Nat. Biotechnol.* **1998**, *16*, 291–294. [CrossRef]
36. Griffiths, A.D.; Tawfik, D.S. Miniaturising the laboratory in emulsion droplets. *Trends Biotechnol.* **2006**, *24*, 395–402. [CrossRef]
37. Chen, W.W.; Balaj, L.; Liau, L.M.; Samuels, M.L.; Kotsopoulos, S.K.; Maguire, C.A.; LoGuidice, L.; Soto, H.; Garrett, M.; Zhu, L.D.; et al. Beaming and droplet digital pcr analysis of mutant idh1 mrna in glioma patient serum and cerebrospinal fluid extracellular vesicles. *Mol. Ther.-Nucleic Acids* **2013**, *2*, e109. [CrossRef]
38. Dressman, D.; Yan, H.; Traverso, G.; Kinzler, K.W.; Vogelstein, B. Transforming single DNA molecules into fluorescent magnetic particles for detection and enumeration of genetic variations. *Proc. Natl. Acad. Sci. USA* **2003**, *80*, 8817–8822. [CrossRef]
39. Kumaresan, P.; Yang, C.J.; Cronier, S.A.; Blazej, R.G.; Mathies, R.A. High-throughput single copy DNA amplification and cell analysis in engineered nanoliter droplets. *Anal. Chem.* **2008**, *80*, 3522–3529. [CrossRef]
40. Chen, J.; Xu, X.; Huang, Z.; Luo, Y.; Tang, L.; Jiang, J.H. BEAMing LAMP: Single-molecule capture and on-bead isothermal amplification for digital detection of hepatitis C virus in plasma. *Chem. Commun.* **2018**, *54*, 291–294. [CrossRef]

41. Abate, A.R.; Chen, C.H.; Agresti, J.J.; Weitz, D.A. Beating Poisson encapsulation statistics using close-packed ordering. *Lab Chip.* **2009**, *9*, 2628–2631. [CrossRef] [PubMed]
42. Lee, D.H.; Park, J.K. Reduction in microparticle adsorption using a lateral interconnection method in a PDMS-based microfluidic device. *Electrophoresis* **2013**, *34*, 3119–3125. [CrossRef] [PubMed]
43. Edd, J.F.; Di Carlo, D.; Humphry, K.J.; Köster, S.; Irimia, D.; Weitz, D.A.; Toner, M. Controlled encapsulation of single-cells into monodisperse picolitre drops. *Lab Chip.* **2008**, *8*, 1262–1264. [CrossRef]
44. Wu, L.; Chen, P.; Dong, Y.; Feng, X.; Liu, B.F. Encapsulation of single cells on a microfluidic device integrating droplet generation with fluorescence-activated droplet sorting. *Biomed. Microdevices* **2013**, *15*, 553–560. [CrossRef]
45. Lagus, T.P.; Edd, J.F. High-throughput co-encapsulation of self-ordered cell trains: Cell pair interactions in microdroplets. *RSC Adv.* **2013**, *3*, 20512–20522. [CrossRef]
46. Ghaemi, R.; Dabaghi, M.; Attalla, R.; Shahid, A.; Hsu, H.H.; Selvaganapathy, P.R. Use of flame activation of surfaces to bond PDMS to variety of substrates for fabrication of multimaterial microchannels. *J. Micromech. Microeng.* **2018**, *28*, 087001. [CrossRef]
47. Kreutz, J.E.; Munson, T.; Huynh, T.; Shen, F.; Du, W.; Ismagilov, R.F. Theoretical design and analysis of multivolume digital assays with wide dynamic range validated experimentally with microfluidic digital PCR. *Anal. Chem.* **2011**, *83*, 8158–8168. [CrossRef]
48. Roberts, C.C.; Rao, R.R.; Loewenberg, M.; Brooks, C.F.; Galambos, P.; Grillet, A.M.; Nemer, M.B. Lab on a Chip Comparison of monodisperse droplet generation in flow-focusing devices with hydrophilic and hydrophobic surfaces. *Lab Chip.* **2012**, *12*, 1540–1547. [CrossRef]
49. Lee, W.; Walker, L.M.; Anna, S.L. Role of geometry and fluid properties in droplet and thread formation processes in planar flow focusing. *Phys. Fluids* **2009**, *21*, 032103. [CrossRef]
50. Chen, I.J.; Wu, T.; Hu, S. A hand-held, power-free microfluidic device for monodisperse droplet generation. *MethodsX* **2018**, *5*, 984–990. [CrossRef]
51. ten Klooster, S.; Sahin, S.; Schroën, K. Monodisperse droplet formation by spontaneous and interaction based mechanisms in partitioned EDGE microfluidic device. *Sci. Rep.* **2019**, *9*, 7820. [CrossRef] [PubMed]
52. Selck, D.A.; Karymov, M.A.; Sun, B.; Ismagilov, R.F. Increased robustness of single-molecule counting with microfluidics, digital isothermal amplification, and a mobile phone versus real-time kinetic measurements. *Anal. Chem.* **2013**, *85*, 11129–11136. [CrossRef] [PubMed]
53. Sun, B.; Shen, F.; McCalla, S.E.; Kreutz, J.E.; Karymov, M.A.; Ismagilov, R.F. Mechanistic evaluation of the pros and cons of digital RT-LAMP for HIV-1 viral load quantification on a microfluidic device and improved efficiency via a two-step digital protocol. *Anal. Chem.* **2013**, *85*, 1540–1546. [CrossRef] [PubMed]
54. Sun, B.; Rodriguez-Manzano, J.; Selck, D.A.; Khorosheva, E.; Karymov, M.A.; Ismagilov, R.F. Measuring fate and rate of single-molecule competition of amplification and restriction digestion, and its use for rapid genotyping tested with hepatitis C viral RNA. *Angew. Chemie Int. Ed.* **2014**, *53*, 8088–8092. [CrossRef]
55. Schoepp, N.G.; Schlappi, T.S.; Curtis, M.S.; Butkovich, S.S.; Miller, S.; Humphries, R.M.; Ismagilov, R.F. Pathogen-specific Phenotypic Antibiotic Susceptibility Test Directly from Clinical Samples in as Fast as 30 Minutes Using Digital LAMP Quantification. *Sci. Transl. Med.* **2017**, *3693*, 1–20.
56. Luo, K.; Chang, W.; Lee, G. An Integrated Array-Based Emulsion Droplet Microfluidic Device for Digital Loop-Mediated Isothermal Amplification (LAMP) Analysis. In Proceedings of the 2016 IEEE 11th Annual International Conference on Nano/Micro Engineered and Molecular Systems (NEMS), Sendai, Japan, 17–20 April 2016; pp. 17–20.
57. Gansen, A.; Herrick, A.M.; Dimov, I.K.; Lee, L.P.; Chiu, D.T. Digital LAMP in a sample self-digitization (SD) chip. *Lab Chip.* **2012**, *12*, 2247. [CrossRef] [PubMed]
58. Xu, P.; Zheng, X.; Tao, Y.; Du, W. Cross-Interface Emulsification for Generating Size-Tunable Droplets. *Anal. Chem.* **2016**, *88*, 3171–3177. [CrossRef]
59. Kim, H.; Choi, I.H.; Lee, S.; Won, D.-J.; Oh, Y.S.; Kwon, D.; Sung, H.J.; Jeon, S.; Kim, J. Deterministic bead-in-droplet ejection utilizing an integrated plug-in bead dispenser for single bead-based applications OPEN. *Sci. Rep.* **2017**, *7*, 46260. [CrossRef]
60. Iwai, K.; Sochol, R.D.; Lin, L. A Bead-in-Droplet Solution Exchange System via Continuous Flow Microfluidic Railing. In Proceedings of the IEEE International Conference on Micro Electro Mechanical Systems (MEMS), Taipei, Taiwan, 20–24 January 2013; pp. 1203–1206. [CrossRef]
61. Wang, C.H.; Lien, K.Y.; Wu, J.J.; Lee, G.B. A magnetic bead-based assay for the rapid detection of methicillin- resistant Staphylococcus aureus by using a microfluidic system with integrated loop-mediated isothermal amplification. *Lab Chip.* **2011**, *11*, 1521–1531. [CrossRef]
62. Price, A.K.; Macconnell, A.B.; Paegel, B.M. Microfluidic Bead Suspension Hopper. *Anal. Chem.* **2014**, *86*, 5039–5044. [CrossRef]
63. Vashist, S.K.; Luong, J.H.T. Antibody Immobilization and Surface Functionalization Chemistries for Immunodiagnostics. In *Handbook of Immunoassay Technologies: Approaches, Performances, and Applications*; Elsevier: Amsterdam, The Netherlands, 2018; pp. 19–46. [CrossRef]
64. Sassolas, A.; Hayat, A.; Marty, J.L. Immobilization of enzymes on magnetic beads through affinity interactions. *Methods Mol. Biol.* **2013**, *1051*, 139–148. [CrossRef] [PubMed]
65. Clark, I.C.; Abate, A.R. Microfluidic bead encapsulation above 20 kHz with triggered drop formation. *Lab Chip.* **2018**, *18*, 3598–3605. [CrossRef] [PubMed]

66. Collins, D.J.; Neild, A.; de Mello, A.; Liu, A.Q.; Ai, Y. The Poisson distribution and beyond: Methods for microfluidic droplet production and single cell encapsulation. *Lab Chip.* **2015**, *15*, 3439–3459. [CrossRef]
67. Shintaku, H.; Kuwabara, T.; Kawano, S.; Suzuki, T.; Kanno, I.; Kotera, H. Micro cell encapsulation and its hydrogel-beads production using microfluidic device. *Microsyst. Technol.* **2006**, *13*, 951–958. [CrossRef]
68. Zilionis, R.; Nainys, J.; Veres, A.; Savova, V.; Zemmour, D.; Klein, A.M.; Mazutis, L. Single-cell barcoding and sequencing using droplet microfluidics. *Nat. Publ. Gr.* **2016**, *12*, 44–73. [CrossRef]
69. Klein, A.M.; Mazutis, L.; Akartuna, I.; Tallapragada, N.; Veres, A.; Li, V.; Peshkin, L.; Weitz, D.A.; Kirschner, M.W. Droplet barcoding for single-cell transcriptomics applied to embryonic stem cells. *Cell* **2015**, *161*, 1187–1201. [CrossRef] [PubMed]
70. Kramer, T.; Rödiger, S. Absolute Quantification of Nucleic Acids on a Planar Droplet Digital PCR Array. 2016, 10–11. Available online: https://www.researchgate.net/publication/303895837_Absolute_Quantification_of_Nucleic_Acids_on_a_Planar_Droplet_Digital_PCR_Array?channel=doi&linkId=575b178108ae414b8e46774e&showFulltext=true (accessed on 5 November 2022). [CrossRef]
71. Rodiger, S.; Schierack, P.; Shroder, C. A Highly Versatile Microscope Imaging Technology Platform for the Multiplex Real-Time Detection of Biomolecules and Autoimmune Antibodies. *Adv. Biochem. Eng. Biotechnol.* **2013**, *133*, 35–74. [CrossRef]
72. Rödiger, S.; Liebsch, C.; Schmidt, C.; Lehmann, W.; Resch-Genger, U.; Schedler, U.; Schierack, P. Nucleic acid detection based on the use of microbeads: A review. *Microchim. Acta* **2014**, *181*, 1151–1168. [CrossRef]
73. Begolo, S.; Zhukov, D.V.; Selck, D.; Li, L.; Ismagilov, R.F. The pumping lid: Investigating multi-material 3D printing for equipment-free, programmable generation of positive and negative pressures for microfluidic applications. *Lab Chip.* **2014**, *14*, 4616–4628. [CrossRef]
74. Weigl, B.; Domingo, G.; LaBarre, P.; Gerlach, J. Towards non- and minimally instrumented, microfluidics-based diagnostic devices. *Lab Chip.* **2008**, *8*, 1999–2014. [CrossRef]
75. Sanchez Noriega, J.L.; Chartrand, N.A.; Valdoz, J.C.; Cribbs, C.G.; Jacobs, D.A.; Poulson, D.; Viglione, M.S.; Woolley, A.T.; Van Ry, P.M.; Christensen, K.A.; et al. Spatially and optically tailored 3D printing for highly miniaturized and integrated microfluidics. *Nat. Commun.* **2021**, *12*, 5509. [CrossRef] [PubMed]
76. Sajeesh, P.; Sen, A.K. Particle separation and sorting in microfluidic devices: A review. *Microfluid Nanofluid.* **2014**, *17*, 1–52. [CrossRef]

micromachines

Article

Surface Modification of 3D Printed Microfluidic Devices for Controlled Wetting in Two-Phase Flow

Chandler A. Warr [1], Nicole G. Crawford [1], Gregory P. Nordin [2] and William G. Pitt [1,*]

[1] Department of Chemical Engineering, Brigham Young University, Provo, UT 84602, USA
[2] Department of Electrical and Computer Engineering, Brigham Young University, Provo, UT 84602, USA
* Correspondence: pitt@byu.edu; Tel.: +1-801-422-2589

Abstract: Microfluidic devices (MFDs) printed in 3-D geometry using digital light projection to polymerize monomers often have surfaces that are not as hydrophobic as MFDs made from polydimethylsiloxane. Droplet microfluidics in these types of devices are subject to droplet adhesion and aqueous spreading on less hydrophobic MFD surfaces. We have developed a post-processing technique using hydrophobic monomers that renders the surfaces of these devices much more hydrophobic. The technique is fast and easy, and involves flowing monomer without initiator into the channels and then exposing the entire device to UV light that generates radicals from the initiator molecules remaining in the original 3-D polymerization. After treatment the channels can be cleared and the surface is more hydrophobic, as evidenced by higher contact angles with aqueous droplets. We hypothesize that radicals generated near the previously printed surfaces initiate polymerization of the hydrophobic monomers on the surfaces without bulk polymerization extending into the channels. The most hydrophobic surfaces were produced by treatment with an alkyl acrylate and a fluorinated acrylate. This technique could be used for surface treatment with other types of monomers to impart unique characteristics to channels in MFDs.

Keywords: 3D printing; microfluidic device; droplet formation; surface energy; post-polymerization processing; wetting

Citation: Warr, C.A.; Crawford, N.G.; Nordin, G.P.; Pitt, W.G. Surface Modification of 3D Printed Microfluidic Devices for Controlled Wetting in Two-Phase Flow. *Micromachines* **2023**, *14*, 6. https://doi.org/10.3390/mi14010006

Academic Editor: Pingan Zhu

Received: 7 November 2022
Revised: 12 December 2022
Accepted: 13 December 2022
Published: 20 December 2022

1. Introduction

Microfluidic devices have been rapidly gaining a foothold in advanced technology applications, including biomedical applications such as diagnostic analysis and personalized medicine [1,2]. Several manufacturing methods have been developed for microfluidic devices (MFD) [3], which we will define herein as devices with flow channels less than 1 mm in characteristic diameter. Both image projection (DLP) stereolithography and multi-jet printing methods have been applied to MFD fabrication, with the former demonstrating significantly higher resolution microfluidic features [4,5]. 3D printing microfluidic devices allows for more complex 3D geometries than some traditional manufacturing methods such as injection molding, but scale-up can be an issue [6]. However, with advancements in the field of 3D printing it has been theorized that by printing many devices at a time (~100) the cost and scalability of this manufacturing method may be commercially viable down the road [7]. While our interest is in medical devices, the same principles of manufacturing and of application apply to MFDs in other technology sectors [8].

One of our main areas of research focuses on forming aqueous droplets in a continuous oil phase [9]. Such droplets have been used as microreactors of pL to nL size in which to interrogate single cells (or particles) in a fluid environment to isolate the cell from other cells and from the external environment. These individualized microreactors have been used for genetic and proteomic analysis of cells [10,11] and of general cell growth when challenged with chemical or biological factors (antimicrobial agents) [12,13]. Often analytical output is conveyed by reporter molecules or reporter reactions from within the droplet. Because these microreactors are used for single cells analysis, it is essential to preclude merging with other droplets. It is also important to eliminate sticking of the droplets to the walls of the channels so the droplets can be conveyed through the MDF to an analytical station. Most often, anti-merging is accomplished by adding surfactants to the oil or aqueous phase, and sometimes by electrostatic charge on the droplets.

In two-phase flow with dispersed aqueous droplets, the oil phase generally consists of fluorocarbons because its high oxygen solubility provides adequate oxygen over time to growing cells [14,15]; however, sometimes hydrocarbon oils can be employed. For droplets in an oil phase, sticking to walls has not been reported to be a large problem, most likely because most MFDs are made by templating silicone rubber in molds etched in silicon [16]. These silicone rubber casts (such as polydimethylsiloxane, PDMS) are bonded to a transparent glass slide or plate using RF plasma treatment [17–19], forming an adhesive bond that is sufficiently tight and leak-proof. The silicone rubber walls and floor of the MFD are very hydrophobic, but the upper surface is hydrophilic glass, which is often treated with a commercial surfactant like Rain-X (ITW Global Brands, Houston, TX, USA) to render it fairly hydrophobic [19,20]. When passing 2-phase droplet suspensions, the silicone rubber and hydrophobized glass are sufficiently hydrophobic that aqueous droplets do not adhere in the presence of hydrocarbon or fluorocarbon oils. Wetting by water is precluded because contact of an aqueous fluid with a hydrophobic surface by displacing a hydrophobic fluid is not energetically favorable and does not occur spontaneously [21]. Thus wetting in 2-phase flow in 2-D silicone rubber MFDs has not been an issue for most aqueous droplets in oil.

Our lab has developed and promoted a novel 3-D printing process using acrylate monomers that produces a surface that is less hydrophobic than MFDs made from silicone rubber. This process was developed using a tetrafunctional PEG-diacrylate monomer to give sufficient flexibility (compliant) to the resulting polymer structures that thin film membranes can remain structurally intact during repeated flexing. The PEG diacrylate monomer used in this printing is amphiphilic. PEG is inherently hydrophilic, and the polymer acrylate sections of the polymerized network are much less hydrophilic. Thus the overall polymer chain structure is amphiphilic and is sufficiently dynamic that when confronted with a hydrocarbon based liquid at an interface, the PEG chains can be buried under the acrylate linkages to present a less hydrophilic surface to hydrocarbon surface. We suspect that similar dynamic rearrangement will bury PEG chains when presented with a fluorocarbon or hydrocarbon liquid at the surface. On the other hand, when presented with an aqueous phase at the interface, the PEG quickly migrates to the surface and creates a hydrophilic interface which interacts favorably with the aqueous phase. In the lab, such phenomenon produces what is called "contact angle hysteresis", in which the surface changes character over time to act more favorably with the liquid it is currently in contact with [22,23]. In a polymeric MFD this phenomenon is manifest by a channel wall being stably wetted by an oil phase (hydrocarbon or fluorocarbon oil) driven by the dynamic burial of hydrophilic groups (like PEG) under hydrophobic polymer groups; in such a case a very brief contact with aqueous droplets will not give time for molecular rearrangement to promote sticking to the wall. However, a prolonged contact (perhaps even shorter than a second) results in dynamic polymer rearrangement and attachment of an aqueous droplet to the surface, and perhaps subsequent spreading of that droplet over seconds or minutes [24].

Our previous publication on MFD design explored forming droplets in our unique annular channel-in-channel (ACC) microfluidic devices printed from various acrylic monomers. In properly designed droplet generators and flow channels, the sticking of droplets to channel walls could be avoided by using more hydrophobic acrylic monomers or by better mechanical design of the droplet generator. However, our previous devices used continual flow (fairly constant flow rate) of at least the continuous phase in order to continually pull off the droplet (dispersed phase) from the orifice of the ACC droplet generator [9]. In that type of flow, adjusting the base monomer was sufficient to form excellent droplets.

One of our current efforts is the development of a digital droplet-on-demand (d-DoD) system in which the flow of the aqueous droplets and the continuous oil phases are intermittent and are driven by digital positive displacement pumps built into the microfluidic device [5,7,9,25–27], as opposed to constant pressure flows or forced flows from syringe pumps. During these start and stop flows, a droplet may contact a wall for more than a few hundred milliseconds. Even more problematic is that during formation of aqueous droplets with digital pumping systems, the aqueous droplet is perched on or in the orifice of the droplet generator. This leads to wetting and spreading of the aqueous phase at the aqueous-oil interface at the orifice of the droplet generator. The pulsatile nature of our d-DoD system produces a pause in time during droplet formation in which the aqueous phase rests at the water/oil junction at the top of the droplet generator (called the pedestal). During each pause of several hundred milliseconds, it appears that the aqueous phase slightly but progressively spreads along the amphiphilic polymer material and progressively increases the wetted area at the top of the pedestal. As this wetted area increases over time the droplet formation arises from a wider base, resulting in satellite droplets forming in the d-DoD system. As seen in Figure 1a, the wetted region on top of the pedestal is much larger than the designed area intended for the release of fluid for droplet formation (Figure 1b). These droplets are often accompanied by more satellite droplets (see Figure 1d) whose number increases over time as the top of the pedestal is progressively wetted by the spreading aqueous phase. Ideally, all particles or cells intended to be encapsulated should be captured within a main droplet, not inside satellite droplets. Production of satellite droplets also decreases the volume of the primary drop. We hypothesized that increasing the hydrophobicity of the 3D-printed PEGDA material would decrease the magnitude (both number and size) of these satellite droplets.

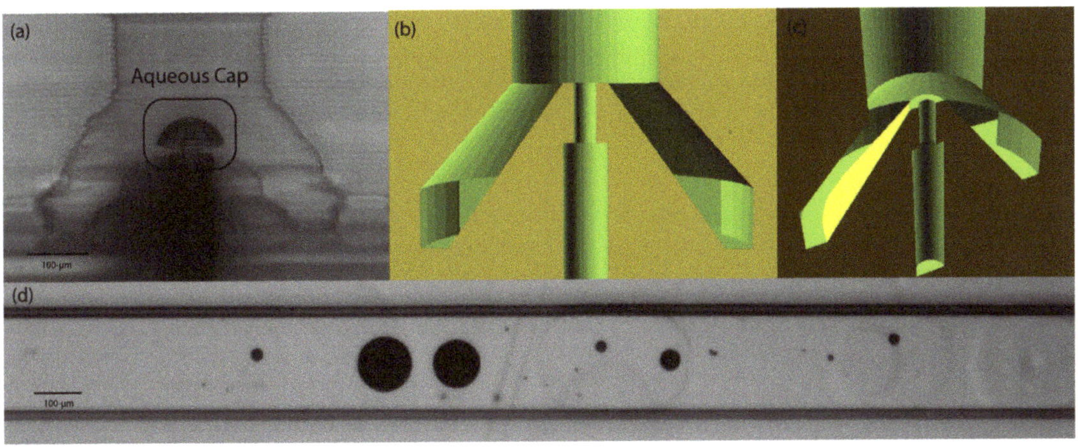

Figure 1. Droplet-on-demand generator printed from PEG-diacrylate monomer. (**a**) Microphotograph of the pedestal of the droplet generator showing the darker aqueous phase spreading over the top of the cone-shaped pedestal. Oil flows up from the outside of the cone, and the aqueous phase flows up the channel in the central core of the cone. (**b**) Schematic showing a side view of the polymerized polymer (in yellow) and the open channels in green forming the cone and pedestal. (**c**) Schematic showing an upward-looking view of the pedestal, which is the light green disk at the top of the darker-green cone. (**d**) Example of droplets and satellites formed when the upper surface of the pedestal is wetted with the aqueous phase.

Initial experimentation to coat the internal microfluidic channels of the 3D-printed PEGDA polymer showed that it is possible to coat the channels using an acrylate monomer with a hydrophobic group attached to it. The current processing protocol for our 3D-printed devices involves a post-exposure which increases the degree of polymerization in the printed material, thus increasing the lifespan of delicate features. This post-exposure is done using a 430 nm light source, which is outside of the effective range of the UV absorber, but still within the range of the photoinitiator. Thus additional polymerization can occur after the initial printing and removal from the printer. The fact that this increases the lifespan of these features and the hardness of the material means that there must be viable photoinitiator and perhaps monomer still present in the polymer [7,25]. It was postulated that some unreacted acrylate groups were also present on the surface of the channel walls and that those may still be available to react with additional monomer. We theorized that this viable photoinitiator and possible unreacted monomers could be used to locally react with a "post-printing-added" acrylate monomer on the surface of the channel walls. Our results show that this is the case and that more hydrophobic surfaces can be attained.

Other published approaches to controlling material and surface properties for parts made by stereolithographic printing include using more than one material in the same print [28,29]. While potentially possible for producing hydrophobic surfaces on a channel by using a hydrophobic print material, this greatly slows down the printing while one material is being swapped out for another material in the printer. For multiple channels with complex geometries, such a process may not be feasible because formation of a vertical channel would require manually swapping monomers during every layer of print.

2. Materials and Methods

2.1. Custom 3D Printer

The 3D printer used in this study was custom made and is described in more detail in other publications [5,7,25,27,30,31]. Briefly, the printer is a digital light projection (DLP) style 3D printer using a Visitech 2560 × 1600 pixel light engine with a 365 nm LED as the projection source and a 100 mm linear Griffin Motion stage as the z-axis control. Additionally, it has several other tip/tilt and focus calibrations built in to bring the custom resin tray in planar focus with the light engine. Custom Python software is used to control calibration parameters and handle printing such that individual printing layers can have their own (or multiple) exposure times depending on the image projected onto the focus plane. This allows for precise control over printing parameters and features.

2.2. Materials

Custom resins were made according to a previously established procedure [30,31] which includes mixing a photopolymerizable acrylate monomer with a UV absorber and photoinitiator. The resin used in this study has been used in several other published studies [5,7,25,27,30,31] and works well to produce consistent features and to print uniformly. It contains a polyethylene glycol diacrylate (PEGDA) monomer, 1% Irgacure 819 as a photoinitiator, and 0.38% avobenzone as a UV absorber. This absorber and photoinitiator combination works well with the 365 nm LED in used in the Visitech light engine. In addition to the 3D printing resin, three other acrylate monomers were used as experimental surface treatments for the study; they are hexanediol diacrylate (HDDA), lauryl acrylate (LA), and tridecafluorooctyl acrylate (FA), which were used as received and whose chemical structures are depicted in Figure 2c. These particular acrylate monomers were chosen because of their hydrophobic chemistry, which potentially could be imparted to the surface of the 3D printed channels. It is worth noting that HDDA is the only difunctional acrylate monomer used in the treatment and has the potential of forming a branched or networked polymer coating. The LA and FA monomers have a single acrylate group with a long hydrophobic tail and are capable only of forming linear polymers attached to the surface.

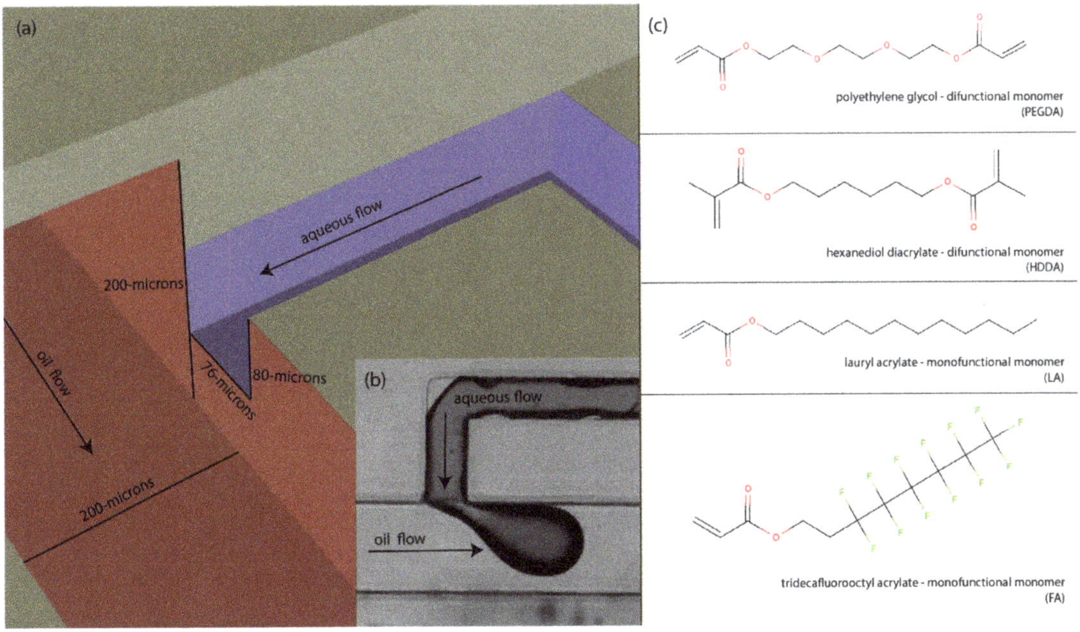

Figure 2. T-junction geometry and monomer chemistry: (**a**) Schematic of cutout cross section of the T-Junction droplet generator with the appropriate dimensions of the oil and aqueous flow channels. (**b**) Micrograph of dyed aqueous fluid forming droplet during operation. In this top-down view the width of the aqueous channel is 76 μm and the width of the oil channel is 200 μm. (**c**) The four photopolymerizable acrylate monomers used in study, including the 3D printing polymer (PEGDA) and the three monomers used for surface treatment (HDDA, LA, FA).

2.3. Optical System

The optical system used to visualize and photograph the MFDs and droplets in this study is a Nikon TE300 inverted microscope equipped with a FLIR Blackfly S USB3 (Teledyne FLIR, Wilsonville, OR, USA) monochrome camera. This camera has high sensitivity and allows computer control of exposure and gain for consistent and comparable image capture.

2.4. 3D Printed T-Junction

A central T-junction droplet generator was chosen for this study because it forces contact between the dispersed aqueous phase and the wall during the formation of droplets and is sensitive to any change in surface hydrophobicity. This droplet generator was chosen over other geometries, such as co-flowing (pinch off), because while these other geometries are more practical for MFD design, they are not considered as sensitive for this type of testing in which we encourage (and even force) the aqueous phase to contact the wall. Throughout this study the T-junction droplet generator geometry was held consistent with dimensions as shown in Figure 2a. Briefly, the continuous oil channels have a square cross section with 200-μm height and width, while the aqueous flow channel has an almost square cross section with a 76-μm width and 80-μm height. These dimensions utilize the 7.6 pixel pitch of the light engine as well as the standard 10-μm layer height used in this custom printer. As seen in Figure 2a, the water outlet has smaller channel dimensions than the continuous oil channel and is centered on the face of the wall. This was done in order to avoid contact of the forming droplet with the top (ceiling) and bottom (floor) surfaces but force direct contact of the forming droplet with the adjacent sidewall of the oil channel. Also of note, the dispersed aqueous phase was dyed to aid in visual differentiation between the two immiscible phases.

2.5. Acrylate Monomer Treatment and Optical Exposure

The acrylate monomer postprint treatment was performed according to the following procedure. (1) The device was removed from the printer and thoroughly washed with isopropyl alcohol (IPA) including all internal microfluidic channels. (2) The hydrophobic monomer was then injected through the relevant fluid channels by hand with a small syringe. (3) The device was then immediately placed under the postexposure station for the UV treatment. The post exposure station consists of a UV postexposure using a 430 nm LED within a 3D printed housing. As mentioned above, this wavelength LED was chosen because it falls within the excitation range of the photoinitiator while being outside of the effective range of the UV absorber, which means that additional polymerization could occur within the polymer (or on the polymer surface) after the initial print is complete. A postprint exposure like this is very common among commercial printers using this same polymerization method. In the experimental design, the treatment dose was varied by changing the exposure time each device had within the post curing station. The maximum treatment dose was initially determined to be a dose sufficient to form walls around a channel, but not cause polymerization within a channel during the normal MFD printing protocol. For example, an optical dose in which during normal printing caused the monomer to not be cleared (by flushing with isopropanol) from the channel, was decreased by 20%; this optical dose was tested again (and decreased again if needed) to iteratively find a dose that would repeatedly clear, and above which the channel would not clear of the monomer. This was the UV dose (time and intensity) used subsequently during postprint treatment with candidate monomers in the channels. In some experiments, multiple treatments were executed sequentially. Since the monomers used in postprint surface treatment did not contain photoinitiator, we did not expect polymerization to occur in the lumen of the channels, but only upon the surface where residual photoinitiator from the original 3D print might reside.

3. Results and Discussion

We hypothesized that it would be possible to use the postulated presence of unreacted initiator at the surface of previously formed channels to polymerize a layer of hydrophobic acrylate monomer. We evaluated this hypothesis by evaluating any change in the surface hydrophobicity by forming water-in-oil droplets in these devices and forcing them against the treated surface. This method of using residual initiator and fresh acrylate monomer to polymerize a surface coating could in theory be used for many other applications to impart particular surface chemistry in MFD channels; however in the study we only investigated the surface hydrophobicity. Several different treatment conditions were examined using three hydrophobic monomers, and then the monomer that best enhanced surface hydrophobicity was used on the ACC droplet generator to investigate how digital droplet generation was influenced.

3.1. Treatment Conditions Using Acrylate Monomers with Residual Initiator

3D printed microfluidic devices were manufactured containing the previously described T-Junction droplet generator followed by a long serpentine channel, a section of which had a narrow channel to force contact between the droplets and the channel wall. Using the treatment protocol described above, some MFDs were put through several cycles of treatment with each of the monomers in order to increase any effect the hydrophobic monomers may have had on the channels. Evaluation involved rating the effectiveness of various treatments on both droplet formation and static (no flow) droplet contact in the narrow channel.

The results are shown in Figure 3, in which the darker phase is the dispersed aqueous phase. Both "formation" and "static" images are shown to reveal how a droplet forms at the T-junction (formation) and how the droplet wets after formation during static (no flow) contact with the channel walls. Micrographs of the flow in the base PEGDA material are shown in column A of Figure 3. No consistent droplet formation was observed in the native

PEGDA material, and downstream of the droplet generator a coflowing phase separated flow was seen as previously reported [9]. Initial testing of the hydrophobic monomer treatments showed that there was indeed a difference in the droplet formation ability of the material after a single treatment. Furthermore, we found that multiple cycles of identical treatment enhanced the hydrophobicity, so three serial treatments were performed with the different hydrophobic monomers. As shown in column B of Figure 3, one treatment of lauryl acrylate was insufficient to allow for consistent droplet formation dramatically different from the PEGDA control. However, when serial treatments were applied, the qualitative hydrophobicity was improved with each subsequent treatment; after the third treatment a non-spreading droplet can be seen forming with the T-junction geometry. Columns C and D of Figure 3 show a similar trend for the results of treatments using hexanediol diacrylate and the fluorinated acrylate. All three individual hydrophobic monomers show progressive improvement for subsequent treatments, and after three photo-treatments all monomer types show comparable droplet formation results.

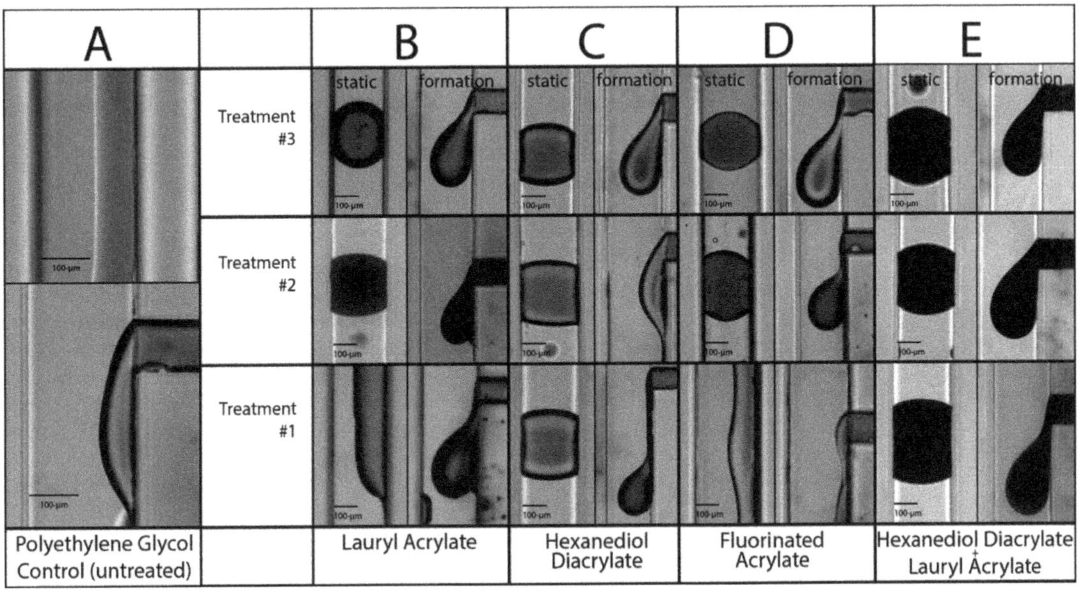

Figure 3. Droplet generation from T-junction with various hydrophobic acrylate coatings; each treatment involved a coating of the acrylate monomer followed by a UV exposure and washing with IPA. Multiple treatments were used to increase the effectiveness of the coatings. Column (**A**): PEG-diacrylate-polymerized T-junction (lower) and serpentine channel (upper) showing 2-phase flow and no droplets formation on untreated polymer. (**B**): Droplet formation on base-polymer treated once (lower), twice (middle) or three times (upper) with lauryl acrylate. (**C**): Droplet formation on base-polymer treated once (lower), twice (middle) or three times (upper) with hexanediol diacrylate. (**D**): Droplet formation on base-polymer treated once (lower), twice (middle) or three times (upper) with fluorinated acrylate. (**E**): Droplet formation on base-polymer treated once with hexanediol diacrylate and then treated once (lower), twice (middle) or three times (upper) with lauryl acrylate.

An additional test condition was performed which used a single hexanediol diacrylate base treatment followed by three standard lauryl acrylate treatments (see column E of Figure 3). Rationale for this treatment was based on the postulate that the HDDA monomer may present additional functional groups on the surface after the first treatment because of its two acrylate groups, compared to the other two monofunctional acrylates tested.

However, after three treatments of the lauryl acrylate following the HDDA base layer, no significant improvement was observed compared the other monomer treatments.

These results show that the method of attaching hydrophobic acrylates via residual UV initiator at the surface increases the hydrophobic quality of the microfluidic channels such that droplet formation becomes practical, as demonstrated with a variety of monomer treatments. It is also worth noting that there is likely some interaction between the continuous oil phase and the hydrophobic monomer that has an effect on droplet formation. For example, a hydrocarbon mineral oil was used as the continuous phase for these tests, which may be more chemically compatible with the lauryl acrylate compared to the fluorinated acrylate; likewise, a fluorinated oil may wet the channel wall better when coated with the fluorinated acrylate.

We propose that a significant benefit of this technique is that it provides a covalent attachment of the modifying molecules to the surface of channels in the MFD. Other methods that involve surfactact coatings or non-reactive processes may have potential to be removed over time. More research is needed to validate the persistence of the surface coating over hours or days of flow. These results suggest that this technique may be useful to covalently attach other functional groups onto the surface of a DLP 3D-printed microfluidic device—molecules such as proteins for biological anchorage, carbohydrates for cell signaling, fluorescent acrylates for visualization, or other such applications. We note that the results presented herein only explored a small subset of possible surface treatments. In addition to other types of monomers, future work can include more variations in post-exposure time, intensity, and wavelength, and perhaps in the initiator concentration.

3.2. *Application in Annular Channel-in-Channel Droplet Generator*

For subsequent investigations on our ACC droplet generator, the three-fold treatment with lauryl acrylate was used because of the ease of use of the monomer (low viscosity), low cost, and rapid curing treatment times. The three-fold surface treatment with FA was applied to the d-DoD system. Figure 4a shows the droplets formed with the native PEGDA polymer without any additional treatment. Many satellite droplets were seen (see Figure 1d), and the top of the pedestal of the droplet generator had an obvious aqueous ring around it that was hypothesized to be creating spurious satellite droplets during drop formation. We hypothesize that the stretched aqueous phase has a larger cross section that breaks into a primary drop and many satellite droplets, rather than a single primary drop.

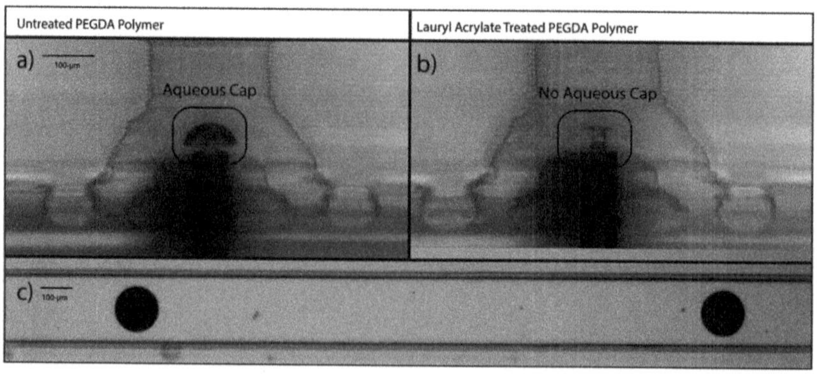

Figure 4. Before and after images of droplet generator. (**a**) Microphotograph of droplet generator made from PEG-diacrylate with no surface treatment after several minutes of droplet generation, showing an aqueous phase in the form of a cap attached to the top of the pedestal. (**b**) Microphotograph of droplet generator made from PEG-diacrylate with three serial treatments of lauryl acrylate, showing no aqueous cap on the pedestal after several minutes of droplet generation. (**c**) Microphotograph of droplets flowing in a serpentine channel, which droplets were formed in the MFD made with three serial treatments of lauryl acrylate.

Figure 4b show the same d-DoD geometry during droplet formation (there is aqueous phase in the core of the cone) after the three treatments of lauryl acrylate post-processing, which shows no aqueous film attachment at the top of the pedestal and much cleaner droplets formed throughout the production of droplets (Figure 4c), indicating the effectiveness of the lauryl acrylate postprocessing treatment in forming better (consistently reproducible) droplets. For example, please compare Figure 4c to Figure 1d. Future work with high speed cameras could be used to confirm this hypothesized formation process.

4. Conclusions

We have successfully investigated and developed a post-printing process that reduces the wetting and adhesion of aqueous droplets in a continuous oil-phase flow in a microfluidic device. The procedure consists of taking a 3D printed microfluidic device made from PEG diacrylate monomers, rinsing the unprinted monomer from the flow channels, refilling the flow channels with a different and more hydrophobic monomer not containing initiator, and then exposing the entire device to UV light to initiate polymerization of the hydrophobic monomer to the channel surfaces. Following exposure the unpolymerized monomer was rinsed from the MFD. Results show that lauryl acrylate, fluorinated acrylate, and hexanediol diacrylate increased the hydrophobicity of the channel surfaces. Repeated surface treatments increased the hydrophobicity further. This same method can have application in applying other surface coatings to MFDs, such as acrylate monomers having functional groups that are fluorescent, that possess unique chemistry or that present charged groups at the surface, that bind or repel targeted chemicals in the flow, or that contain proteins or nucleic acid groups for binding of cells or DNA to the surface. This surface treatment has great potential to provide unique chemistry in MFDs.

Author Contributions: Conceptualization, C.A.W., G.P.N. and W.G.P.; experimental work, C.A.W. and N.G.C.; writing—original draft preparation, C.A.W. and W.G.P.; writing—review and editing, all authors. All authors have read and agreed to the published version of the manuscript.

Funding: This research was partly funded by the National Institutes of Health, grant numbers R15GM123405-02 and R01AI116989 and the Department of Chemical Engineering at Brigham Young University.

Data Availability Statement: There is no numeric data associated with this study. Photographs can be obtained by contacting the authors: chandlerwarr@gmail.com or pitt@byu.edu.

Acknowledgment: We acknowledge and appreciate the assistance of Parker Johns in preparing samples for this study.

Conflicts of Interest: The authors declare no conflict of interest relating to this research and publication. One of the authors (G.P.N.) owns shares in Acrea 3D, a company commercializing microfluidic 3D printing.

References

1. Eduati, F.; Utharala, R.; Madhavan, D.; Neumann, U.P.; Longerich, T.; Cramer, T.; Saez-Rodriguez, J.; Merten, C.A. A microfluidics platform for combinatorial drug screening on cancer biopsies. *Nat. Commun.* **2018**, *9*, 2434. [CrossRef] [PubMed]
2. Li, Y.; Cherukury, H.; Labanieh, L.; Zhao, W.; Kang, D.K. Rapid Detection of beta-Lactamase-Producing Bacteria Using the Integrated Comprehensive Droplet Digital Detection (IC 3D) System. *Sensors* **2020**, *20*, 4667. [CrossRef] [PubMed]
3. Naderi, A.; Bhattacharjee, N.; Folch, A. Digital Manufacturing for Microfluidics. *Annu. Rev. Biomed. Eng.* **2019**, *21*, 325–364. [CrossRef] [PubMed]
4. Gong, H.; Bickham, B.P.; Woolley, A.T.; Nordin, G.P. Custom 3D printer and resin for 18 mum × 20 mum microfluidic flow channels. *Lab Chip* **2017**, *17*, 2899–2909. [CrossRef] [PubMed]
5. Sanchez Noriega, J.L.; Chartrand, N.A.; Valdoz, J.C.; Cribbs, C.G.; Jacobs, D.A.; Poulson, D.; Viglione, M.S.; Woolley, A.T.; Van Ry, P.M.; Christensen, K.A.; et al. Spatially and optically tailored 3D printing for highly miniaturized and integrated microfluidics. *Nat. Commun.* **2021**, *12*, 5509. [CrossRef] [PubMed]
6. Nguyen, T.; Chidambara, V.A.; Andreasen, S.Z.; Golabi, M.; Huynh, V.N.; Linh, Q.T.; Bang, D.D.; Wolff, A. Point-of-care devices for pathogen detections: The three most important factors to realise towards commercialization. *TrAC-Trends Anal. Chem.* **2020**, *131*, 116004. [CrossRef]
7. Gong, H.; Woolley, A.T.; Nordin, G.P. 3D printed high density, reversible, chip-to-chip microfluidic interconnects. *Lab Chip* **2018**, *18*, 639–647. [CrossRef]

8. Lapizco-Encinas, B.H.; Zhang, Y.V. Microfluidic systems in clinical diagnosis. *Electrophoresis* **2022**, 1–29. [CrossRef]
9. Warr, C.A.; Hinnen, H.S.; Avery, S.; Cate, R.J.; Nordin, G.P.; Pitt, W.G. 3D-Printed Microfluidic Droplet Generator with Hydrophilic and Hydrophobic Polymers. *Micromachines* **2021**, *12*, 91. [CrossRef]
10. Agnihotri, S.N.; Ugolini, G.S.; Sullivan, M.R.; Yang, Y.; De Ganzo, A.; Lim, J.W.; Konry, T. Droplet microfluidics for functional temporal analysis and cell recovery on demand using microvalves: Application in immunotherapies for cancer. *Lab Chip* **2022**, *22*, 3258–3267. [CrossRef]
11. Zhang, X.; Li, T.; Liu, F.; Chen, Y.; Yao, J.; Li, Z.; Huang, Y.; Wang, J. Comparative Analysis of Droplet-Based Ultra-High-Throughput Single-Cell RNA-Seq Systems. *Mol. Cell* **2019**, *73*, 130–142.e135. [CrossRef] [PubMed]
12. Hsieh, K.; Mach, K.E.; Zhang, P.; Liao, J.C.; Wang, T.H. Combating Antimicrobial Resistance via Single-Cell Diagnostic Technologies Powered by Droplet Microfluidics. *Acc. Chem. Res.* **2022**, *55*, 123–133. [CrossRef] [PubMed]
13. Postek, W.; Garstecki, P. Droplet Microfluidics for High-Throughput Analysis of Antibiotic Susceptibility in Bacterial Cells and Populations. *Acc. Chem. Res.* **2022**, *55*, 605–615. [CrossRef] [PubMed]
14. Shemesh, J.; Ben Arye, T.; Avesar, J.; Kang, J.H.; Fine, A.; Super, M.; Meller, A.; Ingber, D.E.; Levenberg, S. Stationary nanoliter droplet array with a substrate of choice for single adherent/nonadherent cell incubation and analysis. *Proc. Natl. Acad. Sci. USA* **2014**, *111*, 11293–11298. [CrossRef]
15. Gach, P.C.; Shih, S.C.; Sustarich, J.; Keasling, J.D.; Hillson, N.J.; Adams, P.D.; Singh, A.K. A Droplet Microfluidic Platform for Automating Genetic Engineering. *ACS Synth. Biol.* **2016**, *5*, 426–433. [CrossRef]
16. Subramanian, B.; Kim, N.; Lee, W.; Spivak, D.A.; Nikitopoulos, D.E.; McCarley, R.L.; Soper, S.A. Surface modification of droplet polymeric microfluidic devices for the stable and continuous generation of aqueous droplets. *Langmuir* **2011**, *27*, 7949–7957. [CrossRef]
17. Taylor, D.; Verdon, N.; Lomax, P.; Allen, R.J.; Titmuss, S. Tracking the stochastic growth of bacterial populations in microfluidic droplets. *Phys. Biol.* **2022**, *19*, 026003. [CrossRef]
18. Li, H.; Torab, P.; Mach, K.E.; Surrette, C.; England, M.R.; Craft, D.W.; Thomas, N.J.; Liao, J.C.; Puleo, C.; Wong, P.K. Adaptable microfluidic system for single-cell pathogen classification and antimicrobial susceptibility testing. *Proc. Natl. Acad. Sci. USA* **2019**, *116*, 10270–10279. [CrossRef]
19. Zhang, P.; Kaushik, A.M.; Hsieh, K.; Li, S.; Lewis, S.; Mach, K.E.; Liao, J.C.; Carroll, K.C.; Wang, T.H. A Cascaded Droplet Microfluidic Platform Enables High-Throughput Single Cell Antibiotic Susceptibility Testing at Scale. *Small Methods* **2022**, *6*, e2101254. [CrossRef]
20. Zhang, P.; Kaushik, A.; Hsieh, K.; Wang, T.H. Customizing droplet contents and dynamic ranges via integrated programmable picodroplet assembler. *Microsyst. Nanoeng.* **2019**, *5*, 22. [CrossRef]
21. Pitt, W.G. The Free Energy of Wetting. *Chem. Eng. Educ.* **1993**, *27*, 184–193.
22. Ruiz-Cabello, F.J.M.; Rodríguez-Valverde, M.A.; Cabrerizo-Vilchez, M.A. Contact Angle Hysteresis on Polymer Surfaces: An Experimental Study. *J. Adhes. Sci. Technol.* **2011**, *1*, 2039–2049. [CrossRef]
23. Gao, L.; McCarthy, T.J. Contact angle hysteresis explained. *Langmuir* **2006**, *22*, 6234–6237. [CrossRef] [PubMed]
24. Tretinnikov, O.N.; Ikada, Y. Dynamic Wetting and Contact Angle Hysteresis of Polymer Surfaces Studied with the Modified Wilhelmy Balance Method. *Langmuir* **1994**, *10*, 1606–1614. [CrossRef]
25. Beauchamp, M.J.; Gong, H.; Woolley, A.T.; Nordin, G.P. 3D Printed Microfluidic Features Using Dose Control in X, Y, and Z Dimensions. *Micromachines* **2018**, *9*, 326. [CrossRef]
26. Gong, H.; Woolley, A.T.; Nordin, G.P. High density 3D printed microfluidic valves, pumps, and multiplexers. *Lab Chip* **2016**, *16*, 2450–2458. [CrossRef]
27. Gong, H.; Woolley, A.T.; Nordin, G.P. 3D printed selectable dilution mixer pumps. *Biomicrofluidics* **2019**, *13*, 014106. [CrossRef]
28. Mannel, M.J.; Weigel, N.; Hauck, N.; Heida, T.; Thiele, J. Combining Hydrophilic and Hydrophobic Materials in 3D Printing for Fabricating Microfluidic Devices with Spatial Wettability. *Adv. Mater. Technol.-Us* **2021**, *6*, 2100094. [CrossRef]
29. Ji, Q.L.; Zhang, J.M.; Liu, Y.; Li, X.Y.; Lv, P.Y.; Jin, D.P.; Duan, H.L. A Modular Microfluidic Device via Multimaterial 3D Printing for Emulsion Generation. *Sci. Rep.* **2018**, *8*, 4791. [CrossRef]
30. Gong, H.; Beauchamp, M.; Perry, S.; Woolley, A.T.; Nordin, G.P. Optical Approach to Resin Formulation for 3D Printed Microfluidics. *RSC Adv.* **2015**, *5*, 106621–106632. [CrossRef]
31. Warr, C.; Valdoz, J.C.; Bickham, B.P.; Knight, C.J.; Franks, N.A.; Chartrand, N.; Van Ry, P.M.; Christensen, K.A.; Nordin, G.P.; Cook, A.D. Biocompatible PEGDA Resin for 3D Printing. *ACS Appl. Bio. Mater.* **2020**, *3*, 2239–2244. [CrossRef] [PubMed]

micromachines

Article

Dynamic Behaviours of Monodisperse Double Emulsion Formation in a Tri-Axial Capillary Device

Yuchen Dai, Haotian Cha, Nhat-Khuong Nguyen, Lingxi Ouyang, Fariba Galogahi, Ajeet Singh Yadav, Hongjie An, Jun Zhang, Chin Hong Ooi and Nam-Trung Nguyen *

Queensland Micro-Nanotechnology Centre, Griffith University, Nathan, QLD 4111, Australia
* Correspondence: nam-trung.nguyen@griffith.edu.au

Abstract: We investigated experimentally, analytically, and numerically the formation process of double emulsion formations under a dripping regime in a tri-axial co-flow capillary device. The results show that mismatches of core and shell droplets under a given flow condition can be captured both experimentally and numerically. We propose a semi-analytical model using the match ratio between the pinch-off length of the shell droplet and the product of the core growth rate and its pinch-off time. The mismatch issue can be avoided if the match ratio is lower than unity. We considered a model with the wall effect to predict the size of the matched double emulsion. The model shows slight deviations with experimental data if the Reynolds number of the continuous phase is lower than 0.06 but asymptotically approaches good agreement if the Reynolds number increases from 0.06 to 0.14. The numerical simulation generally agrees with the experiments under various flow conditions.

Keywords: double emulsion; core–shell droplet; microfluidics; tri-axial capillary; computational fluid dynamics; dripping regime

Citation: Dai, Y.; Cha, H.; Nguyen, N.-K.; Ouyang, L.; Galogahi, F.; Yadav, A.S.; An, H.; Zhang, J.; Ooi, C.H.; Nguyen, N.-T. Dynamic Behaviours of Monodisperse Double Emulsion Formation in a Tri-Axial Capillary Device. *Micromachines* **2022**, *13*, 1877. https://doi.org/10.3390/mi13111877

Academic Editor: Pingan Zhu

Received: 13 October 2022
Accepted: 28 October 2022
Published: 31 October 2022

1. Introduction

Double emulsions core–shell droplets are referred to as dispersed droplets containing a smaller one as the core. Double emulsions are highly desired for applications in drug delivery, food science, controlled release of substances, etc., because they can encapsulate a cargo in the core. Due to its significance, various designs for the formation of double emulsions have been extensively studied. Double emulsions are performed either simultaneously or successively and categorised accordingly into one-step and two-step processes. The former is simple and robust [1], while the latter is more efficient and controllable [2]. The one-step approach is the focus of our present work. One-step formation of double emulsions has been achieved with microfluidic devices. The working principle relies on the so-called Rayleigh-Plateau instability, where a liquid jet becomes unstable with its wavelength larger than the circumference. The jet consequently breaks up into segments with minimum surface area, forming a spherical shape [3,4]. Utada et al. [5] connected a glass cylindrical capillary with a square glass one and forced all liquid phases through an exit downstream. Varying the dimension of the outlet orifice can adjust the size of the droplets. The team also employed the breakup mechanism of a single emulsion to explain the size distribution of the double ones and defined an effective capillary number to classify the formation process as dripping and jetting regimes. Nie et al. [6] developed a co-flow microfluidic device in polyurethane elastomer to form double emulsions. The sizes of the core and the outer droplets were tuned with the flow rates of the three liquid phases. In general, double emulsion droplets with diameters from ten to hundreds of micrometres can be formed [1,5,7,8]. However, larger core–shell droplets with diameters ranging from hundreds of micrometres to several millimetres are also required for applications such as cell manipulations [9,10], macrocapsules [11], and microreactors [12]. Producing double emulsion droplets in this size range using typical microfluidic devices requires a significantly low flow rate of the outer phase, usually resulting in a drop accumulation issue because the gravity or buoyancy effect dominates the inertial drag. To prevent this problem, capillary devices were oriented vertically and scaled up to the so-called "millifluidics" [13]. Note that the term "microfluidics" is still valid for droplet smaller than a millimetre even if the device dimension is larger than that [14,15]. Shao et al. [15] experimentally investigated double emulsions in a dual-coaxial capillary with a 2.2 mm diameter tube as the outer channel. Adjusting the relative position of the ends of the inner and middle capillaries can switch the operation between two-step and one-step process. Their results demonstrated that the dynamic effect of the inner phase on the size of shell droplet is insignificant in the two-step set up, but drastically affecting it in the one-step process. Additionally, the team found that the mismatch in formation frequencies of core and shell droplets in the two-step process can be avoided with the one-step device. Schmit et al. [16] utilised a pendant drop method to produce millimetre-size double emulsions in a capillary tube with a 4 mm outer diameter. Their experimental results quantitatively demonstrated the effects of the inner and middle phase flow rates on the number as well as the size of the core in each droplet.

Though it is ultimately essential to obtain experimental data on double emulsions, numerical simulations are also of great importance for understanding and optimising the operation parameters for droplet formations. The velocity and pressure fields can be readily visualised, which are not always accessible with experimental methods. Zhou et al. [17] simulated double emulsions in a capillary device using a diffuse-interface framework. The calculation domain was simplified as a 2-dimensional axisymmetric geometry. The interfacial thickness and position were determined by a phase-field variable. By applying a similar method, Park and Anderson [18] successfully predicted the dripping and jetting regimes as well as their transition for the capillary device reported by Utada et al. [5]. Vu et al. [19] numerically investigated double emulsions in a capillary device using the front-tracking method proposed by Tryggvason et al. [20]. The results showed that the jetting mode can be promoted by increasing the Reynolds number (*Re*) and the Webber number (*We*). Fu et al. [21] established a model based on the ternary Lattice Boltzmann method to simulate the one-step process. Herrada et al. [22] carried out numerical simulations to validate their linear stability analysis on double emulsions in a capillary device. The team adopted the volume-of-fluid (VOF) method to simulate multiphases and interfaces. The effects of viscosity ratio, flow rate ratio, and interfacial tension ratio on the one-step process were also quantitatively demonstrated using the VOF method [23,24]. Azarmanesh et al. [25] simulated both one-step and two-step processes using the VOF method and found that the Capillary number (*Ca*) of the inner phase can significantly affect the outer droplet size in the two-step process, while the *Ca* of the outer phase plays a significant role in one-step counterpart. More recently, Yang et al. [26] numerically investigated the deformation behaviours of one-step double emulsions using VOF method. Examining the streamlines in the channel, the team discovered that the core moving forward relative to the shell was caused by the large vortex passing through the core–shell interface.

The dripping regime is well known to produce a monodisperse droplet size distribution while the jetting regime leads to a polydisperse counterpart. For most applications, a monodisperse emulsion is preferred. As a result, a simple force balance might be conducted to elucidate the formation mechanism and to predict the size distribution for various flow conditions. In general, the forces exerted on a double emulsion are the same as those on a single emulsion. These forces are kinetic, drag, interfacial tension, buoyancy, and Laplace pressure forces [27]. Note that the Laplace pressure was commonly neglected in symmetric and axisymmetric flows but has to be considered in a T-junction configuration due to the significant difference between the head and tail curvatures [28,29]. In addition, the drag force may have various forms while the others remain the same. The deviations were mainly caused by the different wall effect corrections when extending the Stokes drag of unbounded flows to confined counterparts [30,31]. Adjusting the continuous phase flow rate can tune the size of the droplets as its changes the drag force. Temperature can affect the surface tension and the viscosity and thus can be used to tune the droplet size [32,33]. Furthermore, to connect the model of the single droplet with the core–shell droplet, an equivalence of droplets volume ratio and input flow rates ratio was usually introduced [13,34]. Note that this hypothesis is valid if each droplet contains a single core.

Although efforts have been dedicated to double emulsion formation with microfluidics, most reported works were mainly based on either experiments or CFD simulation. Only a few can provide a relatively simple analytical model. Often, results from experiments, theory, and numerical simulation are inconsistent, preventing effective optimisation of devices and operation parameters for the generation of double emulsion. To our best knowledge, no past studies concurrently examined experiments, CFD simulations, and analytical models under the same condition. We propose in the present study a simple experimental setup, a vertically oriented tri-axial co-flow capillary device, to generate core–shell droplets in a one-step process. We investigate the mechanism of the double emulsion formation both experimentally and numerically. Furthermore, we develop and validate a simple but effective analytical model for predicting the droplet size distribution with the wall effect and according to various flow conditions.

2. Experimental Materials and Methods

2.1. Design and Fabrication of the Capillary Device

The tri-axial needles were purchased from Ramé-Hart instrument Co. (Succasunna, NJ, USA). The device structure and dimensions are detailed in Figure 1A and Table 1, with an inner needle tip extension $\Delta z = 0.1$ mm. A straight circular channel was connected with the outlet of the needles. The outlet channel was made of poly-dimethylsiloxane (PDMS) moulded on a straight wire with the same diameter as that of the needles. The inlets of the needles are individually connected with three syringes via tubing. Figure 1B,C show the schematic diagram and the images of the experimental setup.

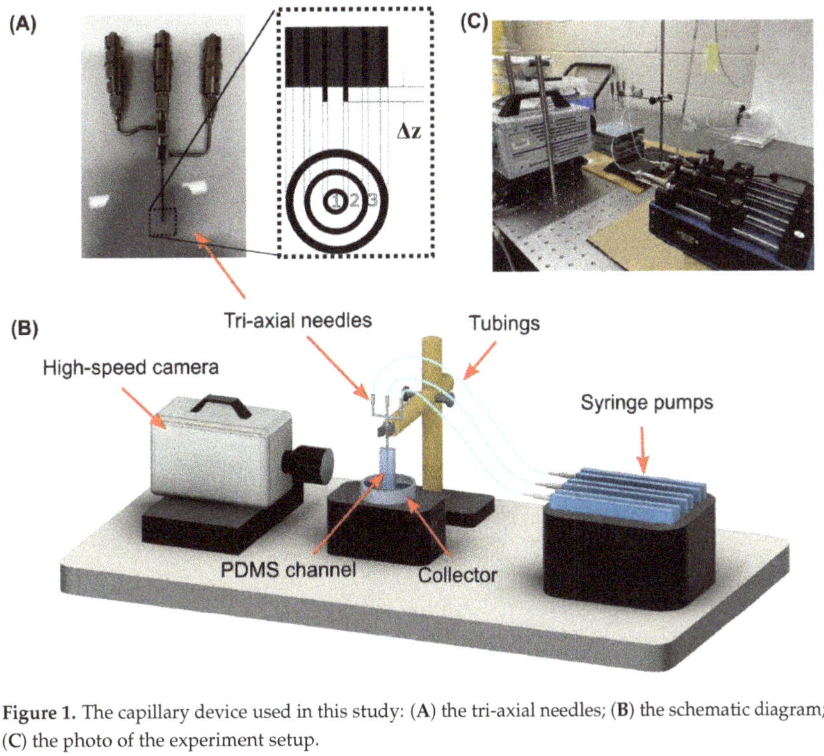

Figure 1. The capillary device used in this study: (**A**) the tri-axial needles; (**B**) the schematic diagram; (**C**) the photo of the experiment setup.

Table 1. The dimensions of the tri-axial needles.

Needle/Phase	Inner Diameter (μm)	Outer Diameter (μm)	Thickness (μm)
1	178	356	89
2	508	813	152
3	1190	1650	229

2.2. Materials

Fluorinated oil (HFE, Novec 7500 3 M, Merck, Darmstadt, Germany) was used as the core dispersed phase, whose density and dynamic viscosity are 1.61 g/mL and 1.31 mPa·s, respectively [8]. The shell dispersed phase was a polymer consisting of 0.06 g ethyl-4(dimethylamino) benzoate (Merck, Darmstadt, Germany), 0.05 g camphorquinone (Merck), and 10 g trimethylolpropane trimethacrylate (TMPTMA, Merck). The corresponding density and dynamic viscosity are 1.07 g/mL and 42 mPa·s, respectively [35]. The continuous phase was 80% glycerol dissolved in 20% DI water, with 0.1% Tween 20, and its density and dynamic viscosity are 1.21 g/mL and 75.42 mPa·s, respectively [36]. The laboratory temperature was maintained at 23 °C.

The interfacial tension σ_{ij} between each pair of two liquid phases were measured using the reverse pendant drop method with an optical tensiometer (Theta Flex from Biolin Scientific, Gothenburg, Sweden). Figure 2A reports the mean contact angles of each pair of two phases, while Figure 2B,C show the images of the pendant drops. The interfacial tensions of the inner and outer interfaces were measured for 10 s for each case with 33 frames per second, and the measured time-averaged mean interfacial coefficients are 3.45 and 8.48 mN/m, respectively.

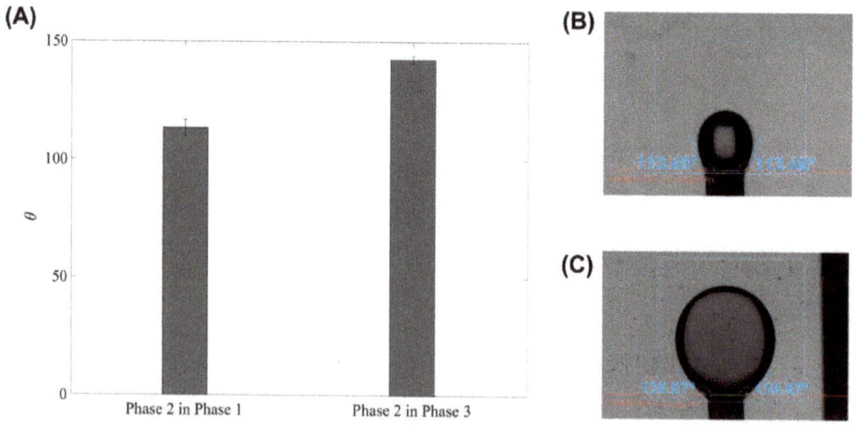

Figure 2. The interfacial tension coefficients measurement: (**A**) the mean contact angles (error bars represent the standard deviations and $n = 7$); (**B**) middle-phase reversed pendant drop in inner phase; (**C**) Middle phase reversed pendant drop in outer phase.

2.3. Experimental Setup and Data Analysis

The three phases were delivered into the capillary device at specific flow rates using syringe pumps (MNT-SPM-100, Welland, Austrlia). The formation and detachment of the droplets from the needle tips under various flow conditions were recorded using a high-speed camera (Photron FASTCAM SA3, San Diego, CA, USA, attached with a 25 mm F2.8 Ultra Macro 2.5-5.0X lens) at 250 frames per second. The open-source software ImageJ 1.53 m (National Institutes of Health USA, Bethesda, MD, USA) was adopted to analyse the captured videos.

3. Numerical Methodology

3.1. Governing Equations

All phases were modelled as incompressible fluids and laminar flows. The continuity and momentum equations are expressed as:

$$\frac{\partial \rho}{\partial t} + \nabla \cdot (\rho \vec{v}) = 0 \tag{1}$$

$$\frac{\partial (\rho \vec{v})}{\partial t} + \nabla \cdot (\rho \vec{v} \vec{v}) = -\nabla \mathrm{p} + \nabla \cdot [\mu (\nabla \vec{v} + \nabla \vec{v}^T)] + \rho \vec{g} + \vec{F} \tag{2}$$

where ρ, t, v, p, μ, g, and F denote density, time, velocity, pressure, dynamic viscosity, gravitational acceleration, and a source term, respectively. The Navier–Stokes equations above are related to the volume fraction of each phase α_i. The density and viscosity in every single computational cell were calculated as follows:

$$(\rho, \mu) = \sum_{i=1}^{3} \alpha_i (\rho_i, \mu_i) \tag{3}$$

The volume of fluid (VOF) method was adopted to model the fluid–fluid interfaces as:

$$\frac{\partial \alpha_i}{\partial t} + \nabla \cdot (\alpha_i \vec{v}) = 0 \tag{4}$$

The sum of volume fractions of three phases within each computational cell follows the constraint:

$$\sum_{i=1}^{3} \alpha_i = 1 \tag{5}$$

The continuous surface force (CSF) model developed by Brackbill et al. [37] was used to model the surface tension via the source term $\vec{F}$ in the momentum equation as:

$$\vec{F} = \sum\nolimits_{pairs(i,j),\ i<j} \sigma_{ij} \frac{\alpha_i \rho_i \kappa_j \nabla \alpha_j + \alpha_j \rho_j \kappa_i \nabla \alpha_i}{\frac{1}{2}(\rho_i + \rho_j)} \tag{6}$$

where κ stands for the interface curvature determined by:

$$\kappa_i = \nabla \cdot \hat{\boldsymbol{n}}_i \tag{7}$$

and $\hat{\boldsymbol{n}}$ is defined as:

$$\hat{\boldsymbol{n}}_i = \frac{\nabla \alpha_i}{|\nabla \alpha_i|} \tag{8}$$

3.2. Computational Domain and Boundary Conditions

As the droplets are formed at the needle outlet, the corresponding circular cross-section channel downstream was selected to be the computational domain. The centreline was regarded as the axis due to the axisymmetric geometry. Boundary conditions at the entrance, exit, and walls were set to be velocity inlet, pressure outlet, and no-slip wall, respectively, Figure 3. The contact angles between each interface and the needle wall were determined based on experimental results in Figure 2B,C because the tri-axial needle and the hook needle used in the interfacial tension measurement are both made of stainless steel. The hydrodynamic entrance length for each needle inlet was estimated by $l_{ent,i} = 0.0575 Re_i D_i$ assuming a laminar pipe flow [38]. The corresponding order of magnitude in this study was calculated as $l_{ent,1} \sim O(1\ \mu\text{m})$, $l_{ent,2} \sim O(0.1\ \mu\text{m})$, $l_{ent,3} \sim O(10\ \mu\text{m})$, respectively. To ensure each flow inlet is fully developed before reaching the expansion region, the inlet lengths for inner, middle, and outer needles in the simulation were fixed, respectively as 300, 200, and 200 µm, much longer than the hydrodynamic entrance lengths. Note that the inner needle inlet length was determined by considering its extension length. Structural meshes were generated, and a grid independence analysis was conducted with three cell sizes: 15 μm, 10 μm, and 5 μm with corresponding grid numbers of 23,652 (coarse), 51,785 (medium), and 204,428 (fine), respectively. The results showed that the medium and fine meshes, under the same input conditions, can model the interfaces with insignificant differences. Since the fine mesh drastically increases the computational resources, the medium one was selected for this study.

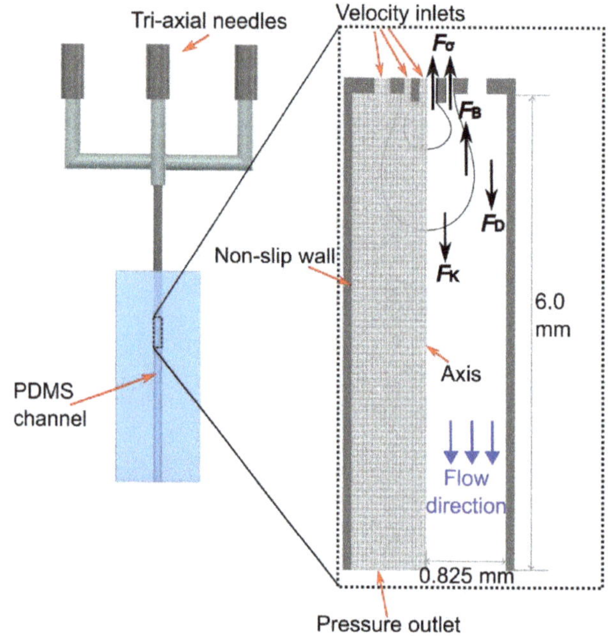

Figure 3. Schematic illustration of the computational domain, structural mesh, and forces exerting on a growing double emulsion droplet.

We performed simulations using the commercially available CFD software package ANSYS 2022. The SIMPLE algorithm was utilised for the pressure-velocity coupling. A PRESTO scheme was selected to calculate the exact value of the pressure term. We implemented the Geo-Reconstruct algorithm to acquire the interface interpolation. The governing equations were discretised by a second-order implicit scheme for the spatial terms and transient formulation. The time step was determined by the Courant–Fredrichs–Lewy number lower than 0.25.

4. Analytical Model

Both experiments and CFD simulations can capture the droplet formation in detail, but building empirical relations based on these methods usually requires relatively intensive case studies. Since we focus on the dripping regime, a simpler analytical model is more convenient for elucidating the droplet breakup mechanism and approximately predicting the droplet size. In the dripping regime, several forces are exerted on the growing droplets along the axial direction: viscous drag F_D and kinetic F_K forces pull them downstream while surface tension F_σ and buoyancy F_B forces act as opposite counterparts, Figure 3.

The assumptions for the analytical model are (i) the continuous flow is in the Stokes regime ($Re_3 << 1$, where $Re_3 = \frac{\rho_3 r_{channel} \overline{v}_3}{\mu_3}$); (ii) the droplet formation is in the dripping regime ($Ca_3 < 1$, where $Ca_3 = \frac{\mu_3 \overline{v}_3}{\sigma_{23}}$); (iii) the shape of droplets is spherical; (iv) laminar and incompressible flow are also presumed as per ansatzes in CFD simulations. The forces involved in the formation process are [13,27,39]:

$$F_K = \rho_1 Q_1 v_1 + \rho_2 Q_2 v_2 \tag{9}$$

$$F_\sigma = \pi d_{o1}\sigma_{12} + \pi d_{o2}\sigma_{23} \tag{10}$$

$$F_B = \frac{4}{3}\pi\left(\frac{d_{drop}}{2}\right)^3 \rho g \tag{11}$$

where Q and d represent volume flow rate input and diameter, respectively; subscripts o and $drop$ denote the outer and the whole droplet. As for the Stokes drag force, we employed the model with the wall effect considered since the diameter of the droplets formed in our device has a similar magnitude as that of the channel [31,40]:

$$F_D = 3\pi\mu_3 d_{drop}\left(\overline{v}_3 K_1 - \overline{v}_{drop} K_2\right) \tag{12}$$

where $\overline{v}_3$ and $\overline{v}_{drop}$ are the mean velocities of the continuous phase and the whole droplet:

$$\overline{v}_3 = \frac{4Q_3}{\pi d_{channel}^2} \tag{13}$$

$$F_D = 3\pi\mu_3 d_{drop}\left(\overline{v}_3 K_1 - \overline{v}_{drop} K_2\right) \tag{14}$$

with K_1 and K_2 being the wall effect correction coefficients as:

$$\begin{aligned} K_1 = 1/\big[1 - 2.10443\lambda + 2.08877\lambda^3 - 0.94813\lambda^5 - 1.372\lambda^6 + 3.87\lambda^8 \\ -4.19\lambda^{10} + O(\lambda^{11})\big] \end{aligned} \tag{15}$$

$$K_2 = K_1\left[1 - \frac{2}{3}\lambda - 0.1628\lambda^3 - 0.4059\lambda^7 + 0.5236\lambda^9 + 1.51\lambda^{10} + O(\lambda^{11})\right] \tag{16}$$

where λ is the dimensionless droplet diameter as follows:

$$\lambda = \frac{d_{drop}}{d_{channel}} \tag{17}$$

Note that Equations (15) and (16) have been proven to show good agreements with the exact solution with $\lambda \leq 0.8$ [31,40]. For larger λ, we may need higher order expansions or the exact solution of Haberman and Sayre [41], which is too long to be expressed here.

By scaling $d_{drop} \sim d_{o2}$, we can obtain the orders of magnitude:

$$F_D \sim O(10^{-5}\ N),\ F_K \sim O(10^{-9}\ N),\ F_\sigma \sim O(10^{-5}\ N),\ F_B \sim O(10^{-7}\ N) \tag{18}$$

Thus, the force balance equation in the current study can be reasonably simplified as:

$$F_D = F_\sigma \tag{19}$$

Now the diameter of the whole droplet d_{drop} can be obtained by numerically solving the algebraic system above. To estimate the size of the core, we employed [13,34]:

$$\left(\frac{d_{core}}{2}\right)^3 = \frac{Q_1}{Q_1 + Q_2}\left(\frac{d_{drop}}{2}\right)^3 \tag{20}$$

5. Results and Discussion

We first examined the formation process with the continuous phase flow rate Q_3 of 250, 500, and 1000 μL/min, while the inner and middle phase flow rates Q_1 and Q_2 were fixed at 3 μL/min and 30 μL/min, respectively. Figure 4 presents the double emulsion formation in the case of Q_1 = 3 μL/min, Q_2 = 30 μL/min, Q_3 = 250 μL/min over a formation period. The Experimental (top row) and the corresponding simulation results (bottom row) are obtained at nine time steps, with an excellent agreement in the droplet formation pattern. However, the total formation time showed a 16% relative deviation. We observed that the core droplet breaks up before the shell does. A satellite droplet also forms after the break-up of the shell droplet. Over the entire formation process, double emulsions are formed in a regular pattern, with each shell droplet containing a core. In addition, as the whole droplet flows downstream, the core droplet moves downwards quicker than the shell one due to the higher density. This phenomenon is slightly overpredicted in the simulation; see the second image in Figure 4. This might also be caused by the fact that the needles are not precisely tri-axial due to the manufacturing error, evidently by the left skewness in the experimental results.

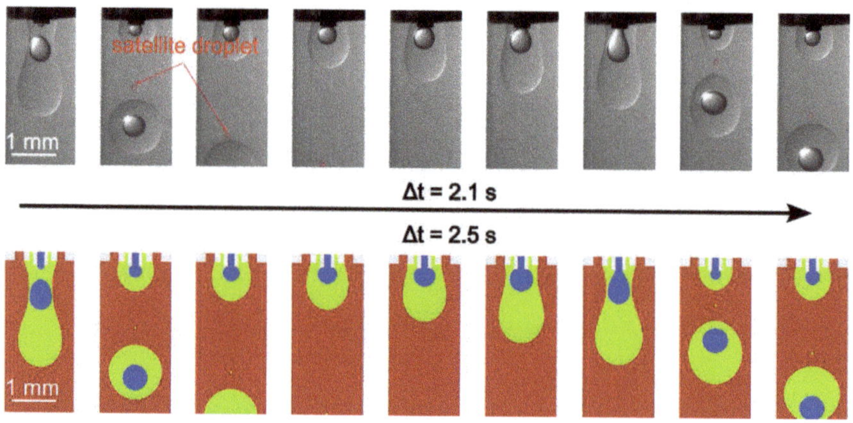

Figure 4. Droplet formation with regular double emulsions at Q_1 = 3 μL/min, Q_2 = 30 μL/min, Q_3 = 250 μL/min.

Furthermore, the other two cases with a continuous phase flow rate Q_3 of 500 and 1000 μL/min, while the inner Q_1 and middle Q_2 phase flow rates were fixed at 3 μL/min and 30 μL/min, respectively, Figure 5. Figure 5A shows that with Q_3 = 500 μL/min, double and single emulsions occur alternately, meaning that only one of every two droplets contains a core. Although the time period still shows a 14% deviation between the experiment and simulation, the flow patterns and droplet sizes were captured with remarkable agreement. Worse conditions can be found with Q_3 is 1000 μL/min, Figure 5B, where only one in three droplets contains a core in the experiment, and one in four droplets contains a core in the simulation. Now the flow patterns show a difference within a period. Since it is necessary to avoid the mismatch in applications, the deviation on the mismatched frequencies is not discussed in detail. These unmatched phenomena can be explained by the mismatch of pinch-off locations. The pinch-off location of the middle phase is stretched downwards due to the increased capillary number of the continuous phase, while the pinch-off location of the inner phase remains near the needle tip. Additionally, we also observed the negligible difference between the size of the double emulsion and the single droplet in both experiments and simulations, meaning that the inner phase flow rate only has a slight effect on the whole droplet size. This observation agrees with previous studies [24,26].

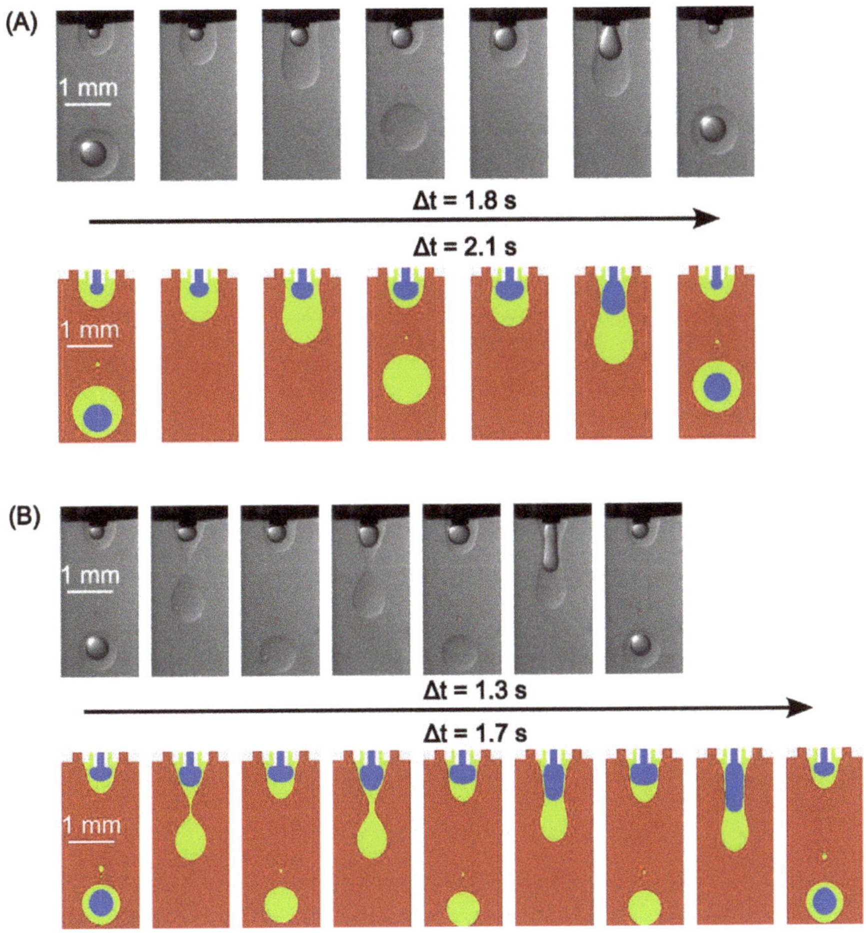

Figure 5. Droplet formation with mismatch at Q_1 = 3 μL/min, Q_2 = 30 μL/min and: (**A**) Q_3 = 500 μL/min; (**B**) Q_3 = 1000 μL/min.

The mismatch phenomenon can be further illustrated along with the velocity streamlines in the case with Q_1 = 3 μL/min, Q_2 = 30 μL/min, Q_3 = 500 μL/min in Figure 6. The blue lines represent the interface between the inner and middle phases, while the red ones stand for the interface between the middle and outer phases. Due to the abrupt expansion of the channel, the velocity profile drastically changes in the vicinity of the needle tips and then develops downstream. Two pairs of vortices, which in fact are two vortex rings due to the axisymmetric geometry, can be observed near the needle tips as a result of the sudden change in the velocity profile, Figure 6A,B. The lower vortex ring formed inside the shell droplet pushes the growing core droplet head backward. The upper vortex ring also acts as an obstacle due to the reverse flow direction along the centreline. As both core and shell droplets grow, Figure 6A–C, the size of the lower vortex ring gradually decreases to null by the shear force of the continuous phase. In contrast, the upper vortex ring increases since the core droplet does not grow beyond the pinch-off location of the shell droplet. After the breakup of the shell droplet, the upper vortex ring crosses the interface between the inner and middle phases, Figure 6D, and is then squeezed by the middle-outer interface and the inner-middle interface. This process forms a new vortex ring from the inner needle tip, Figure 6E. After the mismatch, the growing core droplet accumulated inside the next shell one due to the fixed flow rate input, Figure 6F. Again, as both droplets grow, the size of the lower vortex ring decreases, but the size of the upper one decreases as well in this period, Figure 6F–H. This is because the accumulated growing core droplet is larger, and the vortex ring inside does not attach to the inner needle tip even before the core breaks up. Consequently, the vortex ring is stretched longer and narrower. Thus, more inertia assists the core droplet heading downstream through the space between the vortex ring and the inner-middle interface, Figure 6H. Finally, the core droplet grows beyond the pinch-off location of the shell droplet and breaks up along with the outer emulsion, Figure 6I,J. The whole process then repeats periodically. A similar explanation can be used for the mismatch in the case with Q_1 = 3 µL/min, Q_2 = 30 µL/min, Q_3 = 1000 µL/min. The difference is that the shell droplet pinches off two or three times before the core droplet accumulates to grow beyond the pinch-off position.

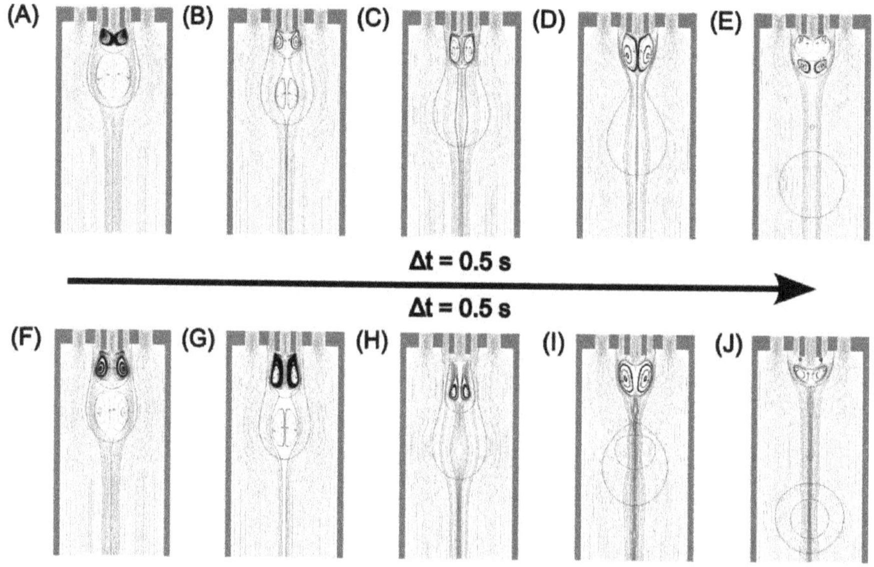

Figure 6. Velocity streamlines near the needle tip at Q_1 = 3 µL/min, Q_2 = 30 µL/min, Q_3 = 500 µL/min: **(A–E)** when the mismatch occurs; **(F–J)** when the core matches with the shell droplet.

To solve the mismatch issue, the core droplet needs to grow beyond the pinch-off location of the shell droplet within each pinch-off period, or $\frac{l_{p2}-\Delta z_{tip1}}{v_{g1}} \leq t_{p2}$. Hence, we define a match ratio of:

$$\xi = \frac{l_{p2} - \Delta z_{tip1}}{v_{g1} t_{p2}} \tag{21}$$

where Δz_{tip1} is the inner needle tip elevation difference; v_{g1} is the core droplet growth velocity estimated as:

$$v_{g1} \approx \frac{Q_1}{\pi\left(\frac{d_{o1}}{2}\right)^2} \tag{22}$$

and the pinch-off time of the shell t_{p2} calculated based on the theoretical model as:

$$t_{p2} = \frac{\frac{4}{3}\pi\left(\frac{d_{drop}}{2}\right)^3}{Q_1 + Q_2} \tag{23}$$

However, to our best knowledge, the pinch-off length of the shell droplet, l_{p2}, cannot be approximated through a simple model. Thus, we measured it and the pinch-off time experimentally based on the averaged value of the upper and lower pinch-off lengths, Figure 7A. Error bars were calculated from standard deviations of the mean (n = 3). We examined Q_3 ranging from 200 to 1200 µL/min with a 100-µL/min interval with the inner phase flow rate Q_1 of 3, 6, 12 µL/min, and the middle phase flow rate Q_2 of 30, 15, 7.5 µL/min, respectively. Note that the corresponding continuous phase Reynolds number, Re_3, and Capillary number, Ca_3, range from 0.02 to 0.14 and from 0.01 to 0.09, respectively. Figure 7B,C show the measured averaged pinch-off length and time against the continuous phase flow rate at different flow rates of the inner and middle phase, respectively. The standard deviations of the mean were calculated from three samples for each case [42].

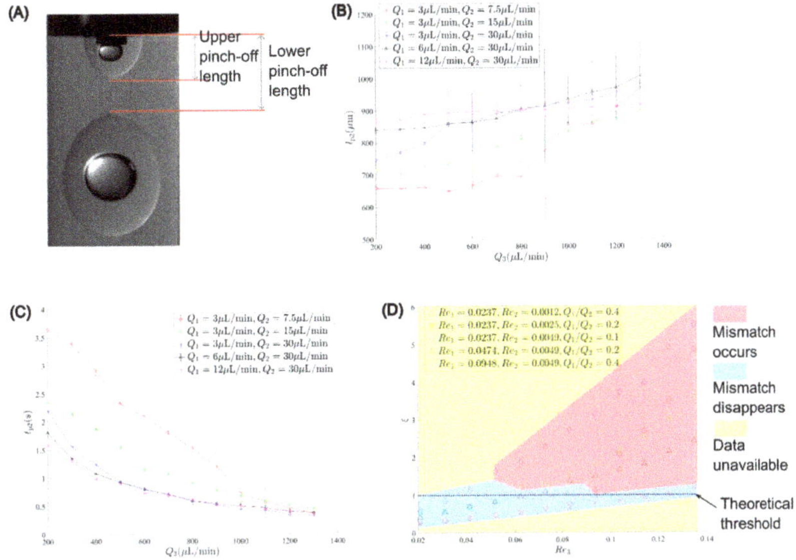

Figure 7. Pinch-off parameters: (**A**) pinch-off length measurement; (**B**) pinch-off length versus the continuous phase flow rate; (**C**) pinch-off time versus the continuous phase flow rate; (**D**) match ratio versus the continuous phase Reynolds number.

With increasing Q_3, the pinch-off length and time of the shell droplet generally increase and decrease, respectively. We also noticed that, when a mismatch occurs, the standard deviation of the pinch-off length drastically increases, while the counterpart of the pinch-off time shows a relatively stable pattern. Additionally, when the sum of Q_1 and Q_2 increases, the pinch-off time for each case decreases and gradually approaches closely as Q_3 increases. In contrast, as the sum of Q_1 and Q_2 increases, the pinch-off length shows a reduction pattern before mismatches occur but a relatively chaotic counterpart due to the mismatches, leading to abrupt increases for different cases. The mismatch issue does not show up in the case of Q_1 = 12 µL/min and Q_2 = 30 µL/min within the Q_3 range in this study, and thus the corresponding line has a negligible change. Equations (21)–(23) indicate that the mismatch issue can be theoretically avoided once the match ratio, ξ, is lower than unity. Increasing Q_1 or decreasing Q_2 can reduce ξ. Increasing Δz_{tip1} can also decrease ξ, but this option was not examined due to the limitation of the tri-axial needles used in this study. Figure 7D shows the operation map with the match ratio versus the continuous phase Reynolds number for various Reynolds numbers of inner and middle phases. The operation map shows the occurrence of mismatches in experiments. We found a very close threshold at $\xi = 1$, beyond which the mismatch occurs. Additionally, the match ratio generally decreases with an increase in the inner-middle phase flow rate ratio. However, even at the same inner-middle phase flow rate ratio $Q_1/Q_2 = 0.4$, the one with a higher total flow rate shows a better performance than the other. Hence, it would be better to increase the inner phase flow rate rather than decrease the middle one.

After resolving the mismatch issue by increasing the inner phase flow rate, we compared the size distribution among experiments, simulations, and theory. The diameters of the droplets in the experiment and simulation were determined through ImageJ by measuring the vertical and horizontal lengths of at least three whole droplets for each case and then averaged as $\bar{d}_i = \frac{a_i+b_i}{2}$, as shown in Figure 8A. Error bars were again calculated from standard deviations of the mean. Figure 8B shows the effect of the continuous phase Reynolds number on the matched droplet size normalized by the channel diameter. As observed, the analytical result on the whole droplet shows a slight deviation from the experimental one when Re_3 ranges from 0.02 to 0.06 and asymptotically approaches good agreement with Re_3 increasing from 0.06 to 0.14. The analytical result of the core droplet diameter shows a better prediction against the experimental data within the range of Re_3 in our current study. This can be explained by the approximated wall effect correction model since increasing Re_3 decreases the dimensionless droplet diameter λ, leading to less influence of the wall effect; see Equations (15) and (16). Furthermore, the reduction in the whole droplet diameter also reduces the deformation effect, which results in a better agreement with the assumption of a spherical droplet. In contrast, the CFD simulation results generally overlap with experimental data, indicating that the current numerical model is valid to capture the double emulsions in the dripping regime through the tri-axial capillary device.

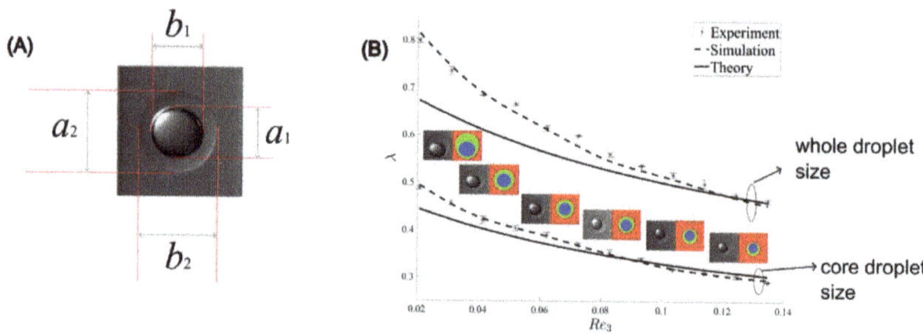

Figure 8. Matched droplets: (**A**) core and shell droplet diameter measurement; (**B**) size distribution comparisons among experiments, simulations, and theory.

6. Conclusions

We experimentally, numerically, and theoretically investigated the formation process of double emulsion in a vertically arranged tri-axial co-flow capillary device. The volume of fluid (VOF) method and continuous surface force (CSF) model were used for the numerical simulation to catch the details of fluid flow. An analytical model based on a simple force balance was built to estimate the droplet size distribution in accordance with the continuous phase Reynolds number. We found that mismatches of core and shell droplets under certain flow conditions can be captured both experimentally and numerically. To quantify the mismatch, we proposed a semi-theoretical model by matching the pinch-off length of the shell droplets with the product of the growth rate of the core and the pinch-off time of the shell. The mismatch issue is expected to be avoided if the match ratio is lower than unity, which was validated with experimental data. Regarding the reduction in the match ratio, we found that increasing the inner phase flow rate shows better performance than reducing the middle one. Considering the wall effect, the analytical model for predicting the size of matched double emulsions showed slight deviations from experiments if the continuous phase Reynolds number is lower than 0.06. However, the behaviour asymptotically approaches good agreement in the Reynolds number range of

0.06 to 0.14. The numerical simulation generally agreed with the experimental data under the investigated flow conditions.

Author Contributions: Conceptualisation, N.-T.N. and Y.D.; methodology, Y.D. and J.Z.; software, Y.D. and H.C.; material, Y.D. and F.G.; formal analysis, Y.D. and N.-K.N.; investigation, Y.D. and L.O.; data curation, Y.D., A.S.Y. and J.Z.; writing—original draft preparation, Y.D.; writing—review and editing, J.Z., H.A. and N.-T.N.; visualisation, Y.D.; supervision, N.-T.N.; project administration, N.-T.N.; funding acquisition, N.-T.N. and C.H.O. All authors have read and agreed to the published version of the manuscript.

Funding: This research was funded by Discovery Project (Grant No. DP220100261), ARC DECRA fellowship (Grant No. DE210100692), and ARC Future Fellowships (FT180100361). This work was partly performed at the Queensland node-Griffith-of the Australian National Fabrication Facility, a company established under the National Collaborative Research Infrastructure Strategy to provide nano and microfabrication facilities for Australian researchers.

Data Availability Statement: The data presented in this study are available on request from the corresponding author.

Conflicts of Interest: The authors declare no conflict of interest.

References

1. Clegg, P.S.; Tavacoli, J.W.; Wilde, P.J. One-step production of multiple emulsions: Microfluidic, polymer-stabilized and particle-stabilized approaches. *Soft Matter* **2016**, *12*, 998–1008. [CrossRef] [PubMed]
2. Ding, S.; Serra, C.A.; Vandamme, T.F.; Yu, W.; Anton, N. Double emulsions prepared by two–step emulsification: History, state-of-the-art and perspective. *J. Control. Release* **2019**, *295*, 31–49. [CrossRef] [PubMed]
3. Taylor, G.I. The formation of emulsions in definable fields of flow. In Proceedings of the Royal Society of London, Series A, Containing Papers of a Mathematical and Physical Character, London, UK, 1 October 1934; pp. 501–523.
4. Tomotika, S. On the instability of a cylindrical thread of a viscous liquid surrounded by another viscous fluid. In Proceedings of the Royal Society of London, Series A—Mathematical and Physical Sciences, London, UK, 1 June 1935; pp. 322–337.
5. Utada, A.S.; Lorenceau, E.; Link, D.R.; Kaplan, P.D.; Stone, H.A.; Weitz, D.A. Monodisperse Double Emulsions Generated from a Microcapillary Device. *Science* **2005**, *308*, 537–541. [CrossRef]
6. Nie, Z.; Xu, S.; Seo, M.; Lewis, P.C.; Kumacheva, E. Polymer Particles with Various Shapes and Morphologies Produced in Continuous Microfluidic Reactors. *J. Am. Chem. Soc.* **2005**, *127*, 8058–8063. [CrossRef] [PubMed]
7. Galogahi, F.M.; Zhu, Y.; An, H.; Nguyen, N.-T. Core-shell microparticles: Generation approaches and applications. *J. Sci. Adv. Mater. Devices* **2020**, *5*, 417–435. [CrossRef]
8. Galogahi, F.M.; Zhu, Y.; An, H.; Nguyen, N.-T. Formation of core–shell droplets for the encapsulation of liquid contents. *Microfluid Nanofluidics* **2021**, *25*, 1–11. [CrossRef]
9. Lewińska, D.; Bukowski, J.; Kożuchowski, M.; Kinasiewicz, J.; Weryński, A. Electrostatic microencapsulation of living cells. *Biocybern. Biomed. Eng.* **2008**, *28*, 69–84.
10. Vadivelu, R.K.; Kamble, H.; Shiddiky, M.J.A.; Nguyen, N.-T. Microfluidic Technology for the Generation of Cell Spheroids and Their Applications. *Micromachines* **2017**, *8*, 94. [CrossRef]
11. Bremond, N.; Santanach-Carreras, E.; Chu, L.-Y.; Bibette, J. Formation of liquid-core capsules having a thin hydrogel membrane: Liquid pearls. *Soft Matter* **2010**, *6*, 2484–2488. [CrossRef]
12. Sreejith, K.R.; Gorgannezhad, L.; Jin, J.; Ooi, C.H.; Takei, T.; Hayase, G.; Stratton, H.; Lamb, K.; Shiddiky, M.; Dao, D.V. Core-Shell Beads Made by Composite Liquid Marble Technology as A Versatile Microreactor for Polymerase Chain Reaction. *Micromachines* **2020**, *11*, 242. [CrossRef]
13. Chaurasia, A.S.; Sajjadi, S. Millimetric core–shell drops via buoyancy assisted non-confined microfluidics. *Chem. Eng. Sci.* **2015**, *129*, 260–270. [CrossRef]
14. Nguyen, N.-T.; Wereley, S.T.; Shaegh, S.A.M. *Fundamentals and Applications of Microfluidics*; Artech House: Norwood, MA, USA, 2019.
15. Shao, T.; Feng, X.; Jin, Y.; Cheng, Y. Controlled production of double emulsions in dual-coaxial capillaries device for millimeter-scale hollow polymer spheres. *Chem. Eng. Sci.* **2013**, *104*, 55–63. [CrossRef]
16. Schmit, A.; Courbin, L.; Marquis, M.; Renard, D.; Panizza, P. A pendant drop method for the production of calibrated double emulsions and emulsion gels. *RSC Adv.* **2014**, *4*, 28504–28510. [CrossRef]
17. Zhou, C.; Yue, P.; Feng, J.J. Formation of simple and compound drops in microfluidic devices. *Phys. Fluids* **2006**, *18*, 092105. [CrossRef]
18. Park, J.M.; Anderson, P.D. A ternary model for double-emulsion formation in a capillary microfluidic device. *Lab Chip* **2012**, *12*, 2672–2677. [CrossRef] [PubMed]

19. Vu, T.V.; Homma, S.; Tryggvason, G.; Wells, J.C.; Takakura, H. Computations of breakup modes in laminar compound liquid jets in a coflowing fluid. *Int. J. Multiph. Flow* **2012**, *49*, 58–69. [CrossRef]
20. Tryggvason, G.; Bunner, B.; Esmaeeli, A.; Juric, D.; Al-Rawahi, N.; Tauber, W.; Han, J.; Nas, S.; Jan, Y.-J. A Front-Tracking Method for the Computations of Multiphase Flow. *J. Comput. Phys.* **2001**, *169*, 708–759. [CrossRef]
21. Fu, Y.; Zhao, S.; Bai, L.; Jin, Y.; Cheng, Y. Numerical study of double emulsion formation in microchannels by a ternary Lattice Boltzmann method. *Chem. Eng. Sci.* **2016**, *146*, 126–134. [CrossRef]
22. Herrada, M.A.; Montanero, J.M.; Ferrera, C.; Gañán-Calvo, A.M. Analysis of the dripping–jetting transition in compound capillary jets. *J. Fluid Mech.* **2010**, *649*, 523–536. [CrossRef]
23. Chen, Y.; Wu, L.; Zhang, L. Dynamic behaviors of double emulsion formation in a flow-focusing device. *Int. J. Heat Mass Transf.* **2015**, *82*, 42–50. [CrossRef]
24. Nabavi, S.A.; Vladisavljević, G.T.; Gu, S.; Ekanem, E.E. Double emulsion production in glass capillary microfluidic device: Parametric investigation of droplet generation behaviour. *Chem. Eng. Sci.* **2015**, *130*, 183–196. [CrossRef]
25. Azarmanesh, M.; Farhadi, M.; Azizian, P. Double emulsion formation through hierarchical flow-focusing microchannel. *Phys. Fluids* **2016**, *28*, 032005. [CrossRef]
26. Yang, Z.; Ma, X.; Wang, S.; Liu, D. Generation and evolution of double emulsions in a circular microchannel. *Chem. Eng. Sci.* **2022**, *255*, 117683. [CrossRef]
27. Umbanhowar, P.B.; Prasad, V.; Weitz, D.A. Monodisperse Emulsion Generation via Drop Break Off in a Coflowing Stream. *Langmuir* **1999**, *16*, 347–351. [CrossRef]
28. Wang, X.; Riaud, A.; Wang, K.; Luo, G. Pressure drop-based determination of dynamic interfacial tension of droplet generation process in T-junction micro-channel. *Microfluid. Nanofluidics* **2015**, *18*, 503–512. [CrossRef]
29. Murshed, S.M.S.; Tan, S.H.; Nguyen, N.-T.; Wong, T.N.; Yobas, L. Microdroplet formation of water and nanofluids in heat-induced microfluidic T-junction. *Microfluid. Nanofluidics* **2008**, *6*, 253–259. [CrossRef]
30. Wu, P.; Luo, Z.; Liu, Z.; Li, Z.; Chen, C.; Feng, L.; He, L. Drag-induced breakup mechanism for droplet generation in dripping within flow focusing microfluidics. *Chin. J. Chem. Eng.* **2014**, *23*, 7–14. [CrossRef]
31. Happel, J.; Brenner, H. *Low Reynolds Number Hydrodynamics: With Special Applications to Particulate Media*; Springer Science & Business Media: New York, NY, USA, 2012.
32. Tan, S.-H.; Murshed, S.M.S.; Nguyen, N.; Wong, T.N.; Yobas, L. Thermally controlled droplet formation in flow focusing geometry: Formation regimes and effect of nanoparticle sus-pension. *J. Phys. D Appl. Phys.* **2008**, *41*, 165501. [CrossRef]
33. Yap, Y.F.; Tan, S.H.; Nguyen, N.-T.; Murshed, S.M.S.; Wong, T.N.; Yobas, L. Thermally mediated control of liquid microdroplets at a bifurcation. *J. Phys. D Appl. Phys.* **2009**, *42*, 065503. [CrossRef]
34. Chang, Z.; Serra, C.A.; Bouquey, M.; Prat, L.; Hadziioannou, G. Co-axial capillaries microfluidic device for synthesizing size-and morphology-controlled polymer core-polymer shell particles. *Lab Chip* **2009**, *9*, 3007–3011. [CrossRef]
35. Teo, A.J.; Malekpour-Galogahi, F.; Sreejith, K.R.; Takei, T.; Nguyen, N. Surfactant-free, UV-curable core–shell microcapsules in a hydrophilic PDMS microfluidic device. *Aip Adv.* **2020**, *10*, 065101. [CrossRef]
36. Takamura, K.; Fischer, H.; Morrow, N.R. Physical properties of aqueous glycerol solutions. *J. Pet. Sci. Eng.* **2012**, *98*, 50–60. [CrossRef]
37. Brackbill, J.; Kothe, D.; Zemach, C. A continuum method for modeling surface tension. *J. Comput. Phys.* **1992**, *100*, 335–354. [CrossRef]
38. Bergman, T.L.; Incropera, F.P.; DeWitt, D.P.; Lavine, A.S. *Fundamentals of Heat and Mass Transfer*; John Wiley & Sons: Hoboken, NJ, USA, 2011.
39. Erb, R.M.; Obrist, D.; Chen, P.W.; Studer, J.; Studart, A.R. Predicting sizes of droplets made by microfluidic flow-induced dripping. *Soft Matter* **2011**, *7*, 8757–8761. [CrossRef]
40. Bohlin, T. On the Drag on a Rigid Sphere Moving in a Viscous Liquid Inside a Cylindrical Tube; Elanders: 1960. Available online: https://jglobal.jst.go.jp/en/detail?JGLOBAL_ID=201602000920293570 (accessed on 12 October 2022).
41. Haberman, W.L.; Sayre, R.M. Motion of Rigid and Fluid Spheres in Stationary and Moving Liquids Inside Cylindrical Tubes. David Taylor Model Basin Washington DC. 1958. Available online: https://apps.dtic.mil/sti/pdfs/AD0206307.pdf (accessed on 12 October 2022).
42. Dai, Y.; Cha, H.; Simmonds, M.J.; Fallahi, H.; An, H.; Ta, H.T.; Nguyen, N.-T.; Zhang, J.; McNamee, A.P. Enhanced Blood Plasma Extraction Utilising Viscoelastic Effects in a Serpentine Microchannel. *Biosensors* **2022**, *12*, 120. [CrossRef]

micromachines

Article

A Portable, Negative-Pressure Actuated, Dynamically Tunable Microfluidic Droplet Generator

Martin Trossbach [1], Marta de Lucas Sanz [1], Brinton Seashore-Ludlow [2] and Haakan N. Joensson [1,*]

1 KTH Royal Institute of Technology & Science for Life Laboratory, 17165 Solna, Sweden
2 Karolinska Institutet & Science for Life Laboratory, 17165 Solna, Sweden
* Correspondence: hakan.jonsson@scilifelab.se

Abstract: Droplet microfluidics utilize a monodisperse water-in-oil emulsion, with an expanding toolbox offering a wide variety of operations on a range of droplet sizes at high throughput. However, translation of these capabilities into applications for non-expert laboratories to fully harness the inherent potential of microscale manipulations is woefully trailing behind. One major obstacle is that droplet microfluidic setups often rely on custom fabricated devices, costly liquid actuators, and are not easily set up and operated by non-specialists. This impedes wider adoption of droplet technologies in, e.g., the life sciences. Here, we demonstrate an easy-to-use minimal droplet production setup with a small footprint, built exclusively from inexpensive commercially sourced parts, powered and controlled by a laptop. We characterize the components of the system and demonstrate production of droplets ranging in volume from 3 to 21 nL in a single microfluidic device. Furthermore, we describe the dynamic tuning of droplet composition. Finally, we demonstrate the production of droplet-templated cell spheroids from primary cells, where the mobility and simplicity of the setup enables its use within a biosafety cabinet. Taken together, we believe this minimal droplet setup is ideal to drive broad adoption of droplet microfluidics technology.

Keywords: droplet microfluidics; plug & play; portable microfluidics; spheroids; microtissues

Citation: Trossbach, M.; de Lucas Sanz, M.; Seashore-Ludlow, B.; Joensson, H.N. A Portable, Negative-Pressure Actuated, Dynamically Tunable Microfluidic Droplet Generator. *Micromachines* **2022**, *13*, 1823. https://doi.org/10.3390/mi13111823

Academic Editor: Pingan Zhu

Received: 23 September 2022
Accepted: 21 October 2022
Published: 25 October 2022

1. Introduction

Compartmentalization of cells or biomolecular samples into monodisperse reaction vessels via droplet microfluidics offers unique benefits to chemical and biological analysis and screening. Key among these benefits are improved throughput, enhanced sensitivity of detection due to the reduced dilution in these small reaction volumes, and the minimization of reagent and sample consumption [1].

Droplet microfluidics, a versatile and high-throughput platform technology, employs pico- to nanoliter-sized aqueous droplets in an immiscible fluorinated oil phase [2]. Fluorinated oils, such as Novec HFE-7500, have several advantageous properties, such as high gas solubility [3] and the availability of biocompatible surfactants [4], making them the preferred choice for biological assays. These biocompatible surfactants usually consist of a fluorophobic-hydrophilic polyethyleneoxide head group and fluorophilic-hydrophobic tails. These amphiphilic characteristics lead to them populating the interface and stabilizing the metastable droplet by lowering the surface tension, preventing coalescence [5]. The droplets are stable over extended incubation periods [6] and at higher temperatures [7,8].

A plethora of applications and assays have been developed using droplet microfluidics. Single cell assays [9,10], droplet PCR [11,12] and viral detection [13], among many others, demonstrate the versatility of the platform and the potential for customizable workflows. Entire microfluidic pipelines are created by combining several unit operations, e.g., combining droplet production, splitting, or sorting, in sequence. Aided by the control offered by microfluidics, a specifically optimized microfluidic device has been developed for virtually every individual unit operation. Moreover, there have been advances in chip design and flow rate management to enable an even wider range of applications [14,15].

Recently, microfluidic droplets have been explored as vehicles for tissue engineering, e.g., in spheroid assembly [16] or gel encapsulation applications [17,18], and are even gaining traction in research studying stem cell differentiation [19,20]. The droplets used in these applications are usually larger than droplets used for single cell studies to allow for encapsulation of sufficient cell numbers [6] or ensure adequate nutrient supply to maintain cell viability for multi-day droplet culture [21]. Addressing the need for scaling, we previously published an automated workflow for high-throughput spheroid production in microfluidic droplets using a liquid handling robot [6], and presented a production optimization pipeline employing a deep neural network to characterize spheroid morphology [22].

There are numerous active and passive methods for producing aqueous droplets in a continuous phase of which flow-focusing droplet generation is most widely used [23,24]. In negative-pressure droplet generation, a pressure source set below ambient pressure is applied at the device outlet to actuate droplet generation. While positive-pressure or positive flow-driven droplet generation using flow-focusing techniques have been extensively studied, including analysis of the impacts of geometry and flow regimes on droplet size [25–29], negative-pressure operation has received less attention. Conceivably, the general tenets still apply, but conditions such as hydraulic resistances become more impactful [30]. Crucially, while the flow rates of the dispersed and continuous phase are directly controllable in positive-pressure or flow-driven droplet generation, they are not independent input parameters for negative-pressure droplet generation.

However, negative-pressure droplet generation has some key advantages over the more conventional positive-pressure-driven methods. Actuated from the collection outlet, these methods allow for continuous access to the oil and sample inlets and for interfacing with liquid handling robots, enabling automation. Compared to syringe pump-driven droplet generation, these advantages are even more pronounced, and with the added advantage of drastically reduced setup and turnaround time. Hundreds of thousands of droplets, each a compartmentalized reaction vessel, can be produced in a matter of minutes while the samples can be agitated in the inlet reservoir, thus preventing particle aggregation, striation, and sedimentation. This is particularly advantageous for sensitive biological samples, such as mammalian cells.

Mammalian cell culture is dependent on sterile conditions and is often sensitive, as some cells, especially primary or stem cells, react negatively to prolonged exposure to lower-than-physiological temperatures [31–33]. In addition, some antibiotics are light-sensitive, which adds to the need for fast experimental processing for steps outside of the incubator [31]. It follows that any cell processing step should be rapid, and the instrument should be deployable in a constrained space, such as a biosafety cabinet, allowing for a fast turnaround time. Naturally, this not only applies for normal cell culture steps, such as freezing, thawing, and upkeep, but for microfluidic operations as well—possibly even more so, as the cells have to resist additional stresses, such as shear forces. Several commercial providers offer relatively compact pressure controller devices that are used extensively within the microfluidic community. However, they still require additional infrastructure, such as an external power supply and pressure source. This might not pose a challenge to research groups focusing on microfluidics, but it adds a layer of complexity and thus a hurdle to overcome for less expert laboratories wanting to adopt microfluidic techniques.

Thus, while a host of operations is available to a specialized lab with appropriate equipment and sufficient expertise—including, but not limited to, generation, incubation, injection, splitting and sorting—the lack of standardization and use of bulky, customized experimental setups limit the dissemination of these powerful techniques to a wider range of laboratories. This creates a disconnect between what is technically possible today and what is easily available and readily used—both limiting advances in research areas that could benefit from microfluidics tools and hampering awareness of biological users' needs in the droplet microfluidics community.

In order to enable non-specialized labs to make use of droplet microfluidics, we compiled and thoroughly characterized a minimal droplet generation setup using only low-cost, commercially available equipment, with a low barrier to entry in mind. Actuation of liquids by exerting variable (negative) pressures at the outlet and inlets of the system enables wide-ranging tuning of droplet sizes and dynamic changes in droplet composition, using a single circuit geometry. The presented setup can produce a range of droplet rates, sizes, and compositions on demand, providing versatility while maintaining a minimal footprint. We further demonstrated the utility of the setup by encapsulating primary hepatocytes in droplets to generate cell spheroids for screening purposes. As shown, this setup addresses the needs of the wider biomedical research community and encourages the adoption of microfluidic workflows and furthers the democratization of healthcare by, e.g., decreasing costs for screening campaigns.

2. Materials and Methods

2.1. Droplet Generation Setups

Three different experimental setups were tested and configured as shown in Figure 1. Briefly, a single pump allows for limited size tuning by adjusting the applied pressure at the collection pump; an additional second pump attached to the continuous phase reservoir increases the accessible droplet size (Figure 1A).

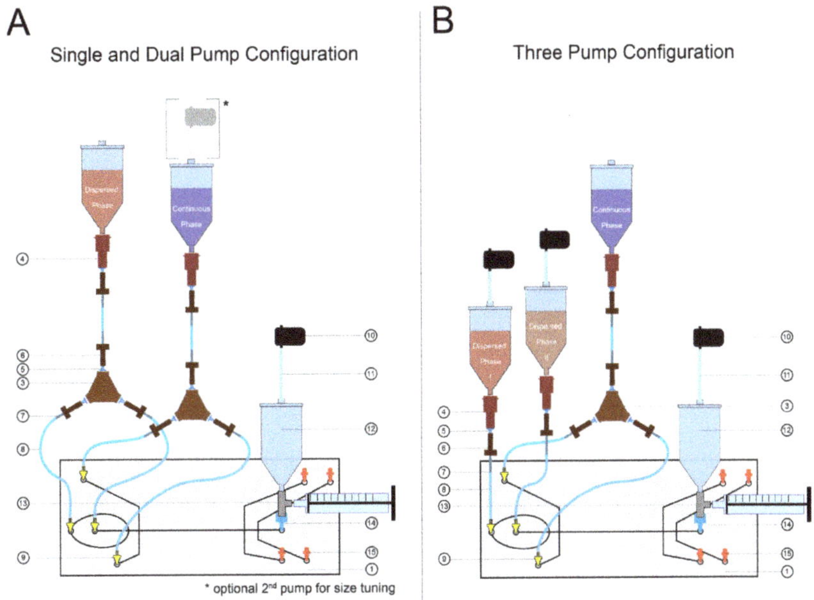

Figure 1. Schematic depictions of the minimal droplet generator setup. (**A**) Illustration of single- and dual-pump configuration. The depiction shows how the components are assembled for operating with one or two pumps; when a second pump is added to the continuous phase reservoir, larger droplets can be generated. (**B**) Schematic depiction of three-pump configuration. The illustration shows how the components are assembled for operating with three pumps. The numbers refer to the part list (see Supplementary Information). Tubing length is customizable, however, we used lengths of 7) 3 cm and 8) 5 cm, with enough overlap to ensure seal. The collection syringe can be any syringe with a Luer lock.

The microfluidic device, the Fluidic 163 (microfluidic ChipShop), features a channel depth of 175 μm, and 140 μm-wide channels for the aqueous and oil phases, respectively. The channel widens to 420 μm 200 μm downstream of the junction (see also Supplemental Figure S1). Attaching two pumps at the dispersed phase reservoirs allows for on-demand tuning of the droplet composition. All necessary microfluidic components are shown, and a detailed list of parts can be found in the Supplementary Material (see Supplemental Figure S2 and Supplemental Table S1). The reservoir container lids connected to pumps were sealed using circular Parafilm cut-outs, prohibiting air leakage. The entire setup is portable (see Supplemental Figure S3) with a compact footprint.

2.2. Pump Characterization

The mp-gas+ membrane pump was connected to a Pressure Unit S (Fluigent) via microfluidic PEEK tubing (Zeus) and the pressure was recorded using OxyGEN software (Fluigent). The frequency response was assessed in 10 Hz increments between 50 and 800 Hz; the amplitude response was assessed in 2 AU increments between 18 and 250 AU.

2.3. Droplet Rate Measurements

Fluorinated oil (Novec HFE-7500, 3M) was supplemented with 2% w/v 008 surfactant (Ran Biotechnologies) for the continuous phase. Fluorescein (Sigma) at a concentration of 20 μM in phosphate buffered saline (PBS, Medicago) was used as the aqueous phase.

The microfluidic chip was connected to tubing as shown in Figure 1A. For the single-pump setup, the mp-gas+ (Bartels Mikrotechnik GmbH) pump at the collection outlet was set to 4.61–13.78 kPa of negative pressure. For the dual-pump setup, the collection pump was set to 13.78 kPa of negative pressure and the mp-gas+ pump attached to the oil inlets was set to 0.3–1.4 kPa of negative pressure.

We focused a 514 nm laser just after the widening of the channel downstream of the nozzle, collected the fluorescence emission using a photomultiplier tube (Hamamatsu), and recorded the fluorescent signal using a custom LabView program (National Instruments).

2.4. Droplet Size Measurements

The microfluidic setup was employed in the same way as described in Section 2.3. To transfer the generated emulsion afterwards, a syringe attached at the Luer T connector is manually aspirated, emptying the collection tank. The emulsion was then imaged using a Ti-E Eclipse (Nikon) and droplet sizes were manually measured using ImageJ [34].

2.5. Flow Rate Measurements

The microfluidic setup was employed in the same way as described in Section 2.3. The pumps were started and the emulsion collected. After 2 min, the pumps were stopped and disconnected to remove any pressure differential. The collected emulsion was broken using an antistatic gun, Zerostat 3 (Milty), according to published protocols [35], and the phases were separated into different tubes. The mass of the samples was measured using a high-precision scale (Ohaus) and their volume was calculated considering the densities of both phases (aqueous phase: 1000 kg/m^3, continuous phase: 1614 kg/m^3).

2.6. Numerical Simulation of Negative Pressure Droplet Generation

To numerically study droplet generation using flow-focusing and negative pressure applied solely at the collection outlet, a 2D approximation of the nozzle region was implemented in COMSOL Multiphysics (COMSOL AB), using the respective densities and dynamic viscosities for water and Novec HFE-7500. Virtual channel boundaries were made using acrylic plastic, and the contact angle was set to superhydrophobic 0 rad to exclude any wetting behavior. The simulation encompassed the first 1.25 s after application of a pressure differential in 5 ms increments. The simulation was carried out on a 2020 M1 16 GB RAM Macbook Pro (Apple).

2.7. Droplet Content Manipulation

We used the same continuous phase as described in Section 2.3. Fluorescein (Sigma) at a concentration of 20 μM in PBS (Medicago) was used as aqueous phase 1 and 15 μm Fluoro-Max Red Dry Fluorescent Particles (ThermoFisher Scientific) dispersed in Milli-Q water were used as aqueous phase 2.

The microfluidic chip was connected to tubing as shown in Figure 1B. The mp-liq pumps (Bartels Mikrotechnik GmbH) connected to the aqueous phase reservoirs were dynamically controlled, we regulated the amplitude parameters of one pump from 24 AU up to 118 AU, while regulating the other from 118 AU down to 24 AU in a corresponding, opposed fashion. A mp-gas+ pump (Bartels Mikrotechnik GmbH) was connected at the collection outlet and set to −13.8 kPa. The emulsion was imaged using a Ti-E Eclipse (Nikon) and ImageJ [34]. Median droplet fluorescence and particle number were manually measured using ImageJ.

2.8. Cell Spheroid Production

A vial of frozen primary hepatocytes (Lonza) was thawed according to the manufacturer's instructions and resuspended in Williams E media with added penicillin-streptomycin (100 U/mL), L-Glutamine (200 mM), insulin (10 µg/mL), 0.1 µM dexamethasone, 5.5 µg/mL transferrin, and 10.95 ng/mL sodium selenite, at a concentration of 7×10^6 cells/mL. An amount of 400 µL of cell-containing media was encapsulated in Novec HFE-7500 (3M), supplemented with 2% *w*/*v* 008 surfactant (RAN Biotechnologies) using a variation of the single-pump configuration (see Supplemental Figure S4) operated at a frequency of 560 Hz and an amplitude of 250 AU, including a 3 mm magnetic stir bar in the cell loading chamber, and we positioned the aqueous reservoir on a rotating magnetic actuator inside a laminar air flow cabinet. Following droplet generation, the emulsion was transferred to a 5 mL syringe and incubated upright in a cell culture incubator with 5% CO_2 and 100% humidity, with a Minisart 5 µm filter (Sartorius) attached to the syringe to allow for gas exchange. Droplet samples were retrieved from the incubation syringe immediately following droplet transfer and after 48 h of incubation. Following 48 h of incubation, spheroids were recovered from the emulsion by filtering on a PTFE membrane (Sartorius) according to a previously published protocol [6]. Spheroids were counted, resuspended in fresh media, and deposited as 20 spheroids per well into a 384 well plate.

2.9. Viability Measurements

A CellTiterGlo (Promega) assay was performed to assess viability. A total of 30 µL of CellTiterGlo was added to 40 µL of standards or spheroids in media and incubated for 1 h at 37 °C and read on an EnSight Multimodal plate reader (Perkin Elmer).

2.10. Albumin Secretion

Albumin levels were measured to quantify albumin secretion. A total of 40 µL of supernatant was frozen at −20 °C until analysis using an Albumin (human) LANCE Ultra TR-FRET Detection Kit on an EnSight plate reader (PerkinElmer) using the manufacturer's protocol.

3. Results

3.1. Pump Characterization

To characterize micropumps that provide actuation in our negative pressure setup, we measured the pressure differential response to the applied membrane frequency and amplitude across the available ranges of frequencies (50–800 Hz) and amplitudes (0–250 AU). When actuated with an amplitude of 150 AU and 250 AU, the micropumps could deliver pressure differentials in the range from −2.8 kPa to −13.7 kPa, with the highest available pressure differential observable occurring with the maximal amplitude of 250 AU and in the frequency range 525–575 Hz (Figure 2A). Notably, the lowest accessible frequencies demonstrated a higher degree of variation in their pressure output. For that reason, and due to considerations with regards to the maximally attainable differential pressure, we chose to examine the frequencies of 560 Hz and 800 Hz more closely.

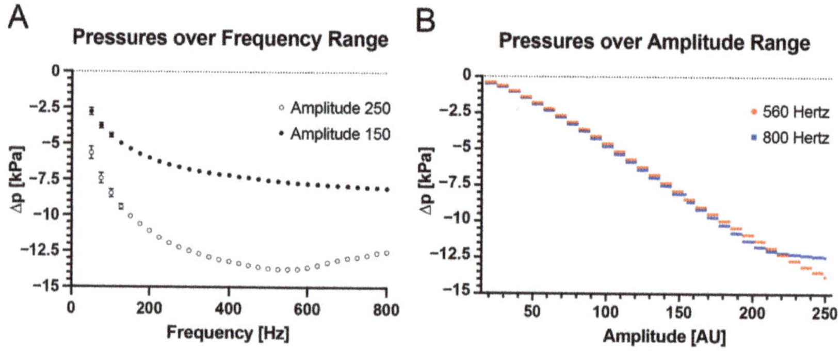

Figure 2. Pump characterization. (**A**) Micropump pressure outputs at fixed amplitudes (150 and 250 AU) for varying frequencies and (**B**) Micropump pressure output at fixed frequencies (560 and 800 Hz) for varying amplitudes. The maximum pressure output was achieved at an amplitude of 250 and between 525 and 575 Hz.

The pressure differential readings from the micropumps demonstrate a stepwise response with a step width of 8 AU for amplitude modulation. Measurable negative pressures were recorded at amplitudes of 17 AU and above for all tested pumps. At 800 Hz, the pressure differential increases at the top end of the amplitude setting were more incremental than at 560 Hz (Figure 2B). We speculate that this might be caused by the high wave frequency, which might limit the travel of the membrane generating the pressure, thereby decreasing the effective amplitude. Based on these findings, we chose to use 560 Hz as the set frequency for the remainder of the experiments and use the amplitude parameter to control pump output. With a set micropump actuation frequency, pressure outputs can be modulated with a pressure step size of 0.5 kPa.

3.2. *Characterization of Droplet Generation*

3.2.1. Single-Pump Operation

In order to gain conceptual insight into the droplet generation behavior of negative pressure actuation applied at the collection outlet with open inlet reservoirs (cf. Figure 1A), we performed numerical simulation using COMSOL Multiphysics. In order to enable simulation within a reasonable amount of time and computational resource use, we limited the simulation to a two-dimensional approximation of the nozzle region. We simulated real-world behavior by setting a surface tension coefficient of 2 mN/m and densities of 1000 kg/m^3 and 1614 kg/m^3, as well as dynamic viscosities of 1 mPa·s and 1.243 mPa·s for the water and fluorinated oil phases, respectively. To avoid wetting phenomena, the contact angle was set to 0 rad.

The results show a trend of decreasing droplet size and increasing droplet frequency with increasing pressure differential applied at the outlet. In other words, the volume of the droplets decreased as the rate increased. Notably, the rate increased faster than the droplet volume decreased, leading to an increased flow rate for the aqueous phase. The oil flow rate was less affected by the higher-pressure differential, resulting in an overall increase in the water-to-oil ratio (see Figure 3).

Using the single-pump configuration, we set out to validate the observations gained from the numerical simulation with the microfluidic setup and noted the same trend: when generating an emulsion using only a single pump attached at the collection outlet, the rate of droplet generation correlated with increasing pressure differential, whereas droplet volume was inversely correlated. Notably, this behavior is consistent with positive pressure generation in the dripping regime, where an increase in flow rates leads to smaller droplets [36]. The oil flow increased as well, but not to the same extent as the aqueous phase flow, leading to an increase in the aqueous-to-oil volume ratio from 0.1 at −4.6 kPa to 0.6 at −13.8 kPa. In this configuration, we were able to generate droplets of 3.1 nL to

5.3 nL at rates of 560 Hz to 114 Hz, respectively (Figure 4A,B). Operating the setup from a dead stop for 2 min, we weighed the phases separately (Figure 4C).

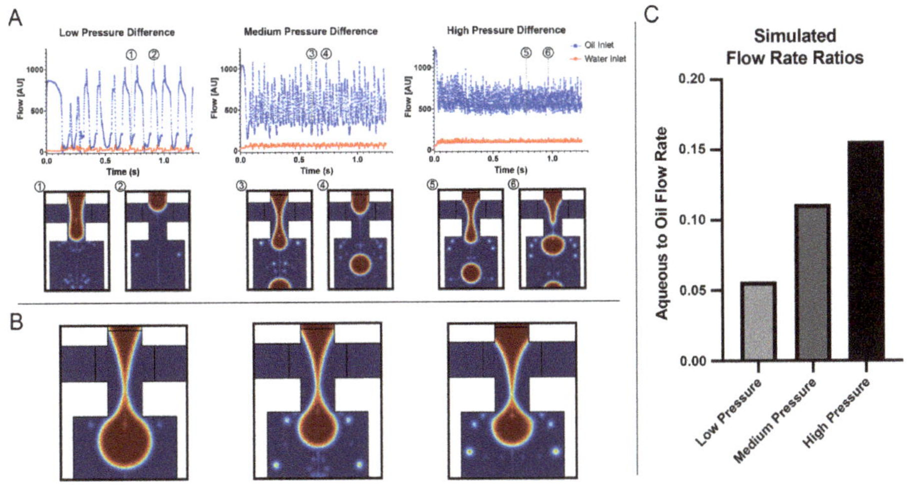

Figure 3. Simulation of microfluidic droplet generation behavior with different applied negative pressures at the outlet. (**A**) shows flows through the respective inlets. The respective flows are oscillating through the droplet break-off, indicating an increase in droplet rate the higher the applied pressure differential. Images (1) and (3) and (5) depict a snapshot of the nozzle at minimal oil flow, (2) and (4) and (6) show the nozzle at maximal oil flow. (**B**) shows the respective droplets immediately prior to break-off. (**C**) shows the dispersed aqueous phase flow normalized to the continuous oil phase flow for the three simulated pressure differentials, indicating an increase in the aqueous-to-oil ratio for increasing negative pressures.

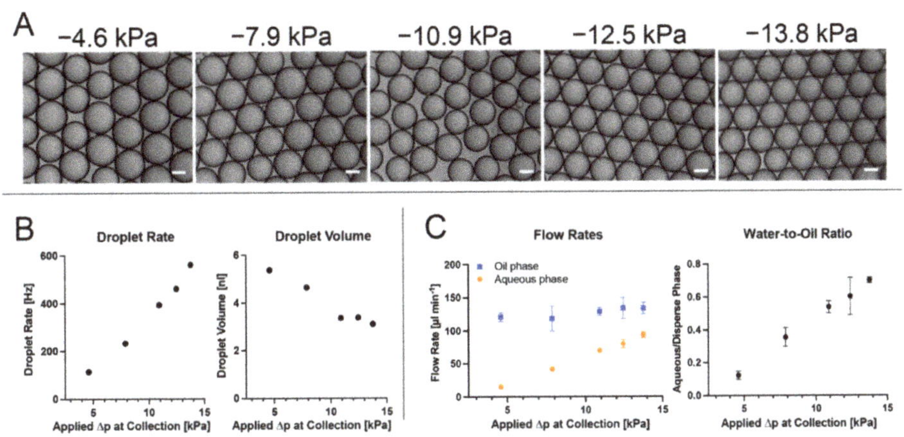

Figure 4. Single−Pump Droplet Generation. (**A**) shows representative images of the resulting emulsions at varying applied pressures at the collection outlet; the white scale bar denotes 100 µm. (**B**) shows droplet rate and volume, the flow rates for both phases, and the water−to−oil ratio for the tested pressures. Notably, the droplet rate increases faster than the droplet volume decreases. (**C**) depicts the flow rate measurements examined by weighing the output after 2 min of operation. Evidently, an increase in the applied negative pressure at the outlet predominantly increases the aqueous phase flow, resulting in an increase in the water−to−oil ratio.

We observed less aqueous phase than expected from droplet rate and droplet volume measurements (for −4.6 kPa, 36 μL expected versus 15 μL measured; for −13.8 kPa, 103 μL expected versus 93 μL measured), which we attributed to the initial equilibration of flows. Due to the higher density of the fluorinated oil, the aqueous phase is pushed back into the reservoir when the system is not pressurized. Only after a negative pressure differential is applied at the outlet, the aqueous phase (re-)enters the chip and moves downstream. This takes longer for lower pressure differentials, accounting for the larger discrepancy of 59% when compared to the 9% discrepancy of the −13.8 kPa sample.

From the channel dimensions of the microfluidic chip, we were able to approximate the relative resistances of the channels when operated as a droplet generation device. Assuming the channel height to be 175 μm, as stated by the manufacturer, we calculated a combined resistance of ~32 kPa m^{-3} s^{-1} for the oil channels and ~30 kPa m^{-3} s^{-1} for the aqueous channels until the junction where both converge.

We assume both phases to be Newtonian, since no large quantities of polymers were dissolved in the aqueous phase. With regards to the fluorinated oil, HFE-7500 has minute non-Newtonian properties, but these should be negligible in our context due to the high shear forces necessary for this behavior to make a significant impact [37]. Consequently, a constant ratio between the two flows should follow if these were the only governing principles. However, we observed diverging behavior, much like in the numerical simulation. Confirming previous findings [22], we measured a greater increase in the flow of the aqueous phase compared to the continuous oil phase with increasing Δp at the outlet. This means that while at an applied pressure difference of −4.6 kPa the ratio of water to oil is 0.11, this increases to 0.61 at −13.8 kPa, reproducing the trend observed in the 2D-simulation. We assume this to be a result of Laplace pressure at the nozzle interface: the effective pressure difference between the aqueous inlet and the water-oil interface $\Delta p_{a\,eff}$ is decreased by the Laplace pressure necessary to balance the surface tension. This means that the derivative of $\Delta p_{a\,eff}$ decreases faster than the derivative of the derivative of the pressure difference from the inlet to the interface at the nozzle Δp_o, or Δp_{nz}, with decreasing applied pressure at the outlet. As the flow rates Q scale with these derivatives, a change in the flow rate ratio is observed (Figure 5).

The Laplace pressure p_{lp} is defined as

$$p_{lp} = \frac{2\gamma}{r} \quad (1)$$

where γ denotes surface tension and r the radius of the interface.

The flow rates Q_a and Q_o for the aqueous and oil phases are dependent on both the pressure differential and the respective resistances:

$$Q_a = \frac{\left(\Delta p_{nz} - \Delta p_{lp}\right)}{R_a} \quad (2)$$

$$Q_o = \frac{\left(\Delta p_{nz} - \Delta p_{Pump}\right)}{R_o} \quad (3)$$

where Δp_{nz} is the pressure difference between the inlet and the aqueous-oil interface at the nozzle, Δp_{lp} is the pressure difference over the interface, R_a is the resistance of the aqueous channel from the inlet until the interface, p_{Pump} is the pressure exerted by the pump, and R_o is the resistance of one oil channel from the inlets up until the interface.

Utilizing Equations (2) and (3), we arrive at Equation (4), describing the ratio of flows:

$$\frac{Q_a}{Q_o} = \frac{\left(\Delta p_{nz} - \Delta p_{lp}\right)}{R_a} \cdot \frac{R_o}{\Delta p_{nz}} = \frac{R_o}{R_a} \cdot \left(1 - \frac{\Delta p_{lp}}{\Delta p_{nz}}\right) \quad (4)$$

Considering that the hydrodynamic resistances of the microfluidic channels do not change, and that both inlets are open and thus subject to atmospheric pressure, and approximating the Laplace pressure to be constant, we can thus conclude that the flow ratio of the aqueous to continuous phase increases with increasing Δp_{nz}, which is a direct result of an increase in Δp_e due to a decrease in p_{Pump}.

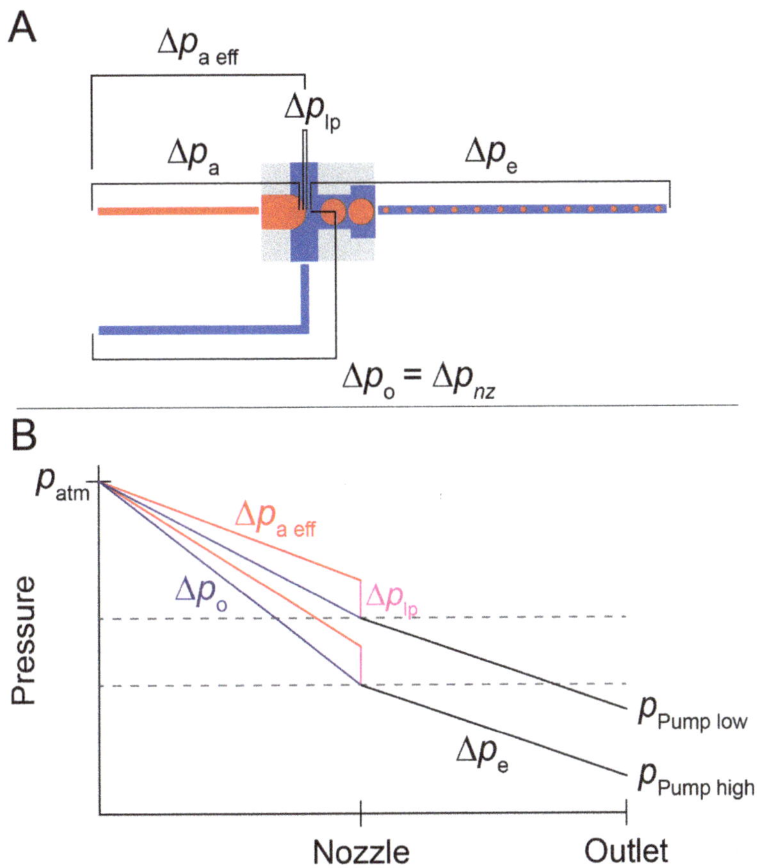

Figure 5. Pressures in the microfluidic system. (**A**) shows a schematic depiction of the microfluidic circuit with the respective pressure differentials and with the two oil channels collapsed into one. Red depicts the aqueous phase, blue the continuous phase. (**B**) represents the idealized pressure along the circuit geometry. Note that Δp_{lp} is not constant in reality, however, the change is minuscule enough to be disregarded for our purposes.

Droplet generation using this single-pump setup ceased with pressure differentials lower than approximately −3 kPa, presumably because of the same Laplace pressure. Therefore, to limit excessive oil consumption, increase aqueous phase throughput, and access larger droplets for, e.g., spheroid production, a different strategy must be pursued.

3.2.2. Dual-Pump Setup for Size Tuning

With a counteracting pump connected to the oil reservoir supplying the oil inlets (cf. Figure 1A), which thereby lowers the effective pressure differential, the oil flow can be modulated. This, in turn, affects the resulting droplet size. Running the collection pump at an amplitude of 250 AU to generate −13.8 kPa, we varied the pressure exerted from the pump connected to the oil reservoir from −0.3 to a maximum of −1.4 kPa, beyond which flow through the oil channels ceased and only aqueous phase was collected. This way, we were able to increase the droplet volume up to 21 nL, or to a diameter of 340 μm. Notably, the flow rate for the aqueous phase remained near constant at 90 μL min^{-1}, while the oil flow rate steadily decreased from 143 μL min^{-1} to 43 μL min^{-1}. In this configuration it appears that the rate of increase in droplet volume is linearly correlated to the decrease in droplet rate, resulting in the observed behavior of near-constant aqueous flow (Figure 6). This is in accordance with previous findings that droplet size in devices with a flow-focusing orifice is governed by geometry and the flow of the continuous phase [29,38].

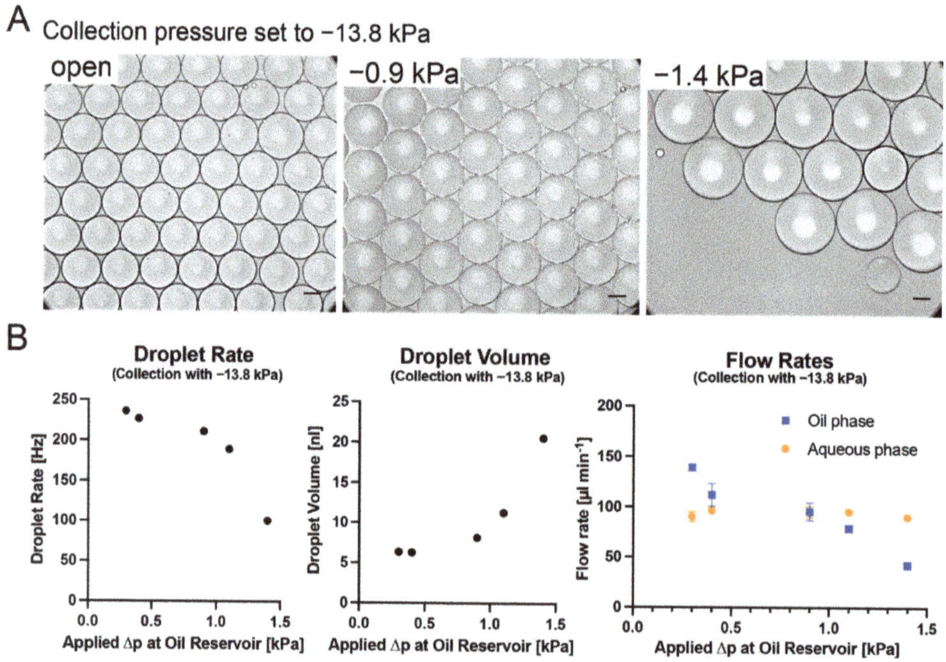

Figure 6. Dual-pump configuration droplet generation. (**A**) shows representative images of the emulsion resulting from −13.8 kPa applied at the outlet, with varying pressures applied to the oil reservoir. Black scale bar denotes 100 μm. (**B**) shows droplet rate and volume and the flow rates for both phases with different pressures applied at the oil reservoir. An increase in the applied negative pressure at the oil reservoir decreases the droplet rate while increasing the droplet volume. The aqueous flow rate is largely unaffected by this, whereas the oil flow rate decreases.

3.2.3. Three-Pump Setup for Droplet Composition Manipulation

To further study the versatility of the setup, we implemented a configuration with one collection pump and a counteracting pump at each aqueous inlet (cf. Figure 1B). We used a suspension of 1.5×10^6 particles as a model for a concentrated mammalian cell culture sample and fluorescein dissolved in PBS as the second aqueous component to be encapsulated. Above amplitude 118 we observed flow from the junction towards the pump-actuated aqueous inlet; thus, we set this as the upper boundary and sequentially adjusted the pump settings from amplitudes 118/0 to 0/118 (see Figure 7A). As a result, we were able to rapidly produce an emulsion with a variety of droplet compositions without changing the sample (see Figure 7B,C).

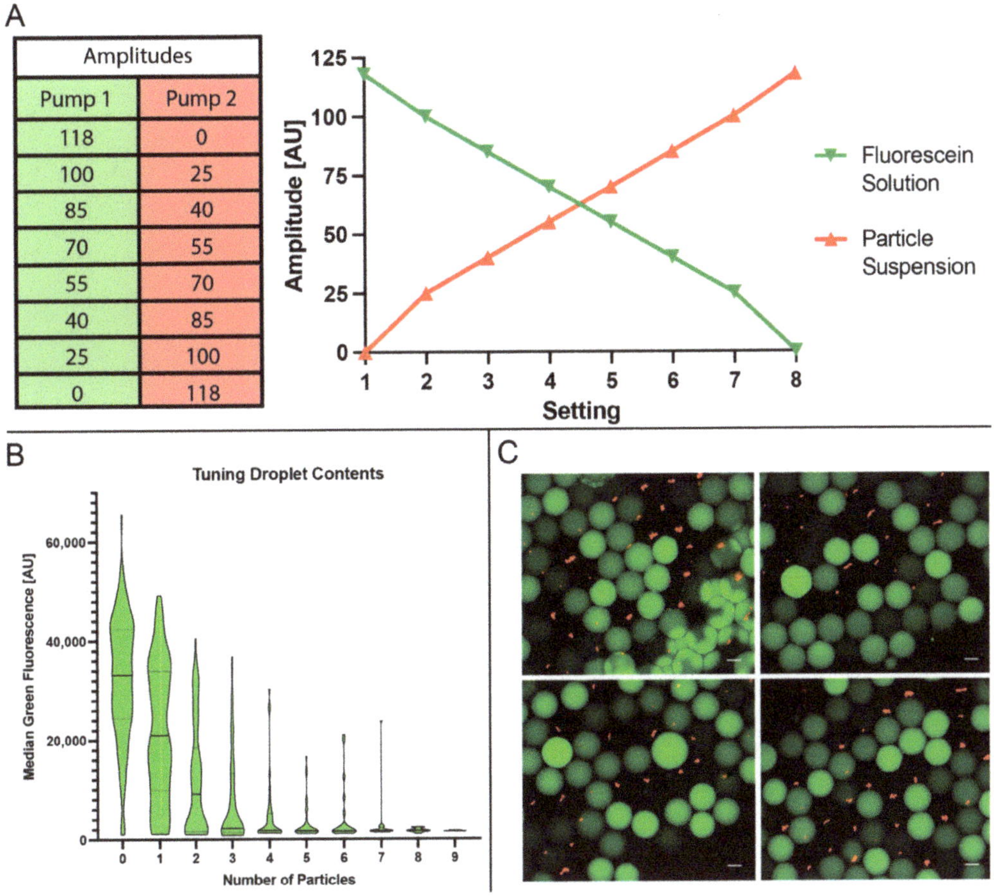

Figure 7. Composition tuning using two pumps at the aqueous inlets. (**A**) illustrates the various settings for dynamic mixing of the two different aqueous samples. (**B**) is a plot of individual droplets' median green fluorescence over their respective particle content. (**C**) shows four representative images of the emulsion used to obtain the results, combining both fluorescent channels in a composite image. White scale bar represents 100 µm.

3.2.4. Single-Pump Setup for Microtissue Production in Biosafety Cabinet

Finally, we intended to verify the practicality of the minimal droplet generator setup for use with a biological sample in the sterile conditions necessary for mammalian cell culture work. Crucially, the minimal droplet generator can be operated by non-experts with very little training, has a minimal footprint, and can be quickly set up. Operating the pump at a frequency of 560 Hz and amplitude of 250 AU (for configuration see Supplemental Figure S4), we encapsulated primary human hepatocytes and successfully demonstrated spheroid assembly after 48 h of droplet incubation (Figure 8A–C). High viability is maintained throughout the incubation period, and subsequent transfer and 120 h incubation in a GrowDex format suggests proliferation (increased ATP) (Figure 8D).

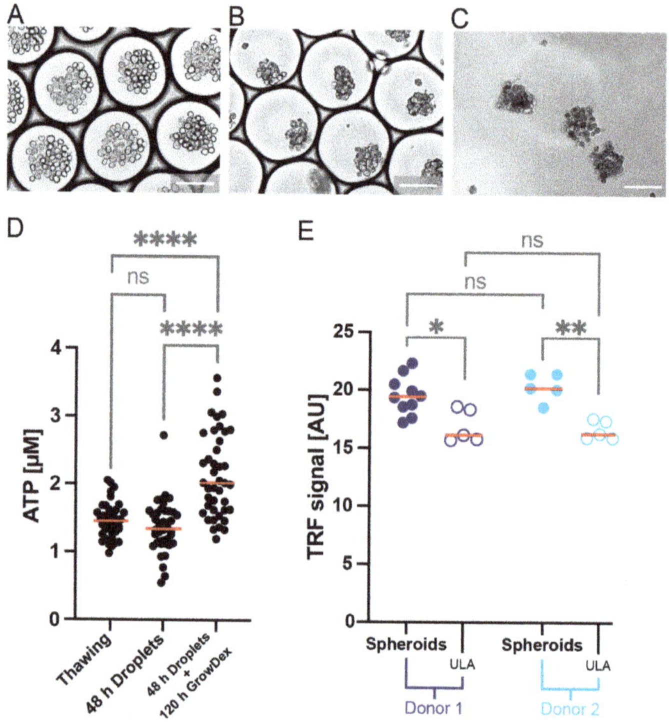

Figure 8. The microfluidic droplet format supports functional cell spheroid formation from primary cells. (**A–C**) show representative brightfield images of primary cell spheroid formation using microfluidic droplets generated with the minimal droplet generation setup. White scale bars indicate 100 µm. (**A**) shows the emulsion directly after encapsulation of primary hepatocytes. (**B**) is an image of the emulsion after 24 h of incubation in a syringe, at 37 °C, 5% CO^2, and 100% humidity. (**C**) shows the spheroids after breaking of the emulsion after 24 h of incubation, demonstrating cohesiveness of the cell assemblies. (**D**) Plot showing ATP concentration as a proxy for viability. Viability of the cells is maintained between thawing and 48-h droplet incubation and breaking of the emulsion. After another 120-h incubation in GrowDex, the viability, as measured by ATP concentration, increases, indicating proliferation. (**E**) compares the functionality of primary hepatocyte spheroids formed in microfluidic droplets encapsulated with the minimal droplet generation setup and spheroids formed in ULA plates from two different sample donors using an albumin quantification assay. Both samples were assessed after 7 days (droplet spheroids were inicubated for 48 h in droplets + 120 h in GrowDex; ULA sample were inicubated for 168 h in ULA plates). In both examined cases, albumin secretion appears to be higher in spheroids assembled in microfluidic droplets. The four stars indicate a p-value of <0.001, two stars a p-value of <0.01, one star a p-value of <0.05, ns indicates insignificance (analyzed with a one-way ANOVA).

Albumin secretion is a hallmark of primary human hepatocyte function and is not retained in all culture models [39]. However, when compared to spheroid formation in an ultra-low attachment (ULA) plate format, microfluidic droplet-derived spheroids exhibit increased albumin secretion as assessed through an albumin quantification assay (Figure 8E). Taken together, these results suggest the microfluidic droplet-derived spheroids are suitable for downstream usage in, e.g., drug or toxicity screens.

4. Conclusions

Here, we characterized and applied a low-cost, minimal footprint droplet generation setup using one to three small membrane pumps. The modular setup is capable of producing monodisperse droplets for a wide range of droplet sizes (see Supplemental Figure S5) and can be dynamically adjusted to vary droplet composition. All parts used are readily available and replaceable. The setup does not require a dedicated microfluidic laboratory infrastructure and can be set up virtually anywhere. All components that come into contact with the sample can be sterilized and the setup as a whole is sufficiently compact to be used in a sterile environment, such as a biosafety cabinet. We demonstrated this by installing the setup in a BSL2 cell culture hood and encapsulating particularly sensitive primary cells for droplet-assisted spheroid formation.

This minimal droplet generator setup represents a step forward in providing access to high-end droplet microfluidics to non-expert users, particularly in the life sciences, with a minimal barrier to entry. The deployment of inexpensive and accessible microfluidics in life science laboratories should yield novel user perspectives to spawn new development and novel technology use cases.

Supplementary Materials: The following supporting information can be downloaded at: https://www.mdpi.com/article/10.3390/mi13111823/s1, Figure S1: Schematic two-dimensional depiction of the nozzle region with the relevant dimensions. The channel depth is 175 µm; Figure S2: Schematic explosion animation of minimal droplet generator setup. The illustration shows all components necessary for repeating our experiments. The numbers refer to the part list (see below). Tubing length is customizable, however, we used lengths of 5 cm of 8) and 3 cm of 7), with enough overlap to ensure no leaking. Not depicted is the collection syringe, which can be any syringe with a Luer lock; Table S1: Parts list of Minimal Droplet Generator Setup. Additionally, a syringe with a Luer tip is needed. Tubings are are interchangeable with other chemically resistant tubing with suitable diameters; Figure S3: The droplet generator has a small footprint and is highly portable. (A) shows three images of the setup set up for droplet generation, depicting the microfluidic device, tubing, syringe and reservoirs, the microcontroller and a 13" Macbook Pro (Apple). (B) shows three images of the setup (minus the Macbook) in a small carrying box; Figure S4: Schematic illustration depicting the used configuration for encapsulation of primary hepatocytes. The numbers refer to the parts list above; Figure S5: Droplet sizes are dependent on applied pressures at the outlets. The measured droplet diameters are monodisperse and decrease with increasing pressure differential in the single pump configuration (gray background). To access larger droplet diameters, a second pump at the oil reservoir might be employed. Red lines signify median, boxes encompass the 25th until the 75th percentile, whiskers indicate 10th until 90th percentile. Dark gray dots show measurements outside of these percentiles.

Author Contributions: Conceptualization, M.T. and H.N.J.; Data curation, M.T.; Formal analysis, M.T.; Funding acquisition, H.N.J.; Investigation, M.T., M.d.L.S. and B.S.-L.; Methodology, M.T.; Project administration, H.N.J.; Resources, B.S.-L. and H.N.J.; Supervision, H.N.J.; Validation, M.T.; Visualization, M.T. and M.d.L.S.; Writing—original draft, M.T. and H.N.J.; Writing—review & editing, M.T., M.d.L.S., B.S.-L. and H.N.J. All authors have read and agreed to the published version of the manuscript.

Funding: This research was funded by the Swedish Innovation agency VINNOVA (Grant number 2018-03338), the Swedish Foundation for Strategic Research SSF (Grant number FFF20-0027), the Knut and Alice Wallenberg Foundation (Grant number 2016.0077), and the Novo Nordisk Foundation.

Institutional Review Board Statement: Not applicable.

Informed Consent Statement: Not applicable.

Data Availability Statement: Not applicable.

Conflicts of Interest: The authors declare no conflict of interest.

References

1. Mashaghi, S.; Abbaspourrad, A.; Weitz, D.A.; van Oijen, A.M. Droplet Microfluidics: A Tool for Biology, Chemistry and Nanotechnology. *Trends Analyt. Chem.* **2016**, *82*, 118–125. [CrossRef]
2. Joensson, H.N.; Andersson Svahn, H. Droplet Microfluidics–a Tool for Single-Cell Analysis. *Angew. Chem. Int. Ed. Engl.* **2012**, *51*, 12176–12192. [CrossRef] [PubMed]
3. Ding, Y.; Howes, P.D.; eMello, A.J. Recent Advances in Droplet Microfluidics. *Anal. Chem.* **2020**, *92*, 132–149. [CrossRef]
4. Holtze, C.; Rowat, A.C.; Agresti, J.J.; Hutchison, J.B.; Angilè, F.E.; Schmitz, C.H.J.; Köster, S.; Duan, H.; Humphry, K.J.; Scanga, R.A.; et al. Biocompatible Surfactants for Water-in-Fluorocarbon Emulsions. *Lab Chip* **2008**, *8*, 1632–1639. [CrossRef]
5. Baret, J.-C. Surfactants in Droplet-Based Microfluidics. *Lab Chip* **2012**, *12*, 422–433. [CrossRef] [PubMed]
6. Langer, K.; Joensson, H.N. Rapid Production and Recovery of Cell Spheroids by Automated Droplet Microfluidics. *SLAS Technol.* **2020**, *25*, 111–122. [CrossRef] [PubMed]
7. Klein, A.M.; Mazutis, L.; Akartuna, I.; Tallapragada, N.; Veres, A.; Li, V.; Peshkin, L.; Weitz, D.A.; Kirschner, M.W. Droplet Barcoding for Single-Cell Transcriptomics Applied to Embryonic Stem Cells. *Cell* **2015**, *161*, 1187–1201. [CrossRef] [PubMed]
8. Zilionis, R.; Nainys, J.; Veres, A.; Savova, V.; Zemmour, D.; Klein, A.M.; Mazutis, L. Single-Cell Barcoding and Sequencing Using Droplet Microfluidics. *Nat. Protoc.* **2017**, *12*, 44–73. [CrossRef]
9. Yao, L.; Shabestary, K.; Björk, S.M.; Asplund-Samuelsson, J.; Joensson, H.N.; Jahn, M.; Hudson, E.P. Pooled CRISPRi Screening of the Cyanobacterium Synechocystis Sp PCC 6803 for Enhanced Industrial Phenotypes. *Nat. Commun.* **2020**, *11*, 1666. [CrossRef]
10. Li, M.; Liu, H.; Zhuang, S.; Goda, K. Droplet Flow Cytometry for Single-Cell Analysis. *RSC Adv.* **2021**, *11*, 20944–20960. [CrossRef]
11. Periyannan Rajeswari, P.K.; Soderberg, L.M.; Yacoub, A.; Leijon, M.; Andersson Svahn, H.; Joensson, H.N. Multiple Pathogen Biomarker Detection Using an Encoded Bead Array in Droplet PCR. *J. Microbiol. Methods* **2017**, *139*, 22–28. [CrossRef] [PubMed]
12. Li, H.; Bai, R.; Zhao, Z.; Tao, L.; Ma, M.; Ji, Z.; Jian, M.; Ding, Z.; Dai, X.; Bao, F.; et al. Application of Droplet Digital PCR to Detect the Pathogens of Infectious Diseases. *Biosci. Rep.* **2018**, *38*, BSR20181170. [CrossRef] [PubMed]
13. Parker, H.E.; Sengupta, S.; Harish, A.V.; Soares, R.R.G.; Joensson, H.N.; Margulis, W.; Russom, A.; Laurell, F. A Lab-in-a-Fiber Optofluidic Device Using Droplet Microfluidics and Laser-Induced Fluorescence for Virus Detection. *Sci. Rep.* **2022**, *12*, 3539. [CrossRef]
14. Wang, J.-H.; Lee, G.-B. Formation of Tunable, Emulsion Micro-Droplets Utilizing Flow-Focusing Channels and a Normally-Closed Micro-Valve. *Micromachines* **2013**, *4*, 306–320. [CrossRef]
15. Hu, B.; Ye, S.; Chen, D.; Xie, B.; Hu, R.; Qiao, Y.; Yu, Y.; Yu, H.; Zheng, X.; Lan, Y.; et al. Tunable and Contamination-Free Injection with Microfluidics by Stepinjection. *Anal. Chem.* **2021**, *93*, 13112–13117. [CrossRef] [PubMed]
16. Chan, H.F.; Zhang, Y.; Ho, Y.-P.; Chiu, Y.-L.; Jung, Y.; Leong, K.W. Rapid Formation of Multicellular Spheroids in Double-Emulsion Droplets with Controllable Microenvironment. *Sci. Rep.* **2013**, *3*, 3462. [CrossRef] [PubMed]
17. Yan, H.; Melin, M.; Jiang, K.; Trossbach, M.; Badadamath, B.; Langer, K.; Winkeljann, B.; Lieleg, O.; Hong, J.; Joensson, H.N.; et al. Immune-modulating Mucin Hydrogel Microdroplets for the Encapsulation of Cell and Microtissue. *Adv. Funct. Mater.* **2021**, *14*, 2105967. [CrossRef]
18. Jang, M.; Koh, I.; Lee, S.J.; Cheong, J.-H.; Kim, P. Droplet-Based Microtumor Model to Assess Cell-ECM Interactions and Drug Resistance of Gastric Cancer Cells. *Sci. Rep.* **2017**, *7*, 41541. [CrossRef]
19. Sart, S.; Tomasi, R.F.-X.; Barizien, A.; Amselem, G.; Cumano, A.; Baroud, C.N. Mapping the Structure and Biological Functions within Mesenchymal Bodies Using Microfluidics. *Sci. Adv.* **2020**, *6*, eaaw7853. [CrossRef]
20. Spitkovsky, D.; Lemke, K.; Förster, T.; Römer, R.; Wiedemeier, S.; Hescheler, J.; Sachinidis, A.; Gastrock, G. Generation of Cardiomyocytes in Pipe-Based Microbioreactor Under Segmented Flow. *Cell. Physiol. Biochem.* **2016**, *38*, 1883–1896. [CrossRef]
21. Periyannan Rajeswari, P.K.; Joensson, H.N.; Andersson-Svahn, H. Droplet Size Influences Division of Mammalian Cell Factories in Droplet Microfluidic Cultivation. *Electrophoresis* **2017**, *38*, 305–310. [CrossRef] [PubMed]
22. Trossbach, M.; Åkerlund, E.; Langer, K.; Seashore-Ludlow, B.; Joensson, H.N. High-Throughput Cell Spheroid Production and Assembly Analysis by Microfluidics and Deep Learning. *bioRxiv* **2022**. [CrossRef]
23. Chong, Z.Z.; Tan, S.H.; Gañán-Calvo, A.M.; Tor, S.B.; Loh, N.H.; Nguyen, N.-T. Active Droplet Generation in Microfluidics. *Lab Chip* **2016**, *16*, 35–58. [CrossRef]
24. Zhu, P.; Wang, L. Passive and Active Droplet Generation with Microfluidics: A Review. *Lab Chip* **2016**, *17*, 34–75. [CrossRef] [PubMed]
25. Anna, S.L.; Bontoux, N.; Stone, H.A. Formation of Dispersions Using "flow Focusing" in Microchannels. *Appl. Phys. Lett.* **2003**, *82*, 364–366. [CrossRef]
26. Chen, X.; Glawdel, T.; Cui, N.; Ren, C.L. Model of Droplet Generation in Flow Focusing Generators Operating in the Squeezing Regime. Microfluid. *Nanofluidics* **2015**, *18*, 1341–1353. [CrossRef]
27. Rahimi, M.; Shams Khorrami, A.; Rezai, P. Effect of Device Geometry on Droplet Size in Co-Axial Flow-Focusing Microfluidic Droplet Generation Devices. Colloids Surf. A Physicochem. *Eng. Asp.* **2019**, *570*, 510–517. [CrossRef]
28. Lashkaripour, A.; Rodriguez, C.; Mehdipour, N.; Mardian, R.; McIntyre, D.; Ortiz, L.; Campbell, J.; Densmore, D. Machine Learning Enables Design Automation of Microfluidic Flow-Focusing Droplet Generation. *Nat. Commun.* **2021**, *12*, 25. [CrossRef]
29. Ibrahim, A.M.; Padovani, J.I.; Howe, R.T.; Anis, Y.H. Modeling of Droplet Generation in a Microfluidic Flow-Focusing Junction for Droplet Size Control. *Micromachines* **2021**, *12*, 590. [CrossRef]

30. Filatov, N.A.; Evstrapov, A.A.; Bukatin, A.S. Negative Pressure Provides Simple and Stable Droplet Generation in a Flow-Focusing Microfluidic Device. *Micromachines* **2021**, *12*, 662. [CrossRef]
31. Helgason, C.D.; Miller, C.L. *Basic Cell Culture Protocols*; Humana Press: New York, NY, USA, 2012; ISBN 9781627031271.
32. Philippeos, C.; Hughes, R.D.; Dhawan, A.; Mitry, R.R. Introduction to Cell Culture. In *Human Cell Culture Protocols*; Mitry, R.R., Hughes, R.D., Eds.; Humana Press: Totowa, NJ, USA, 2012; pp. 1–13, ISBN 9781617793677.
33. Fujita, J. Cold Shock Response in Mammalian Cells. *J. Mol. Microbiol. Biotechnol.* **1999**, *1*, 243–255.
34. Edelstein, A.D.; Tsuchida, M.A.; Amodaj, N.; Pinkard, H.; Vale, R.D.; Stuurman, N. Advanced Methods of Microscope Control Using μManager Software. *J. Biol. Methods* **2014**, *1*, e10. [CrossRef] [PubMed]
35. Karbaschi, M.; Shahi, P.; Abate, A.R. Rapid, Chemical-Free Breaking of Microfluidic Emulsions with a Hand-Held Antistatic Gun. *Biomicrofluidics* **2017**, *11*, 044107. [CrossRef] [PubMed]
36. Köster, S.; Angilè, F.E.; Duan, H.; Agresti, J.J.; Wintner, A.; Schmitz, C.; Rowat, A.C.; Merten, C.A.; Pisignano, D.; Griffiths, A.D.; et al. Drop-Based Microfluidic Devices for Encapsulation of Single Cells. *Lab Chip* **2008**, *8*, 1110–1115. [CrossRef] [PubMed]
37. Bazile, J.-P.; Aguilar, F.; Galliero, G.; Muñoz-Rujas, N.; Montero, E.; Boned, C. High Pressure Viscosity Measurements of the Hydrofluoroether Fluid HFE-7500. In Proceedings of the PPEPPD 2016, Porto, Portugal, 22–26 May 2016; Volume 36.
38. Chen, J.-S.; Jiang, J.-H. Droplet Microfluidic Technology: Mirodroplets Formation and Manipulation. *Chin. J. Anal. Chem.* **2012**, *40*, 1293–1300. [CrossRef]
39. Bachmann, A.; Moll, M.; Gottwald, E.; Nies, C.; Zantl, R.; Wagner, H.; Burkhardt, B.; Sánchez, J.J.M.; Ladurner, R.; Thasler, W.; et al. 3D Cultivation Techniques for Primary Human Hepatocytes. *Microarrays* **2015**, *4*, 64–83. [CrossRef]

micromachines

MDPI

Article

Asymmetric Jetting during the Impact of Liquid Drops on Superhydrophobic Concave Surfaces

Chengmin Chen [1,2,3], Hongjun Zhong [3], Zhe Liu [2], Jianchun Wang [1,2], Jianmei Wang [1,2], Guangxia Liu [1,2], Yan Li [1,2] and Pingan Zhu [4,*]

1 Energy Institute, Qilu University of Technology (Shandong Academy of Sciences), Jinan 250100, China
2 School of Energy and Power Engineering, Qilu University of Technology (Shandong Academy of Sciences), Jinan 250100, China
3 Jinan Key Laboratory of High-Performance Industrial Software, Jinan Institute of Supercomputing Technology, Jinan 250100, China
4 Department of Mechanical Engineering, City University of Hong Kong, Hong Kong 999077, China
* Correspondence: pingazhu@cityu.edu.hk; Tel.: +852-344-22316

Abstract: The impact of liquid drops on superhydrophobic solid surfaces is ubiquitous and of practical importance in many industrial processes. Here, we study the impingement of droplets on superhydrophobic surfaces with a macroscopic dimple structure, during which the droplet exhibits asymmetric jetting. Systematic experimental investigations and numerical simulations provide insight into the dynamics and underlying mechanisms of the observed phenomenon. The observation is a result of the interaction between the spreading droplet and the dimple. An upward internal flow is induced by the dimple, which is then superimposed on the horizontal flow inside the spreading droplet. As such, an inclined jet is issued asymmetrically into the air. This work would be conducive to the development of an open-space microfluidic platform for droplet manipulation and generation.

Keywords: droplet impact; superhydrophobic surface; asymmetric jetting; droplet manipulation

Citation: Chen, C.; Zhong, H.; Liu, Z.; Wang, J.; Wang, J.; Liu, G.; Li, Y.; Zhu, P. Asymmetric Jetting during the Impact of Liquid Drops on Superhydrophobic Concave Surfaces. *Micromachines* **2022**, *13*, 1521. https://doi.org/10.3390/mi13091521

Academic Editor: Giampaolo Mistura

Received: 30 July 2022
Accepted: 6 September 2022
Published: 14 September 2022

1. Introduction

Understanding the dynamics of impacting droplets on nonwetting surfaces is of both scientific and technological importance, such as spraying crops with pesticides [1], spray cooling of hot surfaces [2], 3D inkjet printing of micro/nanostructures [3], shedding virus-laden aqueous droplets away for the anti-pathogen purpose [4], and manipulating droplets with open-space microfluidics [5]. Droplet impact is ubiquitous on solid surfaces with various features, including inclined surface [6], rough surface [7], micro-channels [8], curved surface [9], micro-cellular surface [10,11], heated surface [12], cold surface [13,14], and nanoparticles-coated surfaces [15], to name a few. Varying the surface properties can significantly alter the behaviors and dynamics of droplet impact, such as splashing, spreading, bouncing, jetting, and bubble encapsulation [16–20].

The jetting phenomenon has been previously identified when liquid droplets impact both hydrophilic [21,22] and hydrophobic [23] surfaces. Previous studies showed that jetting is induced by the collapse of the air cavity formed by the deformation of the drop at impact [24–26]. Detailed studies on the mechanism of jetting formation can be found in several recent works [27–29]. Apart from flat hydrophobic surfaces, jetting also occurs on superhydrophobic surfaces with tailored structures, such as ice-leaf-inspired grooved superhydrophobic surfaces [30], superhydrophobic surfaces with anisotropic surface patterning [31], superhydrophobic copper meshes [32], oblique surfaces [24] and artificial dual-scaled superhydrophobic surfaces [33]. The previously reported jetting is symmetric, which is vertical to solid surfaces. However, asymmetric jetting that is inclined to solid surfaces is yet to be observed.

Here, we report asymmetric jetting when water droplets eccentrically impact a macro-sized dimple on superhydrophobic surfaces. The jetting velocity depends on the Weber number (We) and the impact position of the droplet, as demonstrated by experimental results. In parallel, numerical simulations reveal the internal flow field, pressure field, and momentum variation of the droplet. The combination of experimental and numerical studies shed light on the dynamics and mechanism of asymmetric jetting during droplet impact.

2. Experimental Methods

The experimental setup is illustrated in Figure 1. Water droplets were generated from blunt syringe tips. The droplet diameter (d_l, as shown in Figure 1b), which depends on the size of the syringe tips, was about 2.40 $\pm$ 0.05 mm. Following its detachment from the syringe tip, the droplet was accelerated by gravity and then impacted the superhydrophobic surface. The impact velocity (v_0) of the droplet was changed by adjusting the height of the syringe tips, as varied from 0.5 m/s to 0.8 m/s. A copper surface was used in the experiments, on which a hemispherical dimple with a diameter of d_{dimple} = 1.2 mm was excavated. The eccentric distance (d, as shown in Figure 1b) stands for the horizontal distance between the center of the droplet and the center of the dimple. The relative eccentric distance (e) was defined as $e = d/d_{\text{dimple}}$. The surface was cleaned with acetone and ethanol and then coated with candle soot for superhydrophobicity. The candle soot was deposited by exposing the surface to the outer flame of a burning candle for 3–5 s to ensure that the surface areas were fully covered by the candle soot particles. The water contact angle was measured by an Optical Surface Analyzer OSA200 (Ningbo NB Scientific Instruments Co., Ltd., Ningbo, China). The apparent contact angle was about 145.3° (Figure S1 in Supplemental Materials), averaged from three measurements. The sliding angle was 8° and the contact angle hysteresis (the difference between the advancing angle and the receding angle) was about 2.7°.

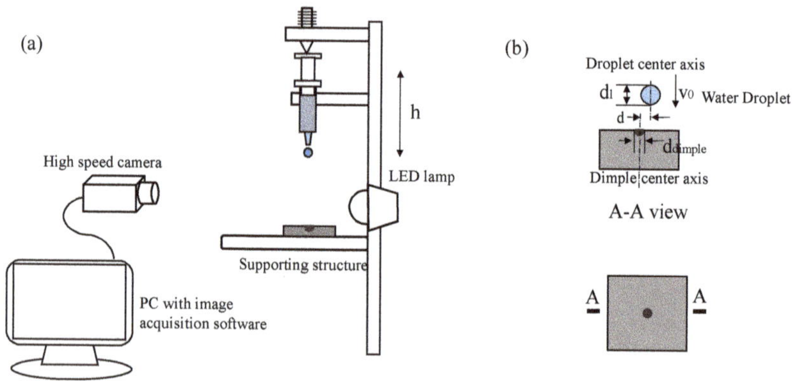

Figure 1. (**a**) Schematic of the experimental setup for droplet impact, (**b**) details of the solid surface.

A high-speed camera (Miro M310, Phantom) was used to record the impact process at 4000 frames per second. The camera was placed parallel to the horizontal surface to obtain the side-view images and videos of impacting droplets. An LED lamp with a diffuser was used for illumination. The recorded images and videos were analyzed by the Phantom camera control software obtained from the camera supplier.

In general, the impact dynamics were characterized by the Weber number defined as the ratio of inertial to surface tension forces (Equation (1)), the Reynolds number defined as the ratio of inertial to viscous forces (Equation (2)), the Bond number defined as the ratio of gravitational to surface tension forces (Equation (3)) and the capillary number defined as the ratio of viscous to surface tension forces (Equation (4)) [34–37]. Here, ρ, v_0, d_l, σ, and

μ are the density, the impact velocity, initial droplet diameter, the surface tension and the viscosity of the liquid, respectively. Water droplets were used in experiments, of which the density is 1000 kg/m^3, the surface tension is 0.072 N/m and the viscosity is 0.001003 Pa·s.

$$\mathrm{We} = \rho {v_0}^2 d_l / \sigma \tag{1}$$

$$\mathrm{Re} = \rho d_l v_0 / \mu \tag{2}$$

$$\mathrm{Bo} = \rho g d_l^2 / 4\sigma \tag{3}$$

$$\mathrm{Ca} = \mu v_0 / \sigma \tag{4}$$

3. Asymmetric Jetting Phenomenon

The behavior of a droplet impacting the dimpled surface showed a remarkable difference from previous observations of droplet impact on flat superhydrophobic surfaces. Figure 2 presents snapshots of a droplet eccentrically impinging the dimpled surface, where the eccentric distance $d = 0.5d_l$, We = 23.41, Re = 1974.10, Bo = 0.20 and Ca = 0.01. The high values of We and Re indicated that inertial forces dominated over capillary and viscous forces during droplet impact, while the low values of Bo and Ca implied that the influences of gravitational and viscous forces were negligible compared with surface tension forces in this study. In Figure 2a, at $t = 0$ ms, the droplet contacted the dimpled surface, with We = 18.97. After contact, the droplet first spread when the time was less than 1.22 ms (Figure 2b) and it fully covered the dimple at $t \sim 1.22$ ms when an inclined jet was issued from the side of the droplet. The jetting angle (θ in Figure 2b and Figure S2 in Supplemental Materials) was about 45° relative to the horizontal plane of the surface. Afterwards, the droplet adopted asymmetric morphology during the spreading and contracting processes. As a result, the bouncing direction of the droplet could be well-controlled by changing the impact position around the dimple, as we identified previously [38].

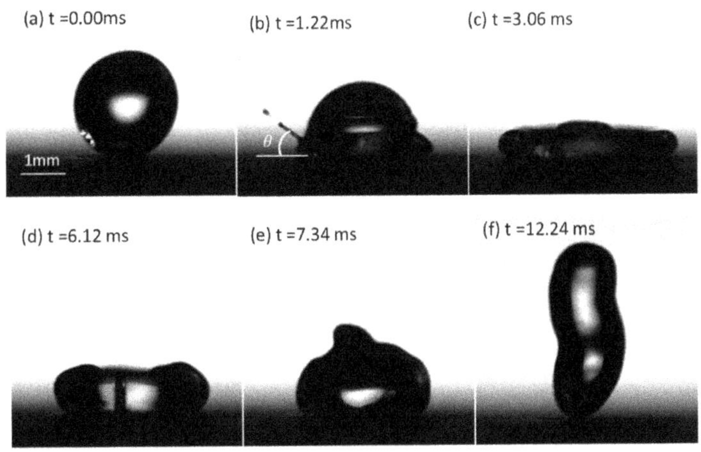

Figure 2. Snapshots showing a drop impacting the dimpled surface at We = 23.41. (**a**) Start of droplet impact at $t = 0.00$ ms. (**b**) Satellite droplets issued at the end of the Jetting. (**c**) Droplet spreading to its maximum diameter. (**d**,**e**) Asymmetric morphology of the droplet during retraction. (**f**) The rebound of the droplet from the surface.

To understand the jetting phenomenon, we plotted the jetting velocity in variation with the Weber number We and relative eccentric distance e in Figure 3. The jetting velocity was determined using image analysis by which the change in the position of the jetting tip was divided by the time interval between two successive frames of the captured video. The jetting velocity increased as the Weber number increased. The highest jetting velocity reached about 4.6 m/s when We = 33.02. At different values of We, the jetting velocity increased at first and then decreased, and the maximum jetting velocity occurred at e = 0.7–0.8 (Figure 3). At We = 15.20 and 18.97, no jetting occurred when e was smaller than ~0.60.

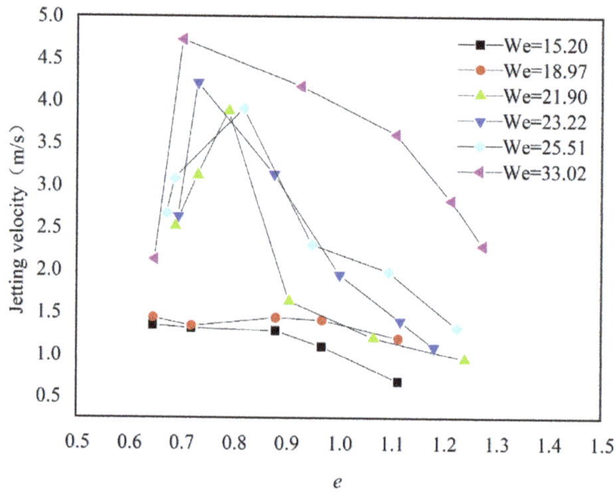

Figure 3. The jetting velocity under different conditions.

Figure 4 shows the jetting angle, which increased as the Weber number increased for We < 30. At a fixed value of We, the jetting angle first increased then decreased with the increase in e. When the values of We were close to each other, a slight difference in the jetting angle was observed, as shown for the datasets with We = 21.90 and 23.22 in Figure 4.

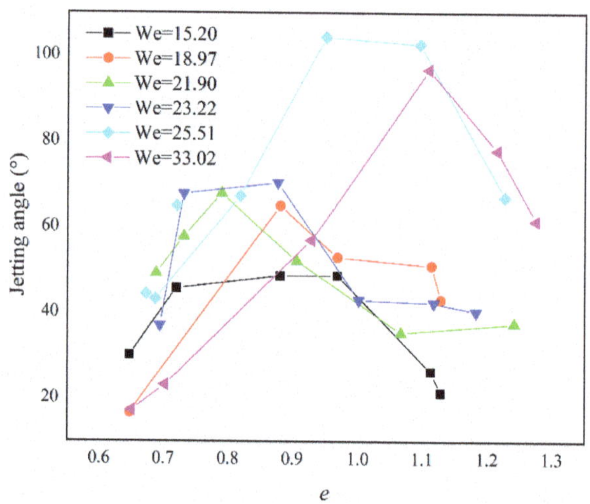

Figure 4. The jetting angle under different conditions.

4. Numerical Simulations of Asymmetric Jetting

To unveil the mechanism of asymmetric jetting, we employed Fluent 2020 to simulate the velocity, momentum, and pressure inside the impacting droplet for a deep understanding of the interaction between liquid droplets and solid surfaces.

4.1. Model Validation

In validating our numerical model, we used the same water droplet properties as indicated in Section 2, and We = 10.23. The mesh step was set to 0.02 mm, as the grid independence study suggested that the simulation was accurate enough when the mesh step is 0.1 mm in size or finer (Figure S3 in Supplemental Materials).

Figure 5 contrasts the shape of the droplet from simulations (Figure 5a–d) and experiments (Figure 5e–h), in which the time was normalized by the capillary time τ_{cap}, as defined in Equation (5). Both numerical and experimental results were consistent with each other. For example, at the normalized time of about 0.2 the droplet covered the dimple, and at the normalized time of about 1.6 the droplets took off from the dimpled surface in both experiment and simulation conditions.

$$\tau_{cap} = \sqrt{\rho d_l^3 / \sigma} \tag{5}$$

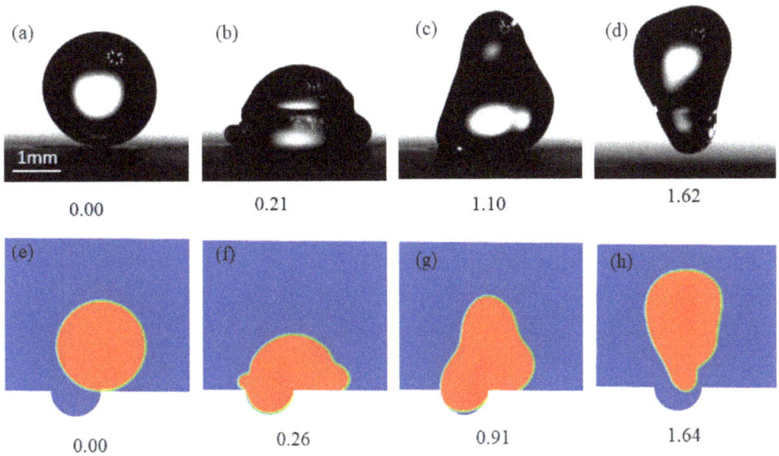

Figure 5. The validation of the simulation model. (**a–d**) Experimental results of a droplet impacting a dimpled surface with the normalized time of 0.00, 0.21,1.10 and 1.62, respectively. (**e–h**) Simulation of a droplet impacting a dimpled surface with the normalized time of 0.00, 0.26, 0.91 and 1.64, respectively.

4.2. Simulation Parameters

To simulate a water droplet with a diameter of 2.4 mm that eccentrically impacts the superhydrophobic dimpled surface, the 3D simulation domain was set as 6 mm × 6 mm × 8 mm (length × width × height). The mesh step was 0.02 mm, which was accurate enough from the grid independence study. The unstructured mesh was used in the dimple region, and the structured mesh was used in the air region. The fluid was an incompressible Newtonian fluid. The surface tension of the droplet was 0.072 N/m. The Weber number of the impacting droplet was 21.90. The water contact angle of the solid surface was set to 180° for the removal of any adhesion between the droplet and solid surface. The volume of fluid (VOF) model was used for tracking the two-phase interface. No-slip boundary condition was applied to the solid surface.

4.3. Simulation Results

4.3.1. The Flow Field

As shown in Figure 3, the jetting velocity varied with the change in *e*. As such, we first investigated the dynamic morphological changes of water droplets impacting the dimpled surface with different *e* at We = 21.9. The simulation results are shown in Figures 5–7.

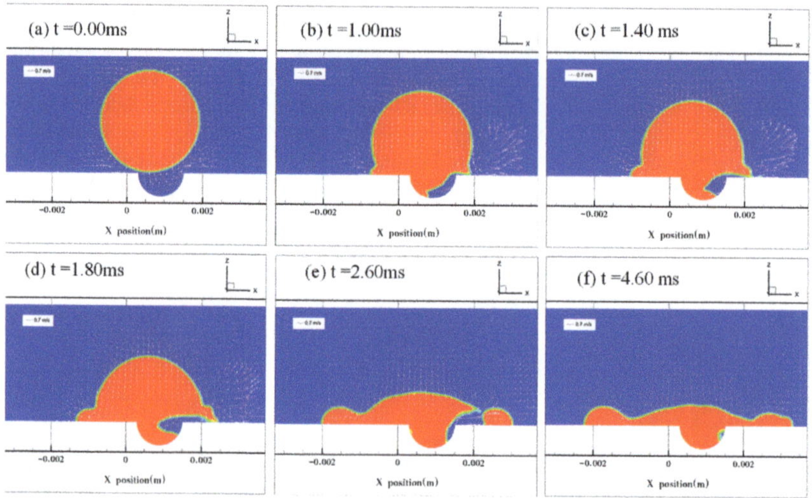

Figure 6. Droplet impacting the dimpled surface at e = 0.25. (**a**) The droplet starts to impact the surface when t = 0.00 ms. (**b–d**) The process of droplet spreading over the dimple, with air trapped inside the dimple. (**b–d**) the air is pushed from the center to the right side of the dimple. (**e**) Air evacuated the dimple at t = 2.60 ms. (**f**) The droplet reaches its maximum spreading diameter.

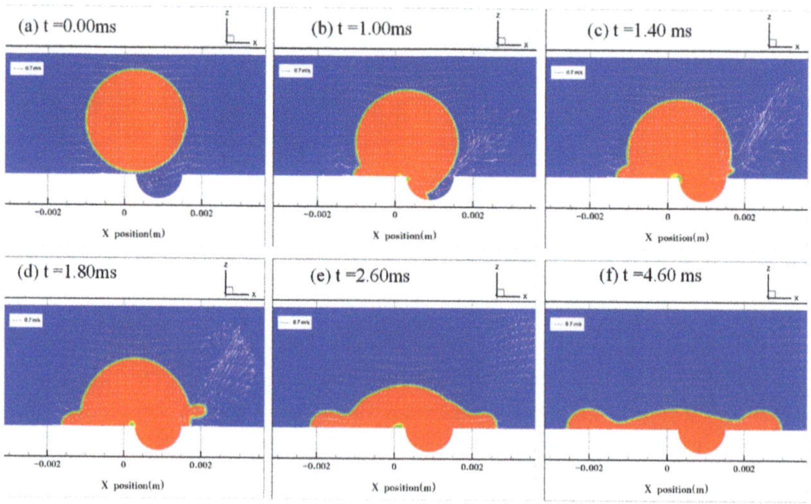

Figure 7. Droplet impacting the dimpled surface at e = 0.5. (**a**) The droplet starts to impact the surface when t = 0.00 ms. (**b**) The edge of the droplet advances along the dimpled surface. (**c**,**d**) The process of the droplet spreading over the dimple. Jetting is formed on the right side of the droplet. (**e**) The diminishing of jetting due to the surface tension force. (**f**) The droplet reaches its maximum spreading diameter.

Figure 6 shows the results at e = 0.25. Figure 6a shows the stage when the droplet made contact with the solid surface at t = 0.00 ms. In Figure 6b, the droplet spreads on the surface and covered the dimple with a small volume of air trapped in the dimple. Because the droplet is moving both downward and outward, the droplet spreading pushed the air to the right side of the dimple. If the droplet coaxially impacts the dimple, the underneath air will be locked in the dimple, and it cannot be evacuated until the droplet rebounds because of the symmetric droplet spreading. In contrast, when the droplet impacts the dimple off the center, air could be squeezed out of the dimple at a high speed during the spreading process (Figure 6c–e), leading to a burst of the droplet's interface (Figure 6e). In Figure 6f, the spreading diameter is maximal. After that, the direction of the flow velocity became reversed, pointing inward and upward, which resulted in bouncing of the droplet. In this case, asymmetric jetting was not observed.

Figures 7 and 8 show the results at $e = 0.5$ and 0.75, respectively. The droplet contacted the solid surface at $t = 0.00$ ms (Figures 7a and 8a). During droplet spreading on the surface and gradual covering of the dimple, the air in the dimple was evacuated at a high velocity (Figures 7b and 8b,c). When the droplet filled the dimple, the motion of the droplet's interface at the right edge of the dimple was governed by the superposition of two types of liquid flow in different directions (Figure 7c,d and Figure 8d): one was the liquid flow to the right in the horizontal direction originating from the rightward droplet spreading and another was the vertically upward liquid flow guided by the curvature of the dimple. The presence of the upward liquid flow induced asymmetric jetting at the edge of the dimple. The jetting angle and velocity were thus determined by the magnitude of the two liquid flow velocities. A larger jetting angle and higher jetting velocity are observed in Figure 8 than in Figure 7 because the magnitude of the upward flow velocity is higher in Figure 8, which is consistent with the experimental results in Figure 3.

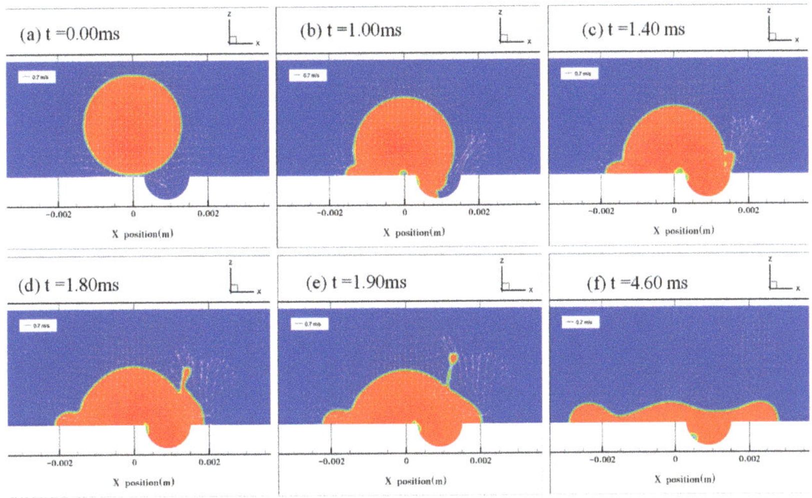

Figure 8. Droplet impacting the dimpled surface at e = 0.75. (**a**) The droplet starts to impact the surface when $t = 0.00$ ms. (**b**) The edge of the droplet advances along the dimpled surface. (**c**) The droplet blankets the dimple. Jetting is formed on the right side of the droplet. (**d**) The jetting position shifts up along the droplet's surface because of the inertia force. (**e**) The breakup of the jet for the formation of satellite droplets. (**f**) The droplet reaches its maximum spreading diameter.

The simulation unveiled the mechanisms for interface bursting and asymmetric jetting. The bursting was caused by the rapid evacuation of the trapped air inside the dimple (Figure 6) and asymmetric jetting was caused by liquid flow in two directions (Figures 7 and 8). Compared with asymmetric jetting, interface bursting occurred at a smaller value of e. As such, a critical value of e was required to trigger the occurrence of asymmetric jetting, consistent with the findings in Figure 3.

4.3.2. The Pressure Distribution inside the Droplet

Figure 9 shows the pressure distribution at $e = 0.25$. The initial pressure inside the droplet was about 200 Pa (Figure 9a). In Figure 8b, the maximum pressure (about 800 Pa) occurred at the dimple's left edge where the droplet first contacted the solid surface. With the spreading of the droplet, the average pressure decreased, and the location of the maximum pressure moved from the left to the right of the dimple. The resulting pressure gradient along the x-direction drove the motion of trapped air from the left to the right inside the dimple (Figure 9b–d). After the droplet's interface burst, the air was evacuated from the dimple, and the pressure decreased, while the maximum pressure still appeared on the right edge of the dimple, which was about 300 Pa (Figure 9e,f). During this process, there was still some air in the position where the maximum pressure appeared, as shown in Figure 6e,f. Figure 9f also suggests that the maximum pressure inside the droplet occurred at the location where the flow velocity was about 0 m/s, and the minimum pressure occurred at the location where the liquid film was the thinnest.

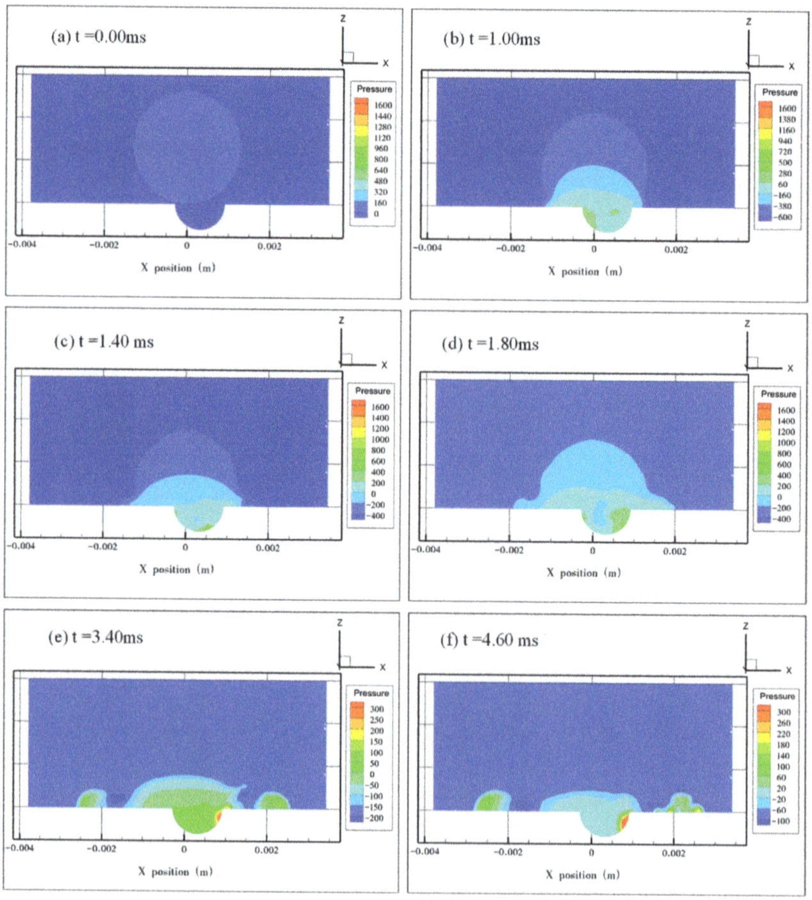

Figure 9. Pressure distribution inside the droplet at $e = 0.25$. (**a**) The initial pressure distribution when the droplet starts to impact the surface at $t = 0.00$ ms. (**b–d**) The maximum pressure moves from the left side to the right side of the dimple along with droplet spreading. (**e**) Pressure distribution after the air is evacuated from the dimple. (**f**) Pressure distribution when the droplet spreads to its maximum diameter.

Figures 10 and 11 show the pressure distribution at $e = 0.5$ and 0.75, respectively. The two cases have a similar tendency in pressure change. At the early stages of droplet spreading, the maximum pressure inside the droplet occurred in the dimple where the air was pushed out. After the jetting was formed, there was no air left in the dimple, and the maximum pressure occurred at the edge of the droplet where the flow velocity was close to 0 m/s. The minimum pressure was located at the position where the liquid film was the thinnest, which is similar to the observation in Figure 9f.

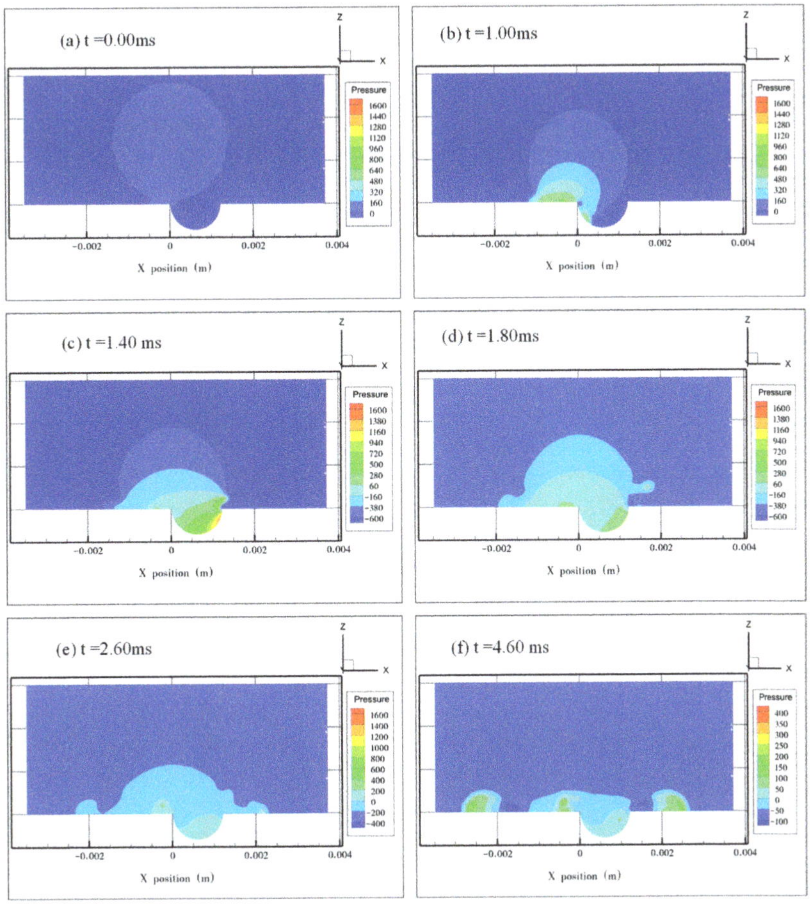

Figure 10. Pressure distribution inside the droplet at e = 0.50. (**a**) The initial pressure distribution when the droplet starts to impact the surface at $t = 0.00$ ms. (**b–e**) The maximum pressure moves from the left side to the right side of the dimple along with droplet spreading. In (**d**), the highest pressure (about 60 Pa) occurs in the tip of the jetting finger. (**f**) Pressure distribution when the droplet spreads to its maximum diameter.

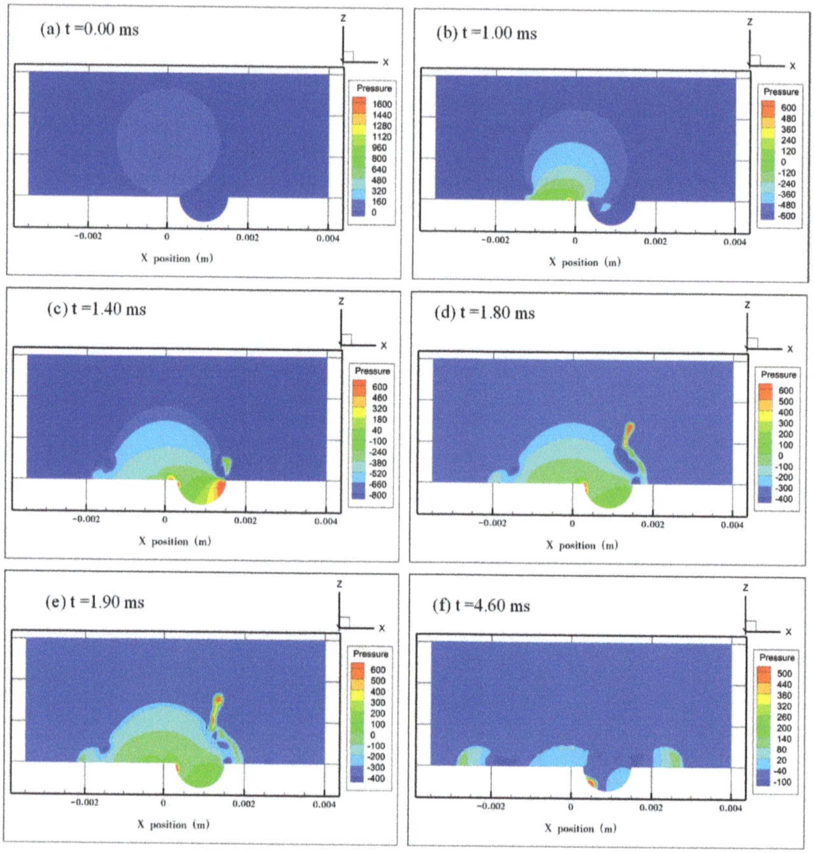

Figure 11. Pressure distribution inside the droplet at e = 0.75. (**a**) The initial pressure distribution when the droplet starts to impact the surface at t = 0.00 ms. (**b**) The maximum pressure occurs on the liquid–solid interface. (**c**) The maximum pressure shifts from the left side to the right side of the dimple with droplet spreading. (**d**,**e**) The maximum pressure occurs in the tip of the jetting finger during the evolution of the jetting. (**f**) Pressure distribution when the droplet spreads to its maximum diameter.

4.3.3. The Momentum Changes in the Droplet

We previously identified the symmetry-breaking of non-specular reflection of impacting droplets and unveiled the momentum difference between droplet impacted on flat and dimpled surfaces [38]. The droplet shape is symmetric at any time instant (see Figure 4), and thus the net momentum is zero in the horizontal direction for the impingement of a droplet on flat surfaces. In sharp contrast, the presence of a dimple affects the internal liquid flow dramatically during droplet impact.

Figure 12 shows the variations in the horizontal momentum (normalized by the initial momentum at droplet impact) with time (normalized by the capillary time). Here, we defined the positive momentum pointing to the right, and vice versa. On the dimpled surface, the magnitude of the horizontal momentum ratio (MHMR) increased at first and then decreased after the liquid reached the bottom of the dimple. The horizontal momentum ratio varied differently with time between the case of interface bursting and asymmetric jetting. The MHMR of the interface bursting case (at e = 0.25) was much smaller than that of the asymmetric jetting case (at e = 0.50 and 0.75), which was attributed to the impounded air (at $e = 0.25$) that prevented the accumulation of horizontal momentum during droplet spreading. For asymmetric jetting, the maximum value of the MHMR appeared later when e increased, because an increase in e indicated a longer distance before the droplet could reach the bottom of the dimple.

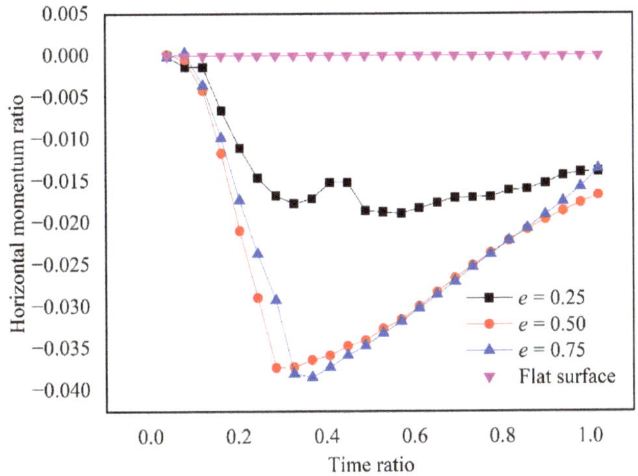

Figure 12. Variations of the horizontal momentum (normalized by the initial momentum at droplet impact) with time (normalized by capillary time).

5. Conclusions

This study identified the asymmetric jetting phenomenon when water droplets impact superhydrophobic surfaces with a macro-sized dimple. The jetting velocity under different conditions was experimentally investigated. Numerical simulations were performed to obtain more details of asymmetric jetting. The following concluding remarks are summarized by combining the results from experimental and numerical studies.

(1) The jetting velocity increases with the increase in the droplet impact velocity. As the eccentric distance increases, the jetting velocity first increases then decreases.

(2) When the eccentric distance is small, interface bursting is induced by the high-speed evacuation of air from the dimple. With the increase in the eccentric distance, asymmetric jetting is triggered by the superposition of liquid flows in the horizontal and vertical directions. The jetting velocity and jetting angle depend on the eccentric distance.

(3) The pressure inside the droplet decreases during droplet spreading. The maximum pressure first occurs at the dimple's left edge where the droplet makes contact with the solid surface, and then moves from the left edge to the right edge of the dimple. After the spreading droplet spans over the dimple, the maximum pressure occurs at the location where the impounded air stays for interface bursting and at the location where the flow velocity is close to 0 m/s for asymmetric jetting.

(4) When the droplet impacts the flat surface, the net horizontal momentum is zero because of the symmetric distribution of fluid flows, whereas the dimple will break down the symmetry of the horizontal momentum, resulting in a non-zero net value. The maximum magnitude of the horizontal momentum appears at the moment when the liquid droplet reaches the bottom of the dimple and is smaller for interface bursting case than asymmetric jetting.

Supplementary Materials: The following supporting information can be downloaded at: https://www.mdpi.com/article/10.3390/mi13091521/s1, Figure S1: Contact angle measurement; Figure S2: The jetting formation process; Figure S3: Grid independence study.

Author Contributions: P.Z. contributed to conceptualization, methodology, writing—review and editing and funding acquisition. C.C. contributed to software application, writing—original draft preparation and funding acquisition. H.Z. contributed to software application. Z.L. contributed to experiments. J.W. (Jianchun Wang) and J.W. (Jianmei Wang) helped with experiment system debugging. G.L. and Y.L. contributed to data analysis. All authors have read and agreed to the published version of the manuscript.

Funding: This research was funded by the Research Grants Council of Hong Kong (ECS 21213621), City University of Hong Kong (9610502 and 7005936), the Open Project of State Key Laboratory of Crystal Materials, Shandong University (KF2014), Key research and development plan in Shandong province (2019GHZ018), National Talent Program (G2022024003L) and Shandong Provincial Key Research and Development Program (Major Technological Innovation Project) (2021CXGC010515).

Institutional Review Board Statement: Not applicable.

Informed Consent Statement: Not applicable.

Data Availability Statement: The data that support the findings of this study are available from the corresponding author upon reasonable request.

Conflicts of Interest: The authors declare no conflict of interest.

References

1. Xue, S.; Xi, X.; Lan, Z.; Wen, R.; Ma, X. Longitudinal drift behaviors and spatial transport efficiency for spraying pesticide droplets. *Int. J. Heat Mass Transf.* **2021**, *177*, 121516. [CrossRef]
2. Wu, H.; Tang, L.; Cen, C.; Lee, C.-F. Effect of droplet size on the jet breakup characteristics of n-butanol during impact on a heated surface. *J. Traffic Transp. Eng.* **2020**, *7*, 320–330. [CrossRef]
3. Zou, W.; Yu, H.; Zhou, P.; Zhong, Y.; Wang, Y.; Liu, L. High-resolution additive direct writing of metal micro/nanostructures by electrohydrodynamic jet printing. *Appl. Surf. Sci.* **2021**, *543*, 148800. [CrossRef]
4. Zhu, P.; Wang, Y.; Chu, H.; Wang, L. Superhydrophobicity preventing surface contamination as a novel strategy against COVID-19. *J. Colloid Interface Sci.* **2021**, *600*, 613–619. [CrossRef]
5. Chen, R.; Jiao, L.; Zhu, X.; Liao, Q.; Ye, D.; Zhang, B.; Li, W.; Lei, Y.; Li, D. Cassie-to-Wenzel transition of droplet on the superhydrophobic surface caused by light induced evaporation. *Appl. Therm. Eng.* **2018**, *144*, 945–959. [CrossRef]
6. Yeong, Y.H.; Burton, J.; Loth, E.; Bayer, I.S. Drop impact and rebound dynamics on an inclined superhydrophobic surface. *Langmuir* **2014**, *30*, 12027–12038. [CrossRef]
7. Tang, C.; Qin, M.; Weng, X.; Zhang, X.; Zhang, P.; Li, J.; Huang, Z. Dynamics of droplet impact on solid surface with different roughness. *Int. J. Multiph. Flow* **2017**, *96*, 56–69. [CrossRef]
8. Shen, Y.; Hu, L.; Chen, W.; Xie, H.; Fu, X. Drop Encapsulated in Bubble: A New Encapsulation Structure. *Phys. Rev. Lett.* **2018**, *120*, 054503. [CrossRef]
9. Liu, Y.; Andrew, M.; Li, J.; Yeomans, J.M.; Wang, Z. Symmetry breaking in drop bouncing on curved surfaces. *Nat. Commun.* **2015**, *6*, 10034. [CrossRef]
10. Zhang, R.; Hao, P.; Zhang, X.; He, F. Dynamics of high Weber number drops impacting on hydrophobic surfaces with closed micro-cells. *Soft Matter.* **2016**, *12*, 5808–5817. [CrossRef]
11. Yang, C.; Cao, W.; Yang, Z. Study on dynamic behavior of water droplet impacting on super-hydrophobic surface with micro-pillar structures by VOF method. *Colloids Surf. A* **2021**, *630*, 127634. [CrossRef]
12. Li, J.; Weisensee, P.B. Low Weber number droplet impact on heated hydrophobic surfaces. *Exp. Therm. Fluid Sci.* **2022**, *130*, 110503. [CrossRef]
13. Moita, A.S.; Moreira, A.L.N. Drop impacts onto cold and heated rigid surfaces: Morphological comparisons, disintegration limits and secondary atomization. *Int. J. Heat Fluid Flow* **2007**, *28*, 735–752. [CrossRef]
14. Chang, S.N.; Qi, H.F.; Zhou, S.; Yang, Y.L. Experimental study on freezing characteristics of water droplets on cold surfaces. *Int. J. Heat Mass Transf.* **2022**, *194*, 123108. [CrossRef]
15. Su, J.; Legchenkova, I.; Liu, C.; Lu, C.; Ma, G.; Bormashenko, E.; Liu, Y. Faceted and Circular Droplet Spreading on Hierarchical Superhydrophobic Surfaces. *Langmuir* **2020**, *36*, 534–539. [CrossRef]
16. Liu, Y.; Moevius, L.; Xu, X.; Qian, T.; Yeomans, J.M.; Wang, Z. Pancake bouncing on superhydrophobic surfaces. *Nat. Phys.* **2014**, *10*, 515–519. [CrossRef]
17. Yarin, A.L. Drop impact dynamics: Splashing, spreading, receding, bouncing. *Annu. Rev. Fluid Mech.* **2006**, *38*, 159–192. [CrossRef]
18. Richard, D.; Clanet, C.; Quere, D. Contact time of a bouncing drop. *Nature* **2002**, *417*, 811. [CrossRef]

19. Han, N.-N.; Sun, B.-M.; He, X. Split of droplets at the nanoscale using mixed-wettability surfaces: A molecular dynamics simulation. *Appl. Surf. Sci.* **2022**, *600*, 154025. [CrossRef]
20. Li, W.; Wang, J.; Zhu, C.; Tian, L.; Zhao, N. Numerical investigation of droplet impact on a solid superhydrophobic surface. *Phys. Fluids* **2021**, *33*, 063310. [CrossRef]
21. Siddique, A.U.; Trimble, M.; Zhao, F.; Weislogel, M.M.; Tan, H. Jet ejection following drop impact on micropillared hydrophilic substrates. *Phys. Rev. Fluids* **2020**, *5*, 063606. [CrossRef]
22. Asai, B.; Tan, H.; Siddique, A.U. Droplet Impact on a Micro-structured Hydrophilic Surface: Maximum Spreading, Jetting, and Partial Rebound. *Int. J. Multiph. Flow* **2022**, *157*, 104235. [CrossRef]
23. Bartolo, D.; Josserand, C.; Bonn, D. Singular jets and bubbles in drop impact. *Phys. Rev. Lett.* **2006**, *96*, 124501. [CrossRef] [PubMed]
24. Guo, J.W.; Zou, S.; Lin, S.J.; Zhao, B.Y.; Deng, X.; Chen, L.Q. Oblique droplet impact on superhydrophobic surfaces: Jets and bubbles. *Phys. Fluids* **2020**, *32*, 122112. [CrossRef]
25. Tsai, P.C.; Pacheco, S.; Pirat, C.; Lefferts, L.; Lohse, D. Drop Impact upon Micro- and Nanostructured Superhydrophobic Surfaces. *Langmuir* **2009**, *25*, 12293–12298. [CrossRef]
26. Thoroddsen, S.T.; Takehara, K.; Nguyen, H.D.; Etoh, T.G. Singular jets during the collapse of drop-impact craters. *J. Fluid Mech.* **2018**, *848*, R3. [CrossRef]
27. Li, J.; Yang, K.; Liang, Y.; Liu, C. Hydrodynamic analysis of the energy dissipation of droplets on vibrating superhydrophobic surfaces. *Int. Commun. Heat Mass Transf.* **2022**, *137*, 106264. [CrossRef]
28. Kibar, A. Experimental and numerical investigations of the impingement of an oblique liquid jet onto a superhydrophobic surface: Energy transformation. *Fluid Dyn. Res.* **2015**, *48*, 015501. [CrossRef]
29. Yamamoto, K.; Motosuke, M.; Ogata, S. Initiation of the Worthington jet on the droplet impact. *Appl. Phys. Lett.* **2018**, *112*, 093701. [CrossRef]
30. Guo, C.F.; Liu, L.; Sun, J.X.; Liu, C.W.; Liu, S.Y. Splashing behavior of impacting droplets on grooved superhydrophobic surfaces. *Phys. Fluids* **2022**, *34*, 052105. [CrossRef]
31. Pearson, J.T.; Bilodeau, D.; Maynes, D. Two-Pronged Jet Formation Caused by Droplet Impact on Anisotropic Superhydrophobic Surfaces. *J.Fluid Eng. T ASME* **2016**, *138*, 074501. [CrossRef]
32. Kumar, A.; Tripathy, A.; Nam, Y.; Lee, C.; Sen, P. Effect of geometrical parameters on rebound of impacting droplets on leaky superhydrophobic meshes. *Soft Matter.* **2018**, *14*, 1571–1580. [CrossRef] [PubMed]
33. Chen, L.Q.; Xiao, Z.Y.; Chan, P.C.H.; Lee, Y.K.; Li, Z.G. A comparative study of droplet impact dynamics on a dual-scaled superhydrophobic surface and lotus leaf. *Appl. Surf. Sci.* **2011**, *257*, 8857–8863. [CrossRef]
34. Liu, C.; Legchenkova, I.; Han, L.; Ge, W.; Lv, C.; Feng, S.; Bormashenko, E.; Liu, Y. Directional Droplet Transport Mediated by Circular Groove Arrays. Part II: Theory of Effect. *Langmuir* **2021**, *37*, 1948–1953. [CrossRef]
35. Liu, C.; Legchenkova, I.; Han, L.; Ge, W.; Lv, C.; Feng, S.; Bormashenko, E.; Liu, Y. Directional Droplet Transport Mediated by Circular Groove Arrays. Part I: Experimental Findings. *Langmuir* **2020**, *36*, 9608–9615. [CrossRef]
36. Josserand, C.; Thoroddsen, S.T. Drop Impact on a Solid Surface. *Annu. Rev. Fluid Mech.* **2016**, *48*, 365–391. [CrossRef]
37. Wu, G.; Chen, S. Simulating the collision of a moving droplet against a moving particle: Impact of Bond number, wettability, size ratio, and eccentricity. *Phys. Fluids* **2021**, *33*, 093313. [CrossRef]
38. Zhu, P.; Chen, C.; Nandakumar, K.; Wang, L. Nonspecular Reflection of Droplets. *Small* **2021**, *17*, 2006695. [CrossRef]

micromachines

Article

Relationship between Onset of Sliding Behavior and Size of Droplet on Inclined Solid Substrate

Yukihiro Yonemoto [1,*], Yosuke Fujii [2], Yoshiki Sugino [2] and Tomoaki Kunugi [3,*]

1 Division of Industrial Fundamentals, Faculty of Advanced Science and Technology, Kumamoto University, 2-39-1, Kurokami, Chuo-ku, Kumamoto 860-8555, Japan
2 Department of Mechanical and Mathematical Engineering, Kumamoto University, 2-39-1, Kurokami, Chuo-ku, Kumamoto 860-8555, Japan
3 College of Energy Engineering, Zhejiang University, 38 Zheda Road, Hangzhou 310027, China
* Correspondence: yonemoto@mech.kumamoto-u.ac.jp (Y.Y.); kunugi.tomoaki.85s@st.kyoto-u.ac.jp (T.K.)

Abstract: Whether a droplet slides or not on inclined solid surface is mainly influenced by a balance between the adhesion force at contact area and the gravitational force exerted on the droplet. Especially as the adhesion force is a key parameter for the determination of the sliding behavior of droplets. The adhesion force is mainly estimated by experimental observation for the sliding motion of the droplet. However, at present it is unknown whether the adhesion force is a constant value regardless of the droplet size or not. In the present study, focused on the onset for sliding of water-ethanol binary mixture droplets on inclined solid surface, experimental investigation on the sliding droplets is performed by considering the droplet volumes ranging from 7 to 600 μL in order to understand the effect of the size of the droplet on the adhesive property. The results are discussed using the existing analytical models. From the results, it is found that the adhesion force increases in the case of large droplet volume, while the force reaches constant value in the case of small droplet volume. This difference is related to the degree of the droplet shape deformation, which leads to a change in the contact angle. Finally, a simple empirical model for the adhesion force including the size effect is proposed.

Keywords: wettability; sliding behavior; contact angle; surface tension; droplet

Citation: Yonemoto, Y.; Fujii, Y.; Sugino, Y.; Kunugi, T. Relationship between Onset of Sliding Behavior and Size of Droplet on Inclined Solid Substrate. *Micromachines* **2022**, *13*, 1849. https://doi.org/10.3390/mi13111849

Academic Editor: Pingan Zhu

Received: 23 September 2022
Accepted: 26 October 2022
Published: 28 October 2022

1. Introduction

The control of liquid on solid substrate which is characterized by wettability is widely seen in industrial and chemical applications, such as coating, inkjet printing and spray cooling [1–3]. For example, in fuel cells [4], liquid water and droplet detachment from electrodes is a fundamental problem for water transport and its management. In a heat exchanger with dropwise condensation [5,6], the removal of droplets from solid surfaces is crucial problem for achieving higher heat transfer. In this kind of system, the dynamic motion of liquid on a solid surface is an important phenomenon where there are many factors related to wettability, adhesion force, flow field around the liquid, and inclination of solid substrate.

There are many studies on the migration of liquid on solid substrates from experimental, numerical, and theoretical point of views. The migration behavior is one of the fundamental behaviors of dynamic wetting. However, in an actual situation, there are many unresolved problems which affect the migration of the liquid, such as internal fluid flow, solid surface condition, external forces, and the treatment of the dynamic contact angle. Yilbas et al. performed experimental and numerical works for the behavior of water droplets on an inclined hydrophobic surface. In their study, rolling behavior was investigated considering the fluid motion in the droplet with PIV (particle image velocimetry) technique. The result reveals the relationship between the fluid velocity in the droplet and the droplet volume [7]. Lv et al. considered the sliding behavior of water droplets on rough surfaces experimentally and analytically [8]. In their study, the effect of the surface microstructure (pillar-structured surface) on the onset of the sliding motion of droplet was discussed and the model for the prediction of the critical sliding angle on rough surface was proposed. Li et al. investigated water droplet detachment characteristics under different gas diffusion layer surfaces in proton exchange membrane fuel cell (PEMFC) experimentally and analytically [9]. In the experiment, the droplet detachment behaviors under air flow rates in the channel were observed. They revealed that the contact angle hysteresis, which is the difference between the advancing and receding contact angles, exhibits the linear relation with respect to the gas flow rate and the contact angle hysteresis of the droplet at the instant of the detachment decreases with the increase of the gas flow rate. In addition to the research mentioned above, there have been many studies focusing on droplet migration behavior, such as the droplet motion sheared by gas stream [10], the pinning characteristics of the droplet on heated or non-heated inclined surfaces [11,12], and modelling for the profile of the droplet on an inclined solid surface [13]. However, the migration behavior of the droplet is not fully understood.

The prediction of the onset for the droplet motion is mainly discussed on the basis of two models. Most popular models are developed by Furmidge [14] and Wolfram and Faust [15], which are mainly developed by considering the balance between external force and the surface tension based on the experimental observation where the sliding droplets on inclined solid surface are investigated. Concretely speaking, in the Wolfram and Faust's (hereafter, called WF) model, the adhesion force per unit length of the contact line is defined as a single constant parameter. Thus, the WF model does not treat the contact angle. On the other hand, the adhesion force in the Furmidge's model is evaluated using the surface tension and the contact angle hysteresis based on the concept of the Young's equation. In a recent study [16], models for the retention force, which were related to the droplet migration behavior, were reviewed and introduced some types of Furmidge's relations. The Furmidge type relation is widely used to understand the migration behavior instead of the WF model. In fact, a simple contact angle measurement of a static droplet is useful to understand and predict its surface conditions. It is mainly used as an indicator for the evaluation of the final determination for the surface preparation, and Furmidge's relation is also used for the surface evaluation [17–19]. In the numerical simulation for the migration behavior [20,21], the values of the contact angle or the contact angle hysteresis where the droplet starts to slide are set as branch conditions. After the onset for the droplet motion, the treatment of the dynamic contact angle is needed, and the relationship between the contact line velocity and the dynamic contact angle is mainly discussed [22]. There are two approaches for the model: one is the molecular kinetic theory [23] and the other is the hydrodynamic theory [24,25]. The molecular kinetic theory is developed by considering the frequency of the random molecular displacements within the three phase zones, which are characterized by the activation free energy. The hydrodynamic theory is developed by considering the relationship between the viscous energy dissipation and the work achieved through the contact line motion. In both models, the contact angle is directly connected to the contact line velocity. However, in an actual situation, the relationship between the contact angle and the contact line velocity is not necessarily a one-to-one correspondence [26] and is not uniquely determined. Thus, at present, there are no versatile numerical models for the dynamic wetting behavior, and the modeling of the dynamic wetting behavior remains an unresolved problem [27].

Considered the dynamic motion of a droplet, the dynamic behavior is basically determined by the force balance among fluid motion, external force, surface tension and adhesion force at the solid–liquid interface, so the contact angle should be secondarily determined by this force balance. In particular, the adhesion force would work as a friction force and become an important factor for the determination of the dynamic contact angle during the dynamic motion. A recent study discusses the adhesion force evaluated by the Furmidge's relation from a viewpoint of the concept of the friction coefficient [28]. If an inherent adhesion force between the liquid and solid surface exists, it would be a constant value regardless of whether a droplet size is large or not. However, it is unknown whether the adhesion force changes depending on the size of the droplet because most previous studies treat a relatively small droplet, such as several dozen μL or less [4,7,8,12,29]. Especially the WF and Furmidge's models treat the same phenomena, so from an engineering point of view, it is important to know the difference in the adhesion forces between two models. Therefore, in the present study, the applicability of the WF and Furmidge's models for the evaluation of adhesion force is considered in the wide range of the droplet volume, and the effect of the droplet size on the adhesion force is investigated. Finally, the relationship between the adhesion forces of two models are discussed and a simple model of the adhesion force, which includes the size effect, is proposed.

2. Evaluation of Adhesiveness

2.1. *Models for Adhesiveness of Droplet*

In the present study, two existing models are used for the evaluation of the adhesiveness of droplets on a solid surface: the WF model [15] and the Furmidge's model [14]. Two existing models mainly describe the phenomena for the sliding behavior of droplets on horizontal or inclined solid substrates and are briefly explained in the following section.

2.2. *Wolfram and Faust's (WF) Model*

In this model, the force balance between the adhesion force resulted from the wetted contact area and the gravitational force of droplet along the solid surface is mainly considered. In the model, the adhesion, E_{w}, is defined as a force exerted on a unit length of the periphery of the contact area. From this concept, the following relation is derived.

$$\rho_{\mathrm{l}} V_0 g \sin\alpha^c = \pi D_0 E_{\mathrm{w}} \tag{1}$$

In Equation (1), ρ_{l}, V_0, g, α^c and D_0 represent the density of liquid, the initial volume of the droplet, gravitational acceleration, the critical inclined angle of the solid surface and the initial contact area diameter of the droplet, respectively. In this model, E_{W} is assumed to be constant. In addition, a contact angle is not considered. According to the concept of Young's equation, the change in the contact angle indicates the change in the surface tension force acting on the contact line, even if the contact area does not change [30]. The contact angle is also an important factor for understanding droplet motion. Therefore, the applicability limit of Equation (1) is unknown if the size of the droplet increases, because the shape of the droplet is assumed as a part of sphere in this model. The evaluation procedure for E_{w} is as follows. By rewriting Equation (1), the relation $\sin\alpha^c = \pi E_{\mathrm{w}} D_0/(\rho_{\mathrm{l}} g V_0)$ is obtained. Here, E_{w} is estimated by fitting the transformed relation to the experimental data where the linear relation of $\sin\alpha_c$ and D_0/V_0 is assumed as $\sin\alpha_c = k\, D_0/V_0$. Here, k is a constant value. Therefore, E_{w} is obtained by the relation $E_{\mathrm{w}} = \rho_{\mathrm{l}} g k/\pi$.

2.3. *Furmidge's Model*

In this model, the shape of the droplet (i.e., wetted contact area) is assumed to be a rectangle. The adhesion is evaluated using advancing and receding contact angles based on the concept for the Young's equation. Then, the relationship is derived by considering the work performed by the gravity and the variation of the adhesion work in the sliding process. Nevertheless, the model is sufficiently able to capture the sliding droplet behavior and is applied to many migration phenomena by an addition of a pre-factor to the original Furmidge's model [31–33]. The Furmidge's model with the pre-factor c_f is as follows:

$$\rho_{\mathrm{l}} V_0 g \sin\alpha^c = c_f l_{\mathrm{width}} \sigma_{\mathrm{lg}} (\cos\theta_{\mathrm{R}} - \cos\theta_{\mathrm{A}}) \tag{2}$$

In Equation (2), c_f, l_{width}, σ_{lg}, θ_{R} and θ_{A} represent the pre-factor, the width of the contact area of the droplet, the surface tension between liquid and gas and the receding and advancing angles, respectively. Equation (2) reduces to the original Furmidge's model when the pre-factor c_f takes unity. In recent models, there are some expressions for the pre-factor c_f

where the Laplace pressure and the parameter for the pinning force are considered [16,34]. However, there is no consensus for the expression of the model. Therefore, the present study mainly focusses on the classical relations of the WF and the original Furmidge's models, as mentioned in the next section, and the pre-factor is discussed in Section 4.3. In a previous study [35], it was revealed that the value of l_{width} is almost constant, which is the same as the initial droplet contact area diameter D_0 until the onset of the droplet sliding. Therefore, in the present study, l_{width} in Equation (2) is treated as D_0.

2.4. Alternative Evaluation for WF and Furmidge's Models

A critical inclined angle in Equations (1) and (2) represents the onset of the droplet sliding motion. Therefore, if one applies the concept of Equation (1) to the Furmidge's model, Equation (2) can be rewritten as follows:

$$\sin \alpha^c = \frac{\sigma_{lg}(\cos\theta_R - \cos\theta_A)}{\rho_l g}\frac{l_{\text{width}}}{V_0} \tag{3}$$

Note, the value of l_{width} can be treated as the initial droplet contact area diameter D_0 until the onset of the sliding motion [35]. Thus, comparing Equation (3) with Equation (1), the value that corresponds to the adhesion force per unit of length defined as E_F can be derived as follows:

$$E_F = \frac{\sigma_{lg}(\cos\theta_R - \cos\theta_A)}{\pi} \tag{4}$$

From Equation (4), it was found that E_F includes the geometrical parameter of the droplet, which is not considered in E_W of Equation (1), because E_W is estimated by fitting Equation (1) to the experimental data, assuming the linear relationship between $\sin\alpha^c$ and D_0/V_0 as mentioned in the Section 2.2. This indicates that E_W may be the averaged value for the adhesion force in the wide range of the droplet volumes. Therefore, the following adhesion $E_W{}'$ is evaluated in addition to Equation (4).

$$E_W{}' = \frac{\rho_l g}{\pi}\left(\sin\alpha^c(V_0)\frac{V_0}{D_0}\right) \tag{5}$$

$E_W{}'$ is evaluated using the experimental data for the onset of the droplet sliding motion in Equation (5). Finally, in the present study, three kinds of the adhesion force for E_W, $E_W{}'$ and E_F are evaluated. Note that the present study does not focus on the morphological effect of the solid surface, such as the surface roughness on the sliding behavior [19,36–38]. In order to consider such a problem, more detailed investigation would be needed, including the definition of the movement of the contact line, because the pinning effect on the contact line motion becomes significant.

3. Experiment

Figure 1 shows the schematic of the experimental apparatus. As shown in Figure 1a, the apparatus mainly consists of the high-speed video camera (HX-5, NAC Image Technology, Ltd., Tokyo, Japan), the rotation stage and the LED light. The solid sample is set on the rotation stage. The droplet is deposited on the solid substrate as shown in Figure 1b. After the deposition, the solid substrate is rotated with a constant angular velocity ω = 0.5 deg sec^{-1}. Then, the droplet motion during the rotation was captured with the high-speed video camera. The geometrical parameters, such as the contact area diameter D, height h and advancing (θ_A) and receding (θ_R) contact angles were measured. In the present study, silicone rubber (SR) was used as the solid substrate. The surface roughness of SR is Ra = 0.02 μm [39]. The SR substrate is a kind of low-surface-energy solid, which enables us to make a stable droplet shape on solid surface with high reproducibility, unlike a high-surface-energy solid, such as a metal [40]. Then, water–ethanol binary mixtures were used for the liquid. The four mixtures with different ethanol mass concentrations were used: 0.072 Nm^{-1} (0 wt%), 0.051 Nm^{-1} (7.7 wt%), 0.038 Nm^{-1} (20.6 wt%) and 0.030 Nm^{-1}

(39.3 wt%). The droplet volumes ranged from 7 to 600 μL. More detailed information on the droplets is listed in Table 1. In this experiment, the temperature and humidity were in the ranges of 20.0–25.0 °C and 50.0–55.0%, respectively. Each experimental condition was performed three times. The contact angles of droplet were measured using commercial software (FAMAS; Kyowa Interface Science Co., Ltd., Saitama, Japan). Figure 2 shows the images of the droplet wettability on SR. The apparent contact angle decreases as the ethanol concentration increases. Here, the droplet volume is 10 μL in each liquid.

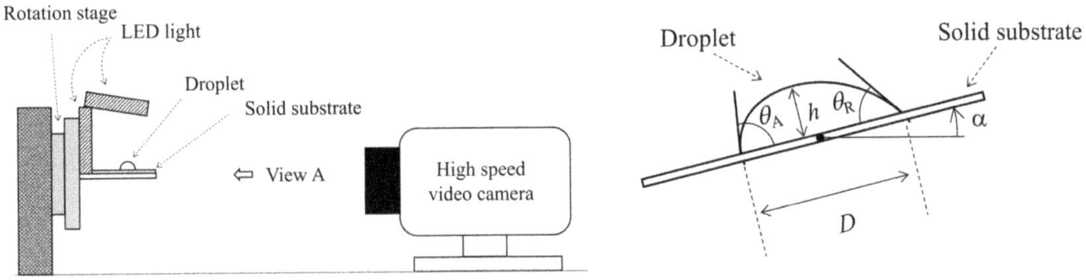

(**a**) Overall view (**b**) View A

Figure 1. Schematic of (**a**) overall view of experimental apparatus and (**b**) view A in (**a**) where the geometrical parameters are defined.

Table 1. Droplet volumes, initial contact area diameter and droplet height in each liquid used in sliding experiment.

Liquids [wt%]	Volumes [μL]	Initial Contact Are Diameter D_0 [m]	Initial Droplet Height h_0 [m]
0	7–600	2.7×10^{-3}–15.5×10^{-3}	1.6×10^{-3}–4.3×10^{-3}
7.7	8–600	3.1×10^{-3}–16.9×10^{-3}	1.6×10^{-3}–3.7×10^{-3}
20.6	10–500	3.7×10^{-3}–18.5×10^{-3}	1.4×10^{-3}–2.7×10^{-3}
39.3	10–400	4.0×10^{-3}–17.6×10^{-3}	1.2×10^{-3}–2.2×10^{-3}

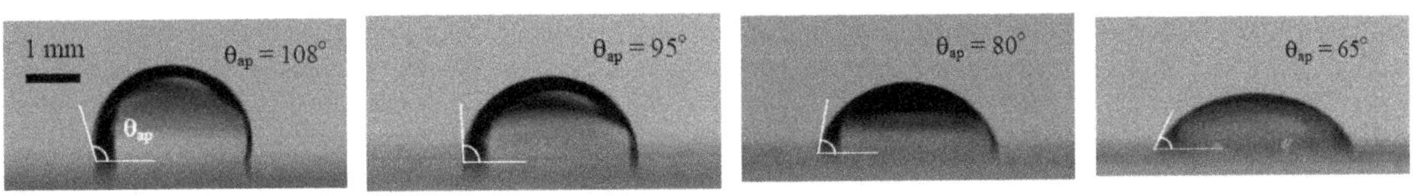

(**a**) 0 wt% (**b**) 7.7 wt% (**c**) 20.6 wt% (**d**) 39.3 wt%

Figure 2. Droplet wettability of each liquid on silicone rubber (SR): (**a**) 0 wt% (0.072 Nm^{-1}), (**b**) 7.7 wt% (0.051 Nm^{-1}), (**c**) 20.6 wt% (0.038 Nm^{-1}) and (**d**) 39.3 wt% (0.030 Nm^{-1}). Droplet volume V_0 is 10 μL.

4. Results and Discussion

4.1. Inclined Angle Dependency of Geometrical Parameters of Droplet

Figure 3 shows the relationship between the inclined angle α, contact area diameter $D(\alpha)$ and the height $h(\alpha)$ of water droplets. The droplet volumes are 10, 100 and 300 μL. In this figure, blue and red points represent the onset of movement of the front and rear contact lines (FCL and RCL), respectively. From Figure 3a, it can be seen that $D(\alpha)$ increases after the front contact line starts to move. On the other hand, in Figure 3b, the droplet height $h(\alpha)$ decreases as α increases. The critical inclined angle where each contact line starts to move becomes small as the droplet volume increases. This obviously indicates that the gravitational force becomes dominant compared with the adhesion force. In fact, the gradient of $dD(\alpha)/d\alpha$ increases and the relationship between $D(\alpha)$ and α exhibits linear as the droplet volume increases.

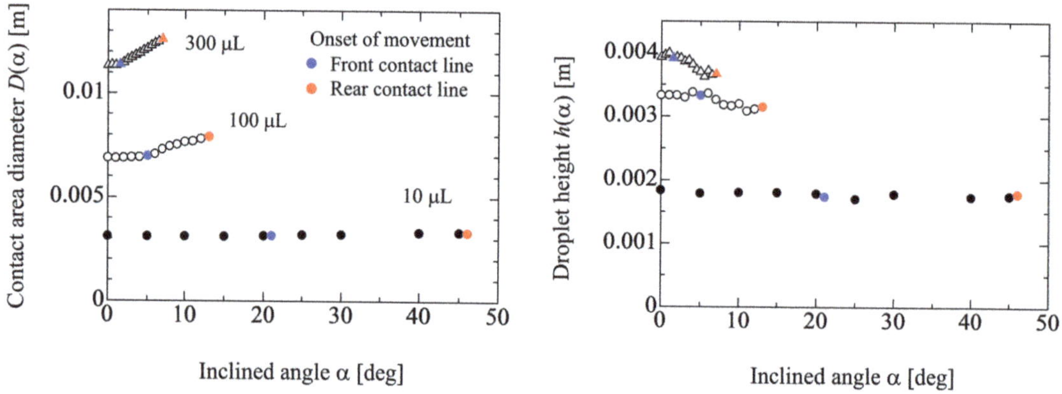

(**a**) Contact area diameter (**b**) Droplet height

Figure 3. Relationship among contact area diameter *D*, height *h* of water droplet (0 wt%) and inclined angle α of solid substrate. The blue and red characters represent the front and rear contact lines movement, respectively.

Figure 4 shows the ethanol concentration dependency of the behaviors for $D(\alpha)$ and $h(\alpha)$. The droplet volume is 100 μL. The changes in $D(\alpha)$ and $h(\alpha)$ against α in Figure 4a,b are qualitatively the same, as shown in Figure 3. For example, in Figure 3a, the critical inclined angle for the movement of the front contact line becomes larger as $D(\alpha)$ increases (i.e., the droplet volume increases). In Figure 4a, the timing of the onset for the movement of the front contact line becomes fast as $D(\alpha)$ increases (i.e., the droplet wettability increases). However, the degree of the difference in the timing is quite different between the results in Figures 3 and 4. This may be understood by considering the relationship between the gravity force and the work of adhesion based on the initial droplet condition. Concretely speaking, the ratio between the gravity force per unit contact line $mg/(\pi D_0)$ and the work of adhesion $\sigma_{lg}(1 + \cos\theta_0)$ is considered for each liquid property. By this ratio, the behavior of the contact line is considered. Figure 5 shows the results for the ratio of two forces for each liquid property in Figures 3 and 4. In Figure 5a, the ratio of two forces becomes large as the droplet volume increases, which indicates the gravity is dominant compared with the wettability. Therefore, the difference in the critical inclined angle for the movement of the front contact line ($\alpha^c{}_{FCL}$) becomes large with respect to the droplet volume. On the other hand, from the result in Figure 5b, it is found that the wettability is dominant compared with the gravity force and the order of three values are similar. Thus, the differences in $\alpha^c{}_{FCL}$ among the three liquid properties are not so large. As to the difference between $\alpha^c{}_{FCL}$ and $\alpha^c{}_{RCL}$ (the critical inclined angle for the movement of the rear contact line), the degree of the deformation for the droplet shape may be related. Concretely speaking, from the results in Figure 4b, the change in the droplet height between the substrate inclined angles at $\alpha^c{}_{RCL}$ and $\alpha^c{}_{FCL}$ in the water case is larger than that in the cases of 20.6 and 39.3 wt%. The deformable case, such as water, easily elongates the contact area diameter. Therefore, in the non-deformable cases, such as 20.6 and 39.3 wt%, the difference between $\alpha^c{}_{RCL}$ and $\alpha^c{}_{FCL}$ is not so large compared with water. In fact, in Figure 4, the averaged values of the difference between $\alpha^c{}_{RCL}$ and $\alpha^c{}_{FCL}$ for 0, 20.6 and 39.3 wt% are 8.7, 2.5 and 2.3 deg, respectively.

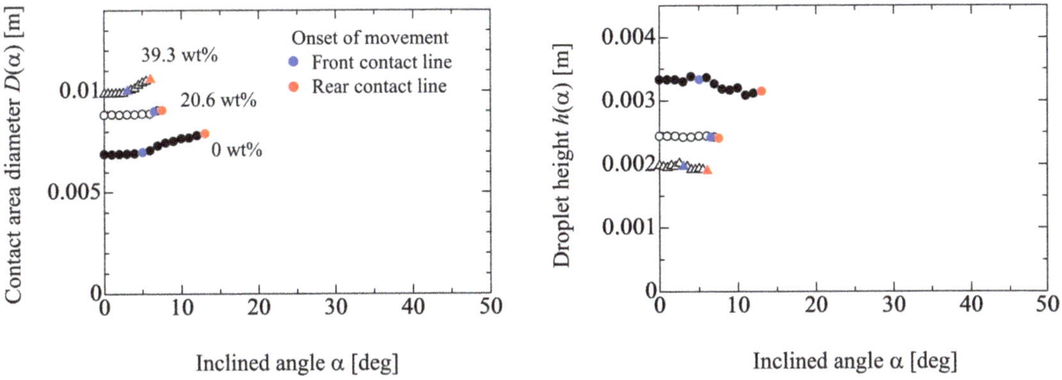

(**a**) Contact area diameter (**b**) Droplet height

Figure 4. Ethanol concentration dependency on the behaviors for D and h. Droplet volume is 100 μL. The blue and red characters represent the front and rear contact lines movement, respectively.

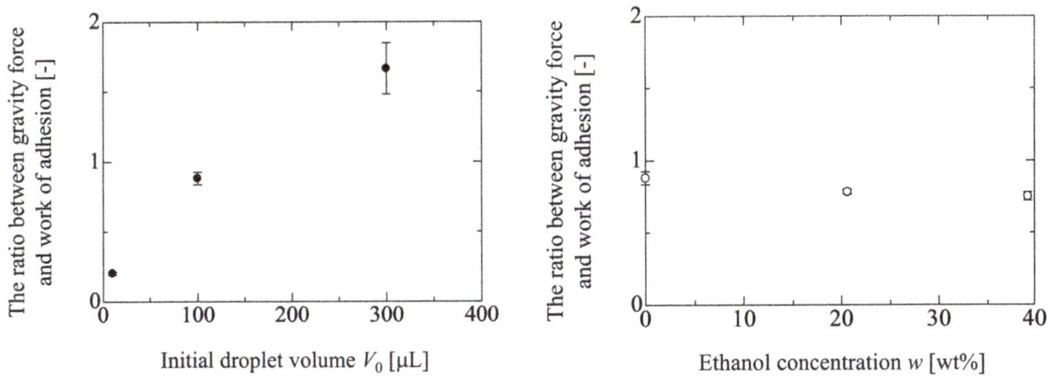

(**a**) 0 wt% (**b**) 100 μL

Figure 5. Ratio between gravity force per unit contact line and the work of adhesion. The ratio is calculated by $mg/(\pi D_0\ \sigma_{lg}\ (1 + \cos\theta_0)$.

Figure 6 shows the changes in the advancing and receding contact angles during the inclination of the solid substrate. The results for water droplets of 10 and 300 μL are depicted in this figure. From these results, the gradients of $|d\theta(\alpha)/d\alpha|$ between the advancing and receding contact angles after the movement of the front contact line are different from each other. This may result from the fact that water exhibits a hydrophobic condition against the SR substrate, which means that the contact line is basically hard to move. In addition, the front contact line moves towards the dry surface and the rear contact line moves towards the wet surface. These conditions may induce the difference in the gradients. Figure 7 shows the effect of the liquid property on the changes in the contact angles. The mass concentrations of ethanol are 0 wt%, 20.6 wt% and 39.3 wt%. The droplet volumes are the same at 100 μL in each case. In the cases of 20.6 wt% and 39.3 wt%, the gradients of $|d\theta(\alpha)/d\alpha|$ between the advancing and receding contact angles are similar to each other. This may result from two cases that are relatively hydrophilic against the solid surface, which indicates that the contact line is easy to move. Therefore, it was thought that the difference in the surface conditions, such as wettability, in these cases does not strongly reflect to the differences in the gradients of $|d\theta(\alpha)/d\alpha|$.

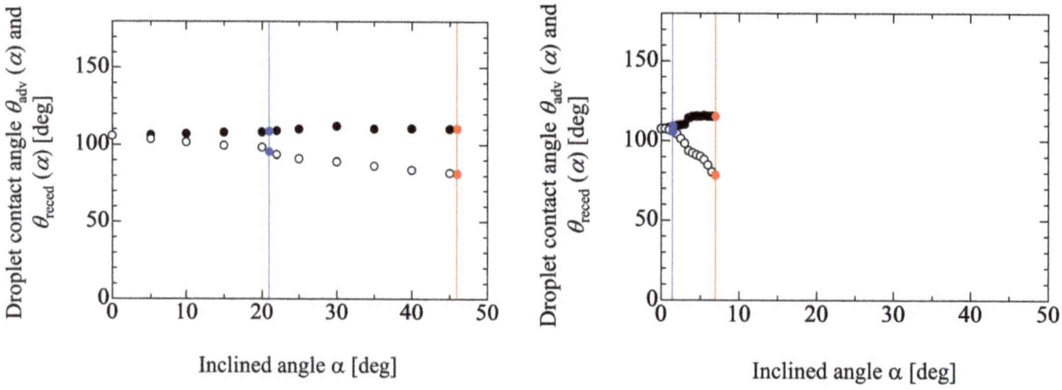

(**a**) 10 μL (**b**) 300 μL

Figure 6. Change in advancing and receding the contact angles of water droplets during the inclination of the solid substrate. The blue and red solid lines represent the front and rear contact lines movement, respectively.

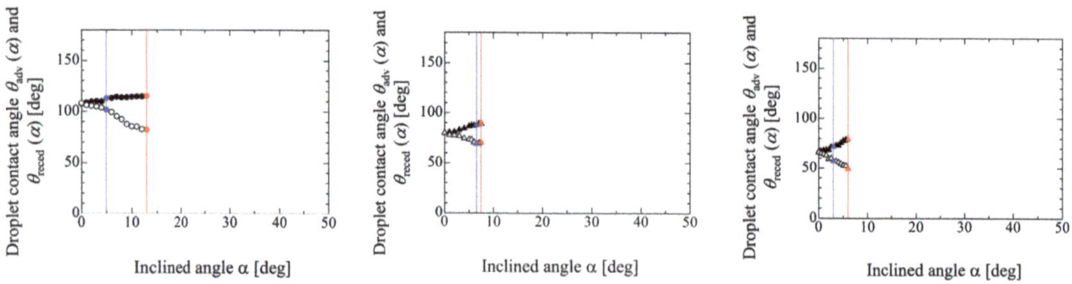

(**a**) 0 wt% (**b**) 20.6 wt% (**c**) 39.3 wt%

Figure 7. Effect of the liquid property on the changes in the contact angles. The blue and red solid lines represent the front and rear contact lines movement, respectively.

4.2. Evaluation of Adhesion Forces

Figure 8 shows the relationship between $\sin\alpha^{c}_{FCL}$ or $\sin\alpha^{c}_{RCL}$ and D_0/V_0 in each liquid property. The open and solid circles represent the critical inclined angles of $\sin\alpha^{c}_{FCL}$ and $\sin\alpha^{c}_{RCL}$, respectively. The red solid line represents the linear fitting by Equation (1) to the experimental data. The critical inclined angles of $\sin\alpha^{c}_{FCL}$ and $\sin\alpha^{c}_{RCL}$ indicate the points where the front and rear contact lines start to move, respectively. Equation (1) is applied to the condition where the droplet moves. For example, as shown in Figure 7a, if the front contact line moves at first where the rear contact line is pinned, after that, the rear contact line starts to move and slide, the open circle is below the solid one. Therefore, the red solid line is used for the fitting. This means that the linear fitting of Equation (1) for the experimental data includes both conditions of $\sin\alpha^{c}_{FCL}$ and $\sin\alpha^{c}_{RCL}$, as seen in Figure 8b–d. From the results of Figure 8, most data show that the front contact line moves before the rear contact line moves. In addition, the differences between the points of front and rear contact line becomes small as the ethanol concentration increases. By fitting Equation (1) to the experimental data shown in Figure 8, the adhesion force of E_W and E_W' can be evaluated for each liquid property. E_F in Equation (4) can be evaluated based on the experimental data for the contact angles of droplet, as shown in Figures 6 and 7.

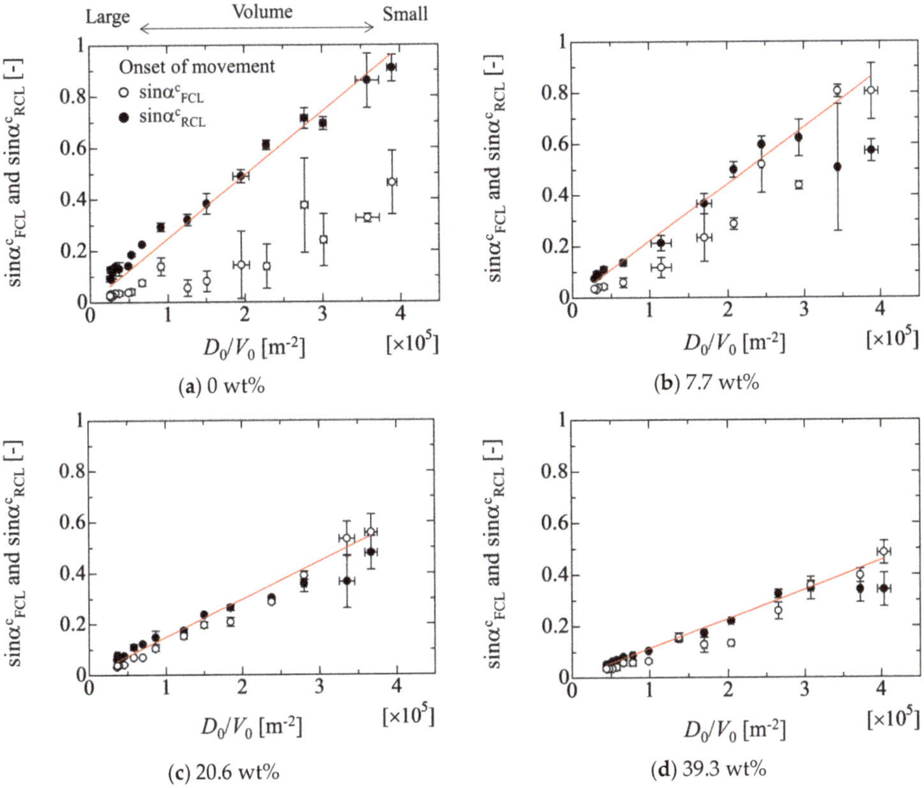

Figure 8. Relationship between critical inclined angles of $\sin\alpha^c{}_{FCL}$ and $\sin\alpha^c{}_{RCL}$ and D_0/V_0 in each liquid property.

Figure 9 shows the droplet size dependency of the adhesion forces in each liquid property. The red solid line indicates the adhesion force of E_W evaluated by the linear fitting approach. Here, the estimated values of E_W are 7.7×10^{-3}, 6.8×10^{-3}, 4.5×10^{-3} and 3.3×10^{-3} $\mathrm{Nm^{-1}}$ for 0 wt%, 7.7 wt%, 20.9 wt% and 39.3 wt%, respectively. The open and solid circles represent the adhesion forces of E_F and E_W', respectively. The results indicate that the values of E_W' and E_F in the case of 0 wt% deviate from the value of E_W as the droplet volume increases. In Figure 9a, the experimental data of E_W' and E_F largely deviate from the linear line of E_W as the droplet volume increases. On the other hand, the deviation of E_W' and E_F from E_W gradually decreases as the ethanol concentration increases. In the case of 39.3 wt%, the values of E_W' and E_F almost coincide with E_W. This volume effect on the adhesion forces may result from the deformation of the droplet shape due to the gravity. The cases from the case (b) to case (d) are basically hydrophilic condition against the SR solid substrate. This means that the droplet height is low and the center of gravity is close to the solid surface. On the other hand, case (a) displays hydrophobic conditions. The center of gravity is far from the solid surface compared with the hydrophilic one. Therefore, the droplet shape easily deforms due to the gravity force in the hydrophobic case.

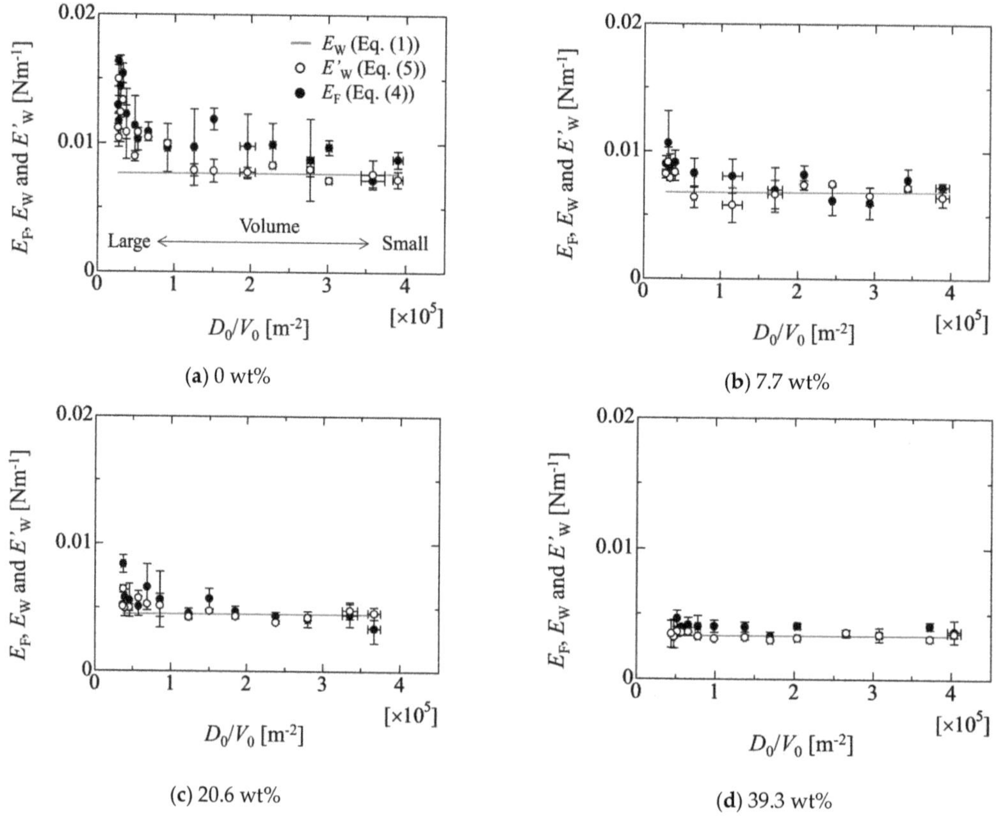

Figure 9. Relationship between adhesion force and D_0/V_0 in each liquid property.

Figure 10a,b represents the images when the front and rear contact line starts to move, respectively. From Figure 10a, it can be seen that the advancing contact angle almost takes the same value as the receding one. However, in Figure 10b, the receding contact angle is smaller than the advancing one, which implies the deformation of the droplet shape. On the other hand, in the case of 39.3 wt%, as shown in Figure 11, the advancing and receding contact angles take similar values in both cases of Figure 11a,b. This means that the droplet deformation almost does not arise in the case of high ethanol concentration. From Figure 9, in addition to the result in Figure 8, at least, it can be seen that the effect of the droplet size on the adhesion force is not so significant if the value of D_0/V_0 is larger than 1.5×10^5 m^{-2}. Here, Figure 12 shows the relationship between the bond number ($\mathrm{Bo} = \rho_l g h_0{}^2/\sigma_{lg}$) evaluated by the initial droplet information and D_0/V_0 in each liquid property. In this figure, this criterion physically means the boundary where the Bo is less than unity depicted by the red dashed line.

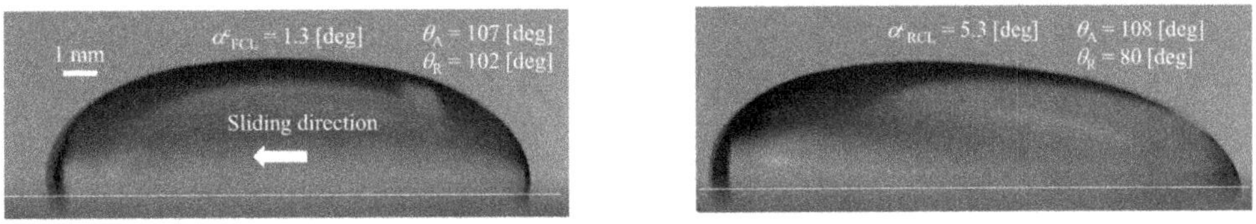

(**a**) FCL movement where RCL is pinned (**b**) FCL and RCL movements

Figure 10. Images for onset of each contact line (CL) of the water droplet (550 µL).

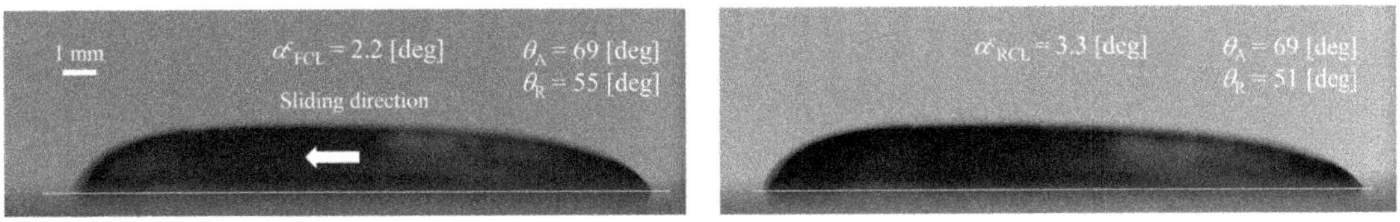

(**a**) FCL movement where RCL is pinned (**b**) FCL and RCL movements

Figure 11. Images for onset of each contact line (CL) of the 39.3 wt% droplet (400 µL).

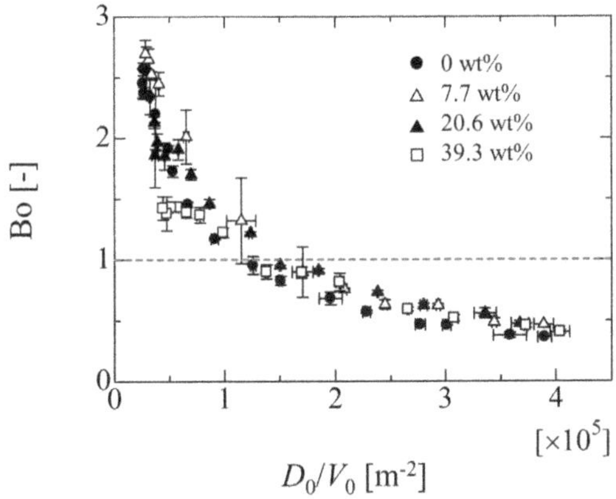

Figure 12. Relationship between Bo number of initial droplet and D_0/V_0.

4.3. *Effect of Droplet Size on Adhesion Force*

From the discussion in the previous section, it can be seen that the droplet size effect on the adhesion force E_F becomes large as the droplet volume increases, as shown in Figure 9. Thus, by focusing on E_W and E_F, the effect of the droplet size on the adhesion force can be considered, as the ratio between E_W, which is constant, and E_F: E_W/E_F. Note that, in this section, it is assumed that the inherent adhesion force between the liquid and solid is expressed by E_W, which is obtained by the linear relationship between sinα_c and D_0/V_0.

Since the result for Bo in Figure 12 exhibits a similar trend in Figure 9, the ratio E_W/E_F may be related to Bo. In particular, from the expression of Equation (2), the ratio E_W/E_F indicates a pre-factor c_f and means the factor which corrects E_F to E_W if the constant E_W is thought as the adhesion forces between liquid and solid. From the results in Figures 9 and 12, a simple relationship can be deduced as follows:

$$\frac{E_W}{E_F} = a\text{Bo} + b \tag{6}$$

Here, from the result in Figure 9, the E_F approaches E_W as the droplet size becomes small (increase of D_0/V_0). This indicates one limit condition in Equation (6) that b is unity because the size effect on the adhesion force becomes small and $E_W \approx E_F$ when Bo→0. Figure 13 shows the relationship between the ratio E_W/E_F and Bo in each liquid case. The trend in the figure exhibits a relatively linear relationship between E_W/E_F and Bo. By fitting Equation (6) to the experimental data in Figure 13, the value a is estimated as $-1.31 \times 10^{-1} \pm 0.42 \times 10^{-1}$.

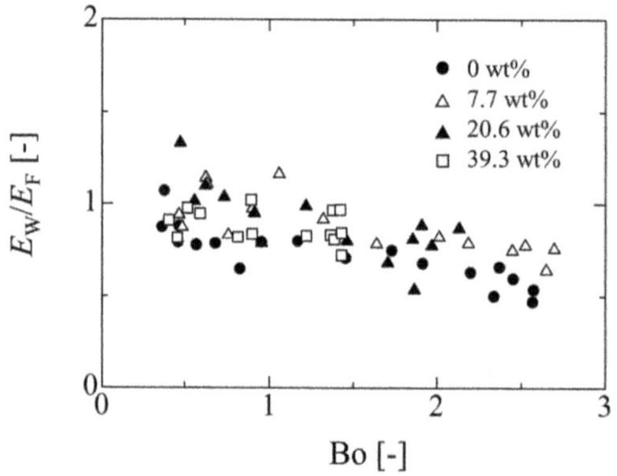

Figure 13. Relationship between Bo number and the ratio of adhesion force E_W and E_F.

The relative errors among the adhesion forces of E_W, $E_F{}^{exp}$ and $E_W{}^{est}$ (Equation (6)) are evaluated in Figure 14. Here, the relative error is calculated by the following relation.

$$e(E) = 100 \times \left| \frac{E_W - E}{E_W} \right| \tag{7}$$

In the figure, the values of $E_F{}^{exp}$ and $E_W{}^{est}$ are substituted into E in Equation (7), and the white and black circles represent the relative error of $e(E_W{}^{est})$ and $e(E_F{}^{exp})$, respectively. The estimated values of $E_W{}^{est}$ by Equation (6) shows relatively good agreement with the linear fitting value of E_W; for example, the relative error for $E_F{}^{exp}$ becomes large as the droplet volume increases in each liquid case. On the other hand, the relative errors for $E_W{}^{est}$ are smaller than that that for $E_F{}^{exp}$. This indicates that the size effect of the droplet on the adhesion force can be well correlated by the simple linear relation of Equation (6). In fact, it is reported that the pre-factor c_f in Equation (2) is related to the size of the droplet [32,33]. However, there is a large discrepancy for the water case in the case of the large droplet. In the water case, the deformation of the droplet is larger than that of other cases. This might relate to the lack of the consideration for the physical conditions, such as a force balance at the contact line and the effect of the droplet surface shape on the adhesion force. Concretely speaking, the models mentioned in the Section 2 are mainly the retention force in the horizontal direction at the contact line. However, forces such as the vertical force at the contact line and the Laplace pressure of the droplet surface are also important factors to determine the droplet conditions. Therefore, such factors must be considered in Equation (6) from a comprehensive point of view in the future.

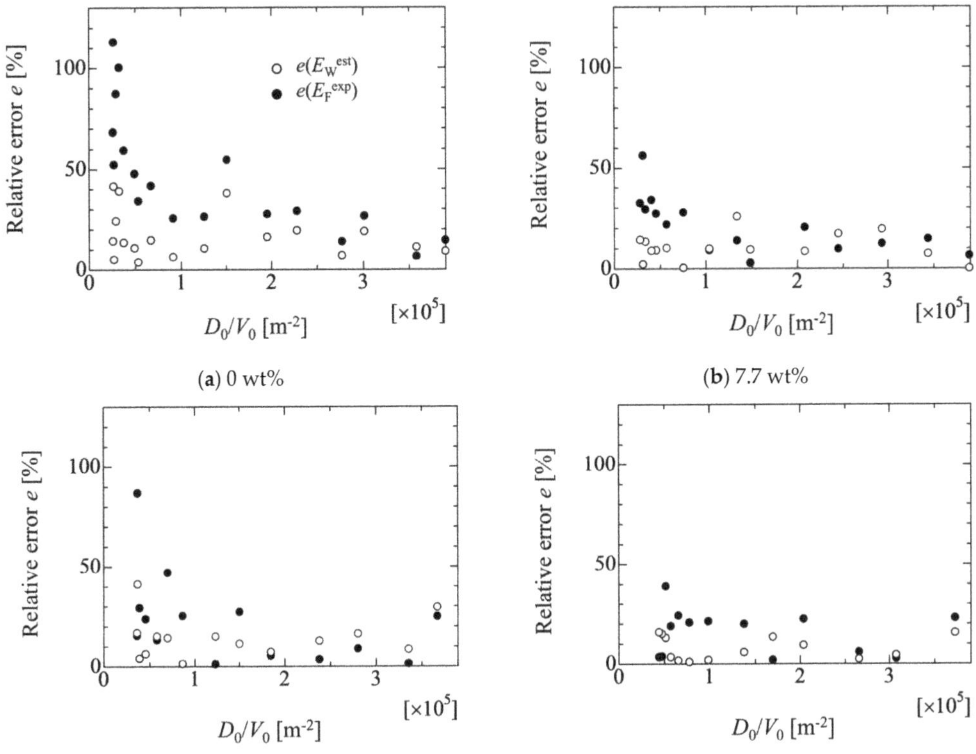

(**a**) 0 wt%; (**b**) 7.7 wt%; (**c**) 20.6 wt%; (**d**) 39.3 wt%

Figure 14. Relative errors of $e(E_W{}^{est})$ and $e(E_F{}^{exp})$ in each liquid case. The values of $e(E_W{}^{est})$ and $e(E_F{}^{exp})$ are calculated by Equation (7) as $100 \times |(E_W - E_W{}^{est})/E_W|$ and $100 \times |(E_W - E_W{}^{exp})/E_W|$, respectively. E_W, $E_F{}^{exp}$ and $E_W{}^{est}$ are estimated by Equations (1), (4) and (6), respectively.

5. Conclusions

The sliding behavior of water–ethanol binary mixture liquids on the silicone rubber was experimentally investigated. The adhesion forces between the liquid and solid surface were evaluated based on the existing models. In particular, the size effect on the adhesion forces were considered by setting the wide range of droplet volumes.

From the results, the critical inclined angle where the front contact line (FCL) starts to move is smaller than that of the rear contact line (RCL) in the case of 0 wt%. However, as the ethanol concentration increases, two values of the critical inclined angles take similar one. Although it was found that the relationship between the critical inclined angle where the droplet starts to move and the D_0/V_0 exhibits almost a linear relation in the high ethanol concentration case, the data for large droplets in the case of 0 wt% deviate from the linear relation. In fact, the adhesion force E_F evaluated using the droplet contact angles (Equation (4)) exhibits larger value than that of the adhesion force of E_W evaluated by a linear fitting approach (Equation (1)). This may result from the deformation of the droplet shape where the deformation of the hydrophobic case is larger than that of the hydrophilic case. This means that the adhesion forces will be influenced by the droplet volume (deformation) if the existing models are used for the evaluation of the adhesion force. From the present study, at least, it was found that the effect of the droplet size on the adhesion force is not so large if the value of D_0/V_0 is larger than 1.5×10^5 m^{-2}. This criterion is the boundary where the Bo of the initial droplet is less than unity. This would become one of the judgement criteria for the appropriate droplet volume for the evaluation of the adhesion force. Furthermore, to consider the effect of the droplet deformation on the adhesion force, the relationship between E_W/E_F and Bo is considered. The result indicates that the E_W/E_F exhibits the good linearity with respect to Bo, which means that the droplet deformation can mainly be considered by the Bo. However, for increased understanding of the larger deformation of the droplet, further detailed investigation is needed by considering the vertical force relation and the Laplace force exerted on the droplet because the relationship used in the present work is only the lateral force relation on the solid surface.

Author Contributions: Y.Y. performed conceptualization data curation, writing—original draft preparation, investigation, writing—reviewing and editing, and conceived methodology. Y.F. and Y.S. performed experimental investigations. T.K. performed conceptualization, writing—reviewing and editing, and conceived methodology. All authors have read and agreed to the published version of the manuscript.

Funding: This study partially supported by JSPS KAKENHI (grant number 17K14590 and 21K03860) and the Natural Science Foundation of Zhejiang Province, China (No. LQ21E090011).

Data Availability Statement: The data presented in this study are available in article.

Conflicts of Interest: The authors declare no conflict of interest.

Nomenclature

The parameters used in the present paper are listed below.

Symbols	
Bo	Bond number [-]
a, b	Parameters in Equation (6) [-]
c_f	Pre-factor in Furmidge's relation [-]
D	Contact area diameter of droplet [m]
E	Adhesion force [Nm^{-1}]
e	Relative error [%]
g	Gravitational acceleration [m s^{-2}]
h	Droplet height [m]
k	Gradient of linear fitting relation for Wolfram and Faust's model [m^2]
l_{width}	Width of the contact area of droplet [m]'
V	Droplet volume [m^3]
Greek Symbols	
α	Inclined angle of solid substrate [deg]
θ	Contact angle [deg]
ρ	Density [kg m^{-3}]
σ	Surface tension [N m^{-1}]
ω	Angular velocity of rotation for solid substrate [deg s^{-1}]
Superscripts	
c	Critical
est	Estimation
exp	Experiment
Subscripts	
0	Initial condition
A	Advancing
ap	Apparent
F	Furmidge's model
FCL	Front contact line
l	Liquid
lg	Liquid–gas interface
R	Receding
RCL	Rear contact line
W	Wolfram model

References

1. Kim, J. Spray cooling heat transfer: The state of the art. *Int. J. Heat Fluid Flow* **2007**, *28*, 753–767. [CrossRef]
2. Hsu, H.-W.; Liu, C.-L. Spray-coating semiconducting conjugated polymers for organic thin film transistor applications. *RSC Adv.* **2014**, *4*, 30145–30149. [CrossRef]
3. Yan, K.; Li, J.; Pan, L.; Shi, Y. Inkjet printing for flexible and wearable electronics. *APL Mater.* **2020**, *8*, 120705-1–120705-22. [CrossRef]
4. Das, P.K.; Grippin, A.; Kwong, A.; Weber, A.Z.; Luo, D.; Lan, L.; Xu, M.; Xu, H.; Li, M.; Wang, L.; et al. Liquid-Water-Droplet Adhesion-Force Measurements on Fresh and Aged Fuel-Cell Gas-Diffusion Layers. *J. Electrochem. Soc.* **2012**, *159*, B489–B496. [CrossRef]
5. Wang, X.; Xu, B.; Chen, Z.; Del Col, D.; Li, D.; Zhang, L.; Mou, X.; Liu, Q.; Yang, Y.; Cao, Q. Review of droplet dynamics and dropwise condensation enhancement: Theory, experiments and applications. *Adv. Colloid Interface Sci.* **2022**, *305*, 102684. [CrossRef]
6. Wang, X.; Xu, W.; Chen, Z.; Xu, B. Dropwise condensation heat transfer on nanostructured superhydrophobic surfaces with different inclinations and surface subcoolings. *Int. J. Heat Mass Transf.* **2021**, *181*, 121898. [CrossRef]
7. Yilbas, B.S.; A-Sharafi, A.; Ali, H.; A-Aqeeli, N. Dynamics of a water droplet on a hydrophobic inclined surface: Influence of droplet size and surface inclination angle on droplet rolling. *RSC Adv.* **2017**, *7*, 48806. [CrossRef]
8. Lv, C.; Yang, C.; Hao, P.; He, F.; Zheng, Q. Sliding of Water Droplets on Microstructured Hydrophobic Surfaces. *Langmuir* **2010**, *26*, 8704–8708. [CrossRef] [PubMed]
9. Li, M.; Li, Y.; Qin, Y.; Yin, Y.; Zhang, J.; Che, Z. Water droplet detachment characteristics on surfaces of gas diffusion layers in PEMFCs. *Int. J. Hydrogen Energy* **2022**, *47*, 10341–10351. [CrossRef]
10. Xie, J.; Xu, J.; Shang, W.; Zhang, K. Mode selection between sliding and rolling for droplet on inclined surface: Effect of surface wettability. *Int. J. Heat Mass Transf.* **2018**, *122*, 45–58. [CrossRef]
11. Park, J.; Kumar, S. Droplet Sliding on an Inclined Substrate with a Topographical Defect. *Langmuir* **2017**, *33*, 7352–7363. [CrossRef] [PubMed]
12. Al-Sharafi, A.; Yilbas, B.S.; Ali, H.; Alaqeeli, N. A Water Droplet Pinning and Heat Transfer Characteristics on an Inclined Hydrophobic Surface. *Sci. Rep.* **2018**, *8*, 3061. [CrossRef] [PubMed]
13. Lv, C.; Shi, S. Wetting states of two-dimensional drops under gravity. *Phys. Rev. E* **2018**, *98*, 042802. [CrossRef]
14. Furmidge, C. Studies at phase interfaces. I. The sliding of liquid drops on solid surfaces and a theory for spray retention. *J. Colloid Sci.* **1962**, *17*, 309–324. [CrossRef]
15. Wolfram, E.; Faust, R. *Chapter 10 in Wetting, Spreading and Adhesion*; Padday, J.F., Ed.; Academic Press: London, UK, 1978; pp. 213–222.
16. Tadmor, R. Open Problems in Wetting Phenomena: Pinning Retention Forces. *Langmuir* **2021**, *37*, 6357–6372. [CrossRef]
17. Qin, Y.; Li, Y.; Zhang, D.; Xu, N.; Zhu, X. Wettability, durability and corrosion properties of slippery laser-textured aluminum alloy surface under water impact. *Surf. Coat. Technol.* **2020**, *394*, 125856. [CrossRef]
18. Peta, K.; Bartkowiak, T.; Galek, P.; Mendak, M. Contact angle analysis of surface topographies created by electric discharge machining. *Tribol. Int.* **2021**, *163*, 107139. [CrossRef]
19. Kuznetsov, G.V.; Islamova, A.G.; Orlova, E.G.; Ivashutenko, A.S.; Shanenkov, I.I.; Zykov, I.Y.; Feoktistov, D.V. Influence o roughness on polar and dispersed components of surface free energy and wettability properties of copper and steel surfaces. *Surf. Coat. Technol.* **2021**, *422*, 127518. [CrossRef]
20. Annapragada, S.R.; Murthy, J.Y.; Garimella, S.V. Prediction of droplet dynamics on an incline. *Int. J. Heat Mass Transf.* **2012**, *55*, 1466–1474. [CrossRef]
21. Goswami, A.; Alen, S.K.; Farhat, N.; Rahman, A. Numerical study of wetting stability and sliding behavior of liquid droplets on microgrooved surfaces. *Colloid Polym. Sci.* **2019**, *297*, 989–1000. [CrossRef]
22. Fernández-Toledano, J.-C.; Blake, T.; De Coninck, J. Taking a closer look: A molecular-dynamics investigation of microscopic and apparent dynamic contact angles. *J. Colloid Interface Sci.* **2021**, *587*, 311–323. [CrossRef]
23. Blake, T.D. The physics of moving wetting lines. *J. Colloid Interface Sci.* **2006**, *299*, 1–13. [CrossRef]
24. Voinov, O.V. Hydrodynamics of wetting. *Fluid Dyn.* **1977**, *11*, 714–721. [CrossRef]
25. Bonn, D.; Eggers, J.; Indekew, J.; Meunier, J.; Rolley, E. Wetting and spreading. *Rev. Mod. Phys.* **2009**, *81*, 739–805. [CrossRef]
26. Yonemoto, Y.; Kunugi, T. Analytical consideration of liquid droplet impingement on solid surfaces. *Sci. Rep.* **2017**, *7*, 2362. [CrossRef]
27. Snoeijer, J.H.; Andreotti, B. Moving Contact Lines: Scales, Regimes, and Dynamical Transitions. *Annu. Rev. Fluid Mech.* **2013**, *45*, 269–292. [CrossRef]
28. McHale, G.; Gao, N.; Wells, G.G.; Barri-Zhang, H.; Ledesma-Aguilar, R. Friction coefficient for droplets on solids: The liq-uid-solid amontons' laws. *Langmuir* **2022**, *38*, 4425–4433. [CrossRef]
29. Kim, Y.H.; Kim, K.; Jeong, J.H. Determination of the adhesion energy of liquid droplets on a hydrophobic flat surface con-sidering the contact area. *Int. J. Heat Mass Transf.* **2016**, *102*, 826–832. [CrossRef]
30. Yonemoto, Y.; Kunugi, T. Experimental and theoretical investigation of contact-angle variation for water-ethanol mixture droplets on a low-surface-energy solid. *Int. J. Heat Mass Transf.* **2016**, *96*, 614–626. [CrossRef]

31. Extrand, C.W.; Kumagai, Y. Liquid Drops on an Inclined Plane: The Relation between Contact Angles, Drop Shape, and Retentive Force. *J. Colloid Interface Sci.* **1995**, *170*, 515–521. [CrossRef]
32. Pierce, E.; Carmona, F.J.; Amirfazli, A. Understanding of sliding and contact angle results in tilted plate experiments. *Colloids Surfaces A: Physicochem. Eng. Asp.* **2008**, *323*, 73–82. [CrossRef]
33. Bouteau, M.; Cantin, S.; Benhabib, F.; Perrot, F. Sliding behavior of liquid droplets on tilted Langmuir–Blodgett surfaces. *J. Colloid Interface Sci.* **2008**, *317*, 247–254. [CrossRef] [PubMed]
34. Kibar, A.; Ozbay, R.; Sarshar, M.A.; Kang, Y.T.; Choi, C.-H. Bubble Movement on Inclined Hydrophobic Surfaces. *Langmuir* **2017**, *33*, 12016–12027. [CrossRef] [PubMed]
35. Yonemoto, Y.; Suzuki, S.; Uenomachi, S.; Kunugi, T. Sliding behavior of water-ethanol mixture droplets on inclined low-surface-energy solid. *Int. J. Heat Mass Transf.* **2018**, *120*, 1315–1324. [CrossRef]
36. Kubiak, K.J.; Mathia, T.G. Anisotropic Wetting of Hydrophobic and Hydrophilic Surfaces–Modelling by Lattice Boltzmann Method. *Procedia Eng.* **2014**, *79*, 45–48. [CrossRef]
37. Prajitno, D.H.; Maulana, A.; Syarif, D.G. Effect of Surface Roughness on Contact Angle Measurement of Nanofluid on Surface of Stainless Steel 304 by Sessile Drop Method. *J. Physics: Conf. Ser.* **2016**, *739*, 012029. [CrossRef]
38. Yonemoto, Y.; Tomimitsu, I.; Shimizu, K.; Kunugi, T. Wettability model for water-ethanol binary mixture droplet on roughened low-surface-energy solids. *Int. J. Multiph. Flow* **2021**, *137*, 103569. [CrossRef]
39. Yonemoto, Y.; Tashiro, K.; Shimizu, K.; Kunugi, T. Predicting the splash of a droplet impinging on solid substrates. *Sci. Rep.* **2022**, *12*, 5093. [CrossRef]
40. de Gennes, P.G.; Brochard-Wyard, F.; Quere, D. *Capillarity and Wetting Phenomena: Drops, Bubbles, Pearls, Waves*; Springer: New York, NY, USA, 2003.

micromachines

Article

An Ultra-Micro-Volume Adhesive Transfer Method and Its Application in fL–pL-Level Adhesive Distribution

Huifang Liu [1,*], Xi Chen [1], Shuqing Wang [2], Shenhui Jiang [1], Ying Chen [1] and Fuxuan Li [1]

1 School of Mechanical Engineering, Shenyang University of Technology, Shenyang 110870, China; chenxi897918099@163.com (X.C.); jshenhui0204@163.com (S.J.); chen1357928@163.com (Y.C.); lfx_19961213@163.com (F.L.)
2 Shenyang Machine Tool (Group) Co., Ltd., Shenyang 110142, China; shuqing_wang@126.com
* Correspondence: hfliu@sut.edu.cn; Tel.: +86-13394210666

Abstract: This study is aimed at addressing the urgent demand for ultra-micro-precision dispensing technology in high-performance micro- and nanometer encapsulation, connection, and assembly manufacturing, considering the great influence of colloid viscosity and surface tension on the dispensing process in micro- and nanometer scale. According to the principle of liquid transfer, a method of adhesive transfer that can realize fL–pL levels is studied in this paper. A mathematical model describing the initial droplet volume and the transfer droplet volume was established, and the factors affecting the transfer process of adhesive were analyzed by the model. The theoretical model of the transfer droplet volume was verified by a 3D scanning method. The relationships between the transfer droplet volume and the initial droplet volume, stay time, initial distance, and stretching speed were systematically analyzed by a single-factor experiment, and the adhesive transfer rate was calculated. Combined with trajectory planning, continuous automatic dispensing experiments with different patterns were developed, and the problems of the transfer droplet size, appearance quality, and position accuracy were analyzed comprehensively. The results show that the average relative deviation of the transfer droplet lattice position obtained by the dispensing method in this paper was 6.2%. The minimum radius of the transfer droplet was 11.7 μm, and the minimum volume of the transfer droplet was 573.3 fL. Furthermore, microporous encapsulation was realized using the method of ultra-micro-dispensing.

Keywords: fL–pL level; transfer droplet volume; microporous encapsulation; ultra-micro-dispensing

Citation: Liu, H.; Chen, X.; Wang, S.; Jiang, S.; Chen, Y.; Li, F. An Ultra-Micro-Volume Adhesive Transfer Method and Its Application in fL–pL-Level Adhesive Distribution. *Micromachines* **2022**, *13*, 664. https://doi.org/10.3390/mi13050664

Academic Editor: Pingan Zhu

Received: 31 March 2022
Accepted: 21 April 2022
Published: 23 April 2022

1. Introduction

MEMS technology is currently important for improving aerospace and military capabilities [1]. Although the development of micro- and nanometer devices is relatively mature, many microelectromechanical systems cannot be applied in practice. Difficulties are related to the high-quality encapsulation, connection, assembly, and integration technology of micro- and nanometer devices. Adhesive micro-distribution technology is an important enabling core technology for the encapsulation, connection, and integration of micro- and nanometer devices. Adhesive micro-distribution technology refers to the use of adhesives to connect micro-sized components [2]. Common dispensing methods include the time/pressure type, piston pump type, transfer printing type, piezoelectric stack type, and piezoelectric ceramic type. Shi proposed a time/pressure-type pL-level micro-volume adhesive transfer method with a minimum dispensing volume of 2 pL, but the viscosity of the applicable adhesive was low [3]. Chen designed a piston pump-type dispensing system with a minimum dispensing volume of 50 μL and a minimum controllable increment of 0.1 μL [4]. Zhang proposed a transfer printing-type dispensing pen. It achieved microporous sealing of 5–20 μm, and the minimum dispensing volume was 4.4 pL [5]. Gu proposed a piezoelectric stack-type adhesive spray system that could spray 143 transfer droplets per second; the droplet volume error was within 3.11%, and the droplet average volume was 0.564 μL [6]. Fan proposed a method to squeeze a capillary tube through a piezoelectric ceramic tube for adhesive transfer, achieving pL-level low-viscosity adhesive transfer, and the volume of the smallest transfer droplet was 8.31 pL [7]. Compared with welding, thermal bonding, and electrostatic bonding, this method does not require high-temperature and high-pressure conditions, while it has the advantages of simple implementation and low stress [8–10]. Therefore, it is widely used in the assembly of microsystems [11]. For example, adhesive transfer technology enables the surface encapsulation of a micro-level electrostatic motor. The micro-accelerometer, which is installed in the automobile airbag, plays a crucial role in driving safety. Using adhesive bonding technology to fix and encapsulate the micro-accelerometer can improve the stability of the device and prolong its service life, thus making driving safer. The adhesive transfer technology proposed in this paper can achieve precise control at the micron level, with a minimum dispensing volume reaching the fL level and the potential for encapsulation in a narrow space. Therefore, it is very advantageous to use this system to fix and encapsulate micro-electrostatic motors and micro-accelerometers. However, too much adhesive at the connection will lead to overflow and pollution, while insufficient adhesive will lead to a decline in connection strength [12]. Obviously, a precise control adhesive transfer volume is very important for the encapsulation, connection, assembly, and integration of micro- and nanometer devices. When using noncontact adhesive spraying for liquid transfer, the inner diameter of the hollow needle must be reduced to a few microns or even smaller, which is very difficult to manufacture [13,14]. Therefore, there are still many problems to be solved in the application of jet methods to adhesive transfer with fL–pL resolution. With the development of MEMS technology, the original liquid transfer technology cannot meet the precision requirements of micro-connections. Therefore, there is an urgent need for an ultra-micro-automatic dispensing technology matching the size and accuracy of micro- and nanometer manufacturing.

The method of transfer printing is to impregnate the liquid through the stamp, and then transfer the impregnated liquid to the base surface through contact. Because the method of transfer printing is based on liquid surface flow rather than internal pipe flow, the effect of high flow resistance on liquid transfer can be reduced, which is promising in the field of micro-volume liquid transfer. Microarray transfer printing technology has been used to generate micro-patterns of various biomolecules and photoresists [15]. The pioneering studies of liquid transfer were conducted by Chadov and Yakhnin [16]. They mainly studied the situation of liquid transfer at different stretching speeds. Three states of liquid transfer were defined, namely, a quasi-static regime, dynamic regime, and transition regime. In addition, Cai's team studied the effect of initial distance on the liquid transfer. They found that, in the process of liquid transfer, when the distance between the donor surface and acceptor surface decreases, the contact area between the liquid and acceptor surface increases [17–19]. Dodds et al. studied the effect of different contact angles on the liquid transfer process [20]. The research showed that there is a lag phenomenon of contact angle in the liquid transfer process, and that the contact angle has an important influence on the liquid transfer volume [21–26]. On the basis of the existing literature, it can be found that most studies did not comprehensively evaluate the relationships between each physical parameter and the adhesive transfer volume. The adhesive transfer volume described in most studies could only reach sub-nanoliter levels, but not fL–pL levels.

This study is aimed at addressing the urgent need for high-precision adhesive transfer technology in the micro- and nanometer encapsulation industry, considering the influence of the adhesive properties on the transfer process, as well as the problem of micro-nozzles being difficult to manufacture. On the basis of the principle of liquid transfer, an adhesive transfer method which can realize fL–pL levels is studied in this paper. A mathematical model describing the initial droplet volume and the transfer droplet volume is established, and the factors affecting the transfer process of adhesive are analyzed by the model. The relationships between the transfer droplet volume and initial droplet volume, stay time, initial distance, and stretching speed are analyzed using a single-factor method. Furthermore, combined with trajectory planning, continuous automatic dispensing experiments with different patterns are carried out, and the research results are applied in microporous encapsulation.

2. Principle of Adhesive Transfer and Modeling of Transfer Quantity

2.1. Adhesive Transfer Principle

The adhesive transfer principle proposed in this paper is to attach the adhesive to the pipetting needle first, and then control the pipetting needle attached with the adhesive to contact the specified base surface. The adhesive is partly transferred from the pipetting needle to the base surface by relying on its own viscosity, and the transferred adhesive forms transfer droplets under the action of surface tension.

The ultra-micro-dispensing process based on the transfer principle is shown in Figure 1. The dispensing process is divided into five stages: adhesive taking, transportation, extrusion, stretching, and return. In the adhesive taking stage, the pipetting needle moves downward within the capillary tube fitted with the adhesive, as shown in Figure 1a. In the transport stage, the pipetting needle passes through the capillary tube. Under the action of viscous force, the adhesive adheres to the surface of the pipetting needle. Under the action of surface tension, the adhesive partly converges at the tip of the pipetting needle to form initial droplets. The pipetting needle with adhesive continues to move downward, as shown in Figure 1b. When the adhesive comes into contact with the base surface, the extrusion stage begins. At this time, the adhesive connects the pipetting needle with the base surface to form a liquid bridge. The liquid bridge is squeezed until the preset initial distance threshold is reached, and the extrusion process ends, as shown in Figure 1c. In the stretching stage, the pipetting needle moves upward, which pulls the liquid bridge off. Some of the adhesive in contact with the liquid bridge is transferred to the base surface, and the transferred adhesive forms transfer droplets under the combined action of surface tension and viscous force, as shown in Figure 1d,e. In the return stage, the pipetting needle moves upward, and the adhesive at the capillary tube mouth moves upward together with the pipetting needle under the combined action of inertial force, viscous force, and surface tension to form a liquid fossa. When the height of the pipetting needle returning to the capillary tube is greater than the height of the liquid fossa, the return stage is over and the system completes an adhesive distribution, as shown in Figure 1f.

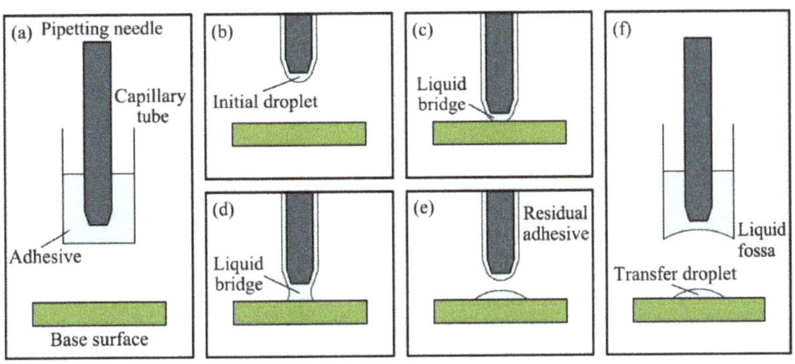

Figure 1. Diagram illustrating the principle of the ultra-micro-dispensing process: (**a**) adhesive taking; (**b**) transportation; (**c**) extrusion; (**d**,**e**) stretch and break; (**f**) return.

2.2. Adhesive Transfer Rate Modeling

In the dispensing process, the adhesive adheres to the surface of the pipetting needle. Through extrusion and stretching, some of the adhesive on the pipetting needle is transferred to the base surface to form transfer droplets, and the residual adhesive on the pipetting needle is brought back to the capillary tube. In this process, the adhesive initially adhered to the surface of the pipetting needle is not completely transferred to the base surface. In order to study the law of adhesive transfer, it is necessary to theoretically calculate and analyze the adhesive transfer rate.

After the pipetting needle passes through the capillary tube, the adhesive adheres to the surface of the pipetting needle under the combined action of viscous force and surface tension. Assuming that the state between the pipetting needle and the adhesive is ideal, the section passing through the axis of the pipetting needle was taken as the research object, as shown in Figure 2.

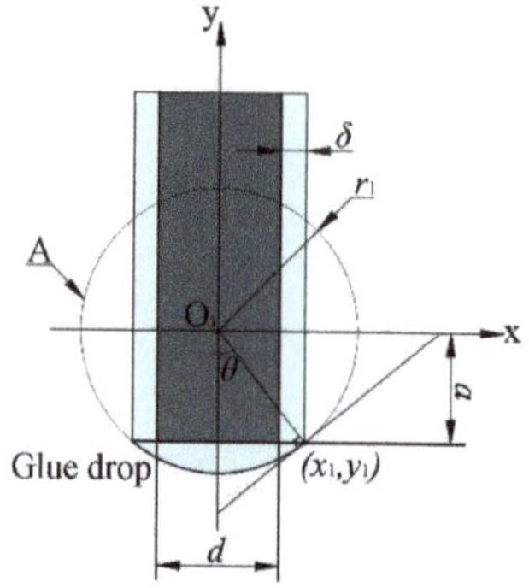

Figure 2. Ideal initial droplet model.

In Figure 2, the diameter of the pipetting needle is d. The thickness of the adhesive adhering to the surface of the pipetting needle is δ. The curved surface radius of the initial droplet is r_1. The contact angle formed between the tip of the pipetting needle and initial droplet is θ. The difference between the circle A's radius and the initial droplet height is a. The intersection formed by the straight line passing through the circle center O_1 and the tangent line passing through circle A is (x_1, y_1). Taking the y-axis as the rotation axis, we can get the initial droplet volume V_1 as follows:

$$V_1 = \int_a^{r_1} \pi x^2 dy = \int_{\cot\theta(\frac{d+2\delta}{2})}^{\csc\theta(\frac{d+2\delta}{2})} \pi(r_1^2 - y_1^2)dy. \tag{1}$$

By integrating Equation (1), the initial droplet volume V_1 can be obtained as follows:

$$V_1 = \frac{\pi(d+2\delta)^3(2+\cos^3\theta - 3\cos\theta)}{24\sin^3\theta}. \tag{2}$$

When the adhesive at the tip of the pipetting needle contacts the base surface, a liquid bridge is formed due to the combined action of the inertial force and surface tension of the adhesive itself. When the pipetting needle moves upward, the liquid bridge is pulled off due to the viscous force of the adhesive. The adhesive left on the base surface is mainly affected by the combined action of mass force, surface tension, and viscous force. Mass force diffuses the adhesive, while surface tension shrinks the adhesive inward. Viscous force holds the adhesive to the base surface. After a period of time, the force of the adhesive reaches equilibrium, and a transfer droplet is finally formed. In an ideal state, the transfer droplet features a spherical crown, as shown in Figure 3.

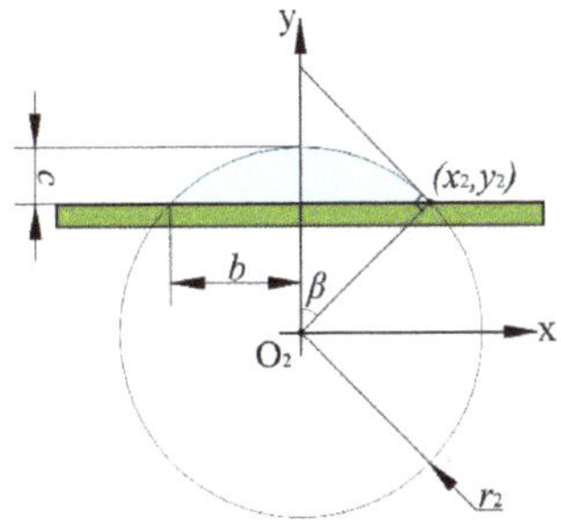

Figure 3. Ideal transfer droplet model.

In Figure 3, the radius of the transfer droplet is b, the height of the transfer droplet is c, the coordinates at the boundary intersection of three phases are (x_2, y_2), the contact angle of the transfer droplet with the base surface is β, and the curved surface radius of the transfer droplet is r_2. Taking the y-axis as the rotation axis, we can determine the transfer droplet volume V_2 as follows:

$$V_2 = \int_{r_2-c}^{r_2} \pi x^2 dy = \int_{\frac{b}{\sin\beta}-c}^{\frac{b}{\sin\beta}} \pi(r_2^2 - y_2^2)dy. \tag{3}$$

By integrating Equation (3), the transfer droplet volume V_2 can be further obtained as follows:

$$V_2 = \pi c^2\left(\frac{b}{\sin\beta} - \frac{c}{3}\right). \tag{4}$$

Therefore, the adhesive transfer rate η can be calculated as follows:

$$\eta = \frac{V_2}{V_1} = \frac{8c^2\sin^3\theta(3b - c\sin\beta)}{\sin\beta(d+2\delta)^3(2+\cos^3\theta - 3\cos\theta)}. \tag{5}$$

It can be seen from Equation (2) that the initial droplet volume is related to the diameter of the pipetting needle, the angle between the tip of the pipetting needle and the initial droplet, and the adhesive thickness. In addition, the diameter of the pipetting needle and the thickness of the adhesive are the main factors affecting the volume of the initial droplet. It can be seen from Equation (4) that the volume of the transfer droplet is related to the height of the transfer droplet, the radius of the transfer droplet, and the contact angle between the transfer droplet and the base surface. It can be seen from Equation (5) that the adhesive transfer rate is related to c, θ, b, β, d, and δ. The size of c and β is mainly related to the wettability of the base surface. The size of θ and δ is mainly related to the pipetting needle material and the viscosity of adhesive. In the process of stretching the liquid bridge, the contact line between the liquid bridge and the base surface shrinks inward. The dimension of the transfer droplets is largely dependent on the size of the contact line, and the wettability of the base surface directly affects the formation of the contact line. A better base surface wettability results in a smaller distance of the contact line moving inward, and the phenomenon of contact angle hysteresis is more obvious. This means that more adhesive is transferred to the base surface, and a larger transfer droplet formed. If the influence of material properties and adhesive properties on the adhesive transfer rate is not considered, the factors affecting adhesive transfer rate are only the radius of the transfer droplet and the diameter of the pipetting needle. Therefore, this paper only analyzes the effects of the transfer droplet radius and the pipetting needle diameter on the adhesive transfer rate.

3. Experimental System and Method

3.1. Development of Experimental System

Through the analysis of the dispensing principle and the adhesive transfer process, the main factors affecting the adhesive transfer rate were preliminarily identified. According to the measurement requirements, the dispensing experimental platform was designed. The ultra-micro-precision automatic dispensing system independently developed and designed is shown in Figure 4. The system consists of a transfer droplet morphology observation and data acquisition module, adhesive transfer module, and drive control module.

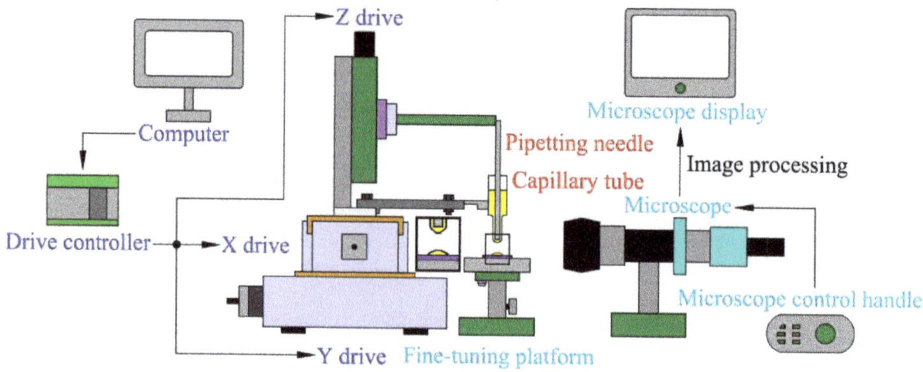

Figure 4. Composition diagram of automatic dispensing system.

The function of the transfer droplet morphology observation and data acquisition module involves auxiliary adjustment for the experimental process, as well as completion of the measurement, annotation, and observation of the transfer droplet. The adhesive transfer module is composed of a pipetting needle, a capillary tube, and a base surface. The pipetting needle is used to complete the adhesion and transfer of the adhesive, the capillary tube is used to store the adhesive, and the base surface is the carrier for the transfer adhesive. The function of the drive control module is to control the movement of the pipetting needle. Under the control of the driving device, the pipetting needle can move the minimum distance up to 1 μm.

The prototypes of the developed dispensing system and the built experimental platform are shown in Figure 5. The transfer droplet morphology observation and data acquisition module is composed of a fine-tuning platform, a microscope, a microscope control handle and a microscope display. The fine-tuning platform and the microscope are an integrated structure. The fine-tuning platform can be adjusted by the fine-tuning knob, and the microscope can be adjusted by the microscope control handle. The magnification of the microscope is between 100× and 1000×, which can meet the observation requirements in most cases. The pipetting needle and capillary tube in the adhesive transfer module are fixed on the Z drive. During the experiment, the adhesive adheres to the pipetting needle, and then the adhesive on the pipetting needle is transferred to the base surface to achieve the dispensing process. The drive control module is composed of a computer, a control panel, a drive controller, and an X/Y/Z drive. The computer is responsible for transmitting the control program to the drive controller, and the motion speed of the drive can be set by the computer. The main function of the control panel is to control the movement trajectory of the pipetting needle and set parameters. The drive controller is responsible for transmitting the motion control signal to the X/Y/Z drive. The X/Y/Z drive executes the move command to drive the pipetting needle to the specified position to complete the transfer of adhesive.

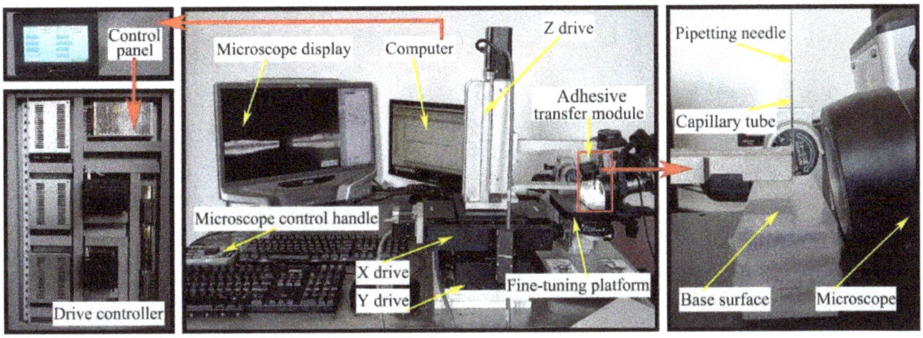

Figure 5. Physical object of experimental platform.

3.2. Material

The materials needed for the experiment mainly include the pipetting needle, adhesive, base surface, and capillary tube. The pipetting needle is a key component in the dispensing experiment, and its diameter is required to reach tens of microns. On the one hand, it is necessary to ensure that the pipetting needle does not bend during processing. On the other hand, the tip of the pipetting needle is required to be flat and smooth. Therefore, the pipetting needle should have the characteristics of high hardness, corrosion resistance, and wear resistance. A tungsten needle was selected as the raw material in this paper. The machining process of the pipetting needle was divided into two stages: electrolysis and grinding. After electrolysis, the pipetting needle had a stepped shape with a tapered tip. Then, high-mesh sandpaper was used to further grind the tip to achieve a flat and smooth effect. An image of the pipetting needle is shown in Figure 6.

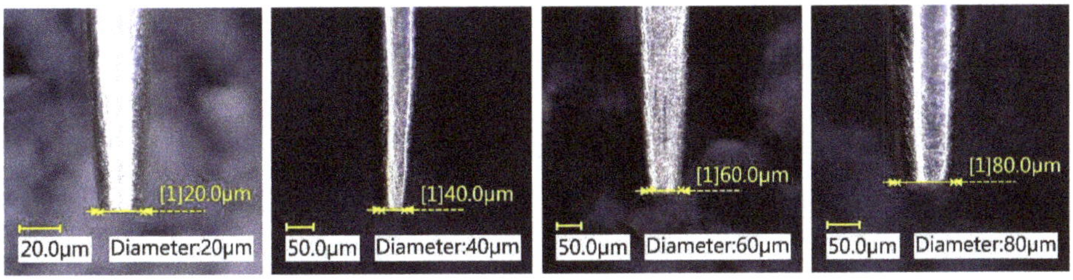

Figure 6. The pipetting needle image.

The adhesive transfer method proposed in this paper is suitable for the transfer of adhesives such as epoxy resin, methyl silicone oil, and polyurethane. Because the purpose of the experiment was to study the influence of the initial droplet volume, stay time, initial distance, and stretching speed on the adhesive transfer rate, the composition and viscosity of adhesive were not taken as the research object. However, experiments using adhesives of different composition and viscosity would yield the same conclusion. Therefore, epoxy resin with a viscosity of 1000 cps was taken as the experimental adhesive.

The base surface was thin glass material with a size of $75 \times 25 \times 1$ mm. An ultrasonic cleaning method was used to ensure that the base surface was clean. The adhesive storage tube was a glass capillary tube with an inner diameter of 0.8 mm and a length of 12 cm.

3.3. Experimental Conditions

When observing the dispensing process through the microscope, it was found that small vibration had a great impact on the experimental results. Therefore, in order to reduce the impact of vibration on the dispensing process, the experimental platform was built on a shock absorption platform. During the experiment, the room was kept relatively sealed, with no dust indoors and a room temperature of 23–25 °C. The experimental process was consistent with the theoretical process mentioned in Section 2.1. The actual adhesive transfer process observed by the microscope is shown in Figure 7.

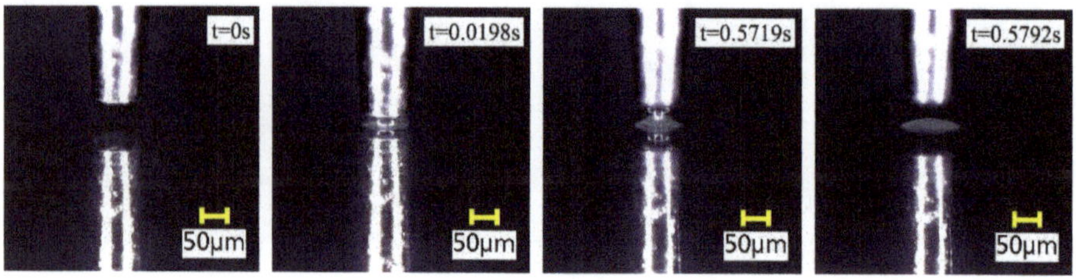

Figure 7. The actual transfer process of the adhesive.

4. Factors Affecting the Volume of Transfer Droplets and Comprehensive Experiments

The purpose of this paper was to study the automatic distribution method of an adhesive at the fL–pL level, as well as analyze the influence of working parameters on the adhesive transfer process. Therefore, a method for evaluating the volume of the transfer droplet was first established. Then, the single-factor method was used to analyze the influence of each factor on the transfer droplet volume and the adhesive transfer rate. Lastly, combined with the trajectory planning method, an application experiment of automatic adhesive dispensing with fL–pL resolution was completed.

4.1. Method for Determining Transfer Droplet Volume

The shape of the transfer droplet was similar to a spherical crown, but it was not a regular spherical crown. Therefore, it was necessary to verify whether Equation (4) could accurately describe the volume of the transfer droplet through experiments.

After analyzing the adhesive transfer rate through the method of theoretical analysis, it was found that the key to studying the adhesive transfer rate was to determine the volume of the transfer droplet. On the one hand, the 3D scanning method could be used to measure the volume of the transfer droplet. This method could obtain an accurate volume, but its efficiency is low. On the other hand, the theoretical calculation method could be used to calculate the transfer droplet volume. This method was more efficient, but its accuracy needed to be verified.

It can be seen from Equation (4) that the volume of the transfer droplet is related to the height of the transfer droplet, the radius of the transfer droplet, and the contact angle between the transfer droplet and the base surface. On the one hand, we used the plane measurement method to measure the above parameters and substituted the measured values into Equation (4) to calculate the transfer droplet volume. The experimental conditions were as follows: a pipetting needle diameter of 60 μm, a stretching speed of 2 mm/s, a stay time of 1 s, and an initial distance of 0 μm. Under these conditions, several dispensing experiments were completed, and 10 transfer droplets were obtained. Figure 8 shows the transfer droplet parameters measured by microscope. The measured height of the transfer droplet was 13 μm, the diameter was 66.5 μm, and the contact angle was 34°. After calculation, we could calculate the transfer droplet volume as 29.3 fL. It can be seen from the figure that the appearance of the transfer droplet was relatively round, and the side of the transfer droplet was similar to a spherical crown. The same method was used to measure the parameters of other transfer droplets, and then the volume of the transfer droplets was calculated. The specific measurement data and calculation results are shown in Table 1. After measurement, it was found that the height of the transfer droplet was 13–15 μm, the radius of the transfer droplet was 31.9–33.3 μm, and the contact angle was 33°–36°. After calculation, we could get that the minimum volume of the transfer droplet was 27.2 pL, the maximum volume was 37.9 pL, and the theoretical average volume of the transfer droplet was 31.5 pL.

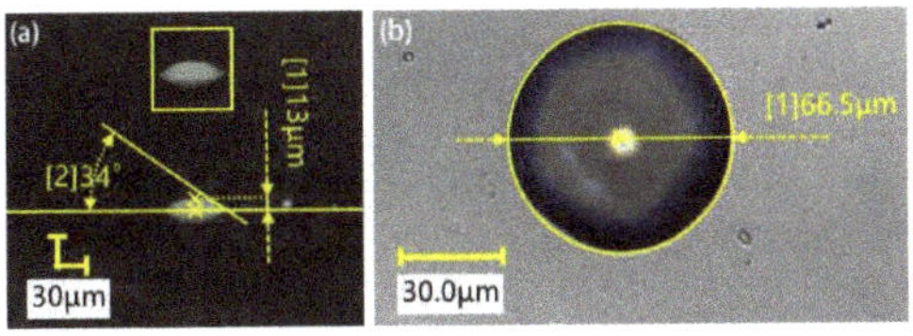

Figure 8. Transfer droplet parameters: (**a**) horizontal direction; (**b**) vertical direction.

Table 1. Measurement data and calculation results.

Number	1	2	3	4	5	6	7	8	9	10
c (μm)	13	13	14	13	15	15	13	14	14	13
b (μm)	33.3	32.4	33.2	31.9	32.1	32.8	33.1	32.9	32.3	32.6
β (°)	34	35	33	35	34	34	36	34	35	34
V_2 (pL)	29.3	27.7	34.7	27.2	37	37.9	27.6	33.4	31.8	28.7

On the other hand, we used 3D scanning measurements to determine the actual volume of the transferred droplet. The principle of 3D scanning measurement is to scan the transfer droplet with a microscope to obtain the original point coordinate data. After the original data are processed, the actual transfer droplet volume can be obtained. Because both the transfer droplet and the base surface have light transmittance, serious deletion can appear during direct scanning. Therefore, it was necessary to spray treatment the base surface with a developer before the dispensing experiment. As the diameter of the developer particles was less than 10 μm, they had no significant effect on the morphology of the transfer droplet. After the developer was completely cured, the experiment was started on the base surface. After the experiment, it was necessary to spray the base surface with the developer again. After the developer was cured, the transfer droplets could be scanned by the 3D scanning method.

The volume of the transfer droplet was measured by the 3D scanning method in three steps: scanning of the original transfer droplet, boundary point processing, and error compensation processing. For example, the transfer droplet in Figure 8 was scanned in 3D to obtain the original image of the transfer droplet, as shown in Figure 9a. It can be seen that there was the problem of a fuzzy boundary around the transfer droplets. Therefore, the point processing method was used to process the transfer droplet image, and the exported image after preliminary processing is shown in Figure 9b. It can be seen that the boundary area of the transfer droplet was defective, which would have led to the deviation of the transferred droplet volume. Hence, the error compensation method was further applied. The repaired 3D export model is shown in Figure 9c. It can be seen that the surface of the transferred droplet was relatively complete without obvious defects.

It can be seen from Figure 9c that the shape of the transfer droplet was not completely regular; thus, the most accurate method to determine the volume of the transfer droplet was to take the coordinate point as calculation unit. Firstly, each calculation unit was integrated, and then all the integration results were added. After error compensation processing, nearly 160,000 calculation units needed to be integrated and summed, requiring a large amount of calculation. Therefore, the integral summation function was used to calculate the volume of the transfer droplet, and the measured volume of this transfer droplet was finally obtained as 28.4 pL. The volume of the remaining transfer droplets was measured using the above method. The error between the measured volume and the theoretical volume is shown in Figure 10a. We found that the measured volume was smaller than the theoretical volume. The average measured volume was 30.4 fL, which is about 96.4% of the theoretical average volume. Two reasons could explain why the measured volume was smaller than the theoretical volume. Firstly, the scan could have had missing parts, whereby error compensation processing would have only supplemented the missing parts with an approximate value close to the actual value. Secondly, the cross-section where the diameter of the transfer droplet was measured could have differed, as shown in Figure 10b. We found that the root of the actual transfer droplet was internal concave. Thus, the internal concave phenomenon became more obvious upon approaching the base plane, which would have made the measured volume smaller than the theoretical volume. After calculation, the average relative error of the transfer droplet volume was 3.8%. After analysis, we found that the theoretical calculation method was accurate, and the level of calculation was small. Therefore, the theoretical calculation method was used to estimate the transfer droplet volume in this paper.

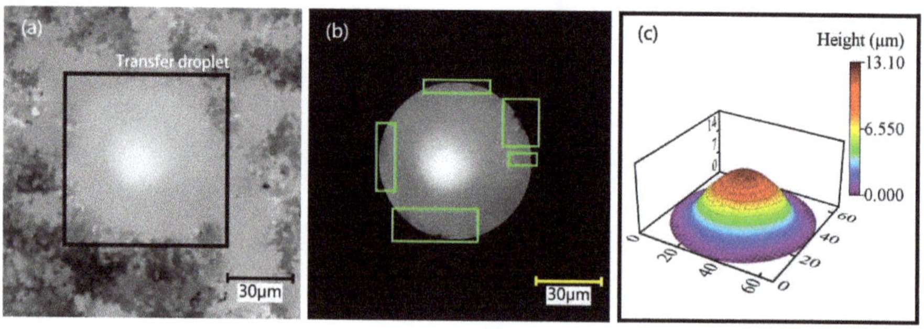

Figure 9. The 3D scanned image: (**a**) original image; (**b**) processed image; (**c**) 3D exported model.

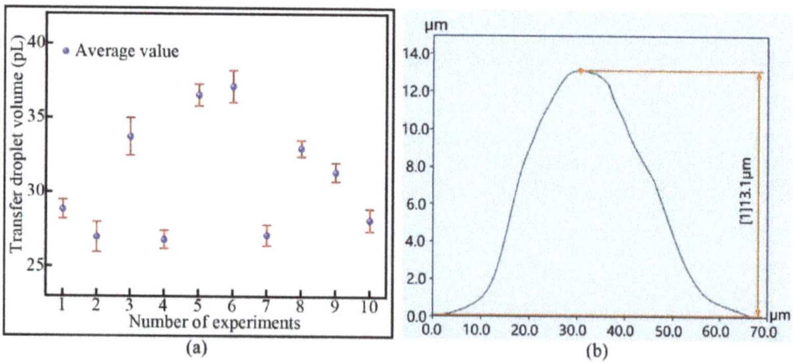

Figure 10. Transfer droplet volume: (**a**) comparison between the measured volume and the theoretical volume; (**b**) cross-section profile of the transfer droplet.

4.2. Analysis of Influencing Factors

According to the conclusion drawn in Section 2.2, we could simplify the research content of this paper by studying the influence of the transfer droplet radius and pipetting needle diameter on the adhesive transfer rate. It can be seen from Equation (4) that the radius of the transfer droplet is the main factor affecting the volume of the transfer droplet. Through observation, we found that the stay time, initial distance, and stretching speed all affected the radius of the transfer droplet. The diameter of the pipetting needle mainly affects the volume of the initial droplet, and the volume of the initial droplet affects the volume of the transfer droplet. Without affecting the experimental results, we could simplify the research content of this paper by studying the influence of the initial droplet volume, stay time, initial distance, and stretching speed on the transfer droplet volume.

The four main influencing factors are defined below. The initial droplet volume refers to the volume of the droplet formed after the tip of the pipetting needle adheres to the adhesive. The stay time refers to the time that the pipetting needle tip remains stationary after contacting the base surface. The initial distance refers to the distance from the tip of the pipetting needle to the base surface. The stretching speed refers to the average speed of stretching the liquid bridge until fracture. Assuming a lack of interaction between factors, the single-factor method was used to analyze the influence of each factor on the transfer droplet volume.

4.2.1. The Effect of Initial Droplet Volume

In the experiment, with other factors unchanged, initial droplets of different diameters were obtained by selecting pipetting needles of different diameters to attach the adhesive. The effect of initial droplet volume on transfer droplet volume was analyzed in detail through experiments. The stay time was set to 1 s, the initial distance was set to 0 μm, the stretching speed was set to 4 mm/s, and the pipetting needle diameters were 20 μm, 40 μm, 60 μm, and 80 μm. Figure 11 shows the transfer droplets observed under the microscope. It can be seen from the figure that the morphology of all transferred droplets was intact without obvious defects.

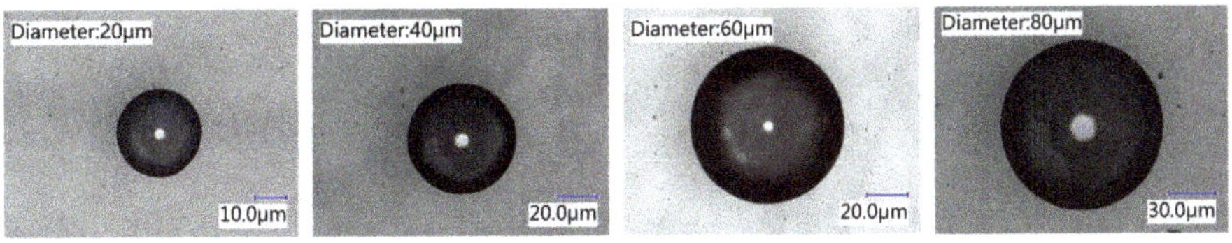

Figure 11. Transfer droplets obtained after changing the diameter of the pipetting needle.

In order to ensure the reliability of the experimental results, after four groups of dispensing experiments were completed with pipetting needles of different diameters, 10 transfer droplets were randomly selected from each group for the measurement of diameter. The measurement results are shown in Figure 12a. It can be seen from the figure that the diameter of the same group of transfer droplets was approximately the same. Figure 12b reflects the relationship between the average diameter of the transfer droplet and the diameter of the pipetting needle. It can be seen from the figure that the transfer droplet diameter increased with the increase in the pipetting needle diameter. The initial droplet volume and the transfer droplet volume were calculated using the method of theoretical calculation, and the adhesive transfer rate was further calculated on this basis. The results are shown in Figure 12c. The results show that the transfer droplet volume and the adhesive transfer rate increased with the increase in the initial droplet volume.

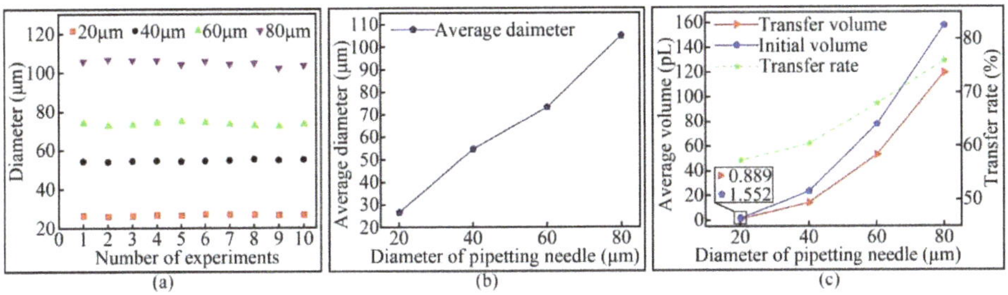

Figure 12. The relationship between the initial droplet volume and the transfer droplet volume: (**a**) the diameter of the transfer droplet; (**b**) the average diameter of the transfer droplet; (c) volume comparison.

A larger diameter of the pipetting needle resulted in a larger initial droplet volume adhering to the tip of the pipetting needle. When the height of the liquid bridge remained unchanged, a larger initial droplet volume resulted in a larger contact area between the liquid bridge and the base surface, as well as a larger diameter of the transferred droplet. As the contact area increased, the adhesion effect of the base surface to the liquid bridge increased, adsorbing more adhesive, and the volume of the transfer droplet increased. Viscous force plays a leading role in the transition regime [16], and large initial droplets exert greater pressure on the base surface. Therefore, the adhesive transfer rate increased under the combined action of viscous force and pressure. Zhu's team also reached the same conclusion [27], where the contact area was also mentioned as the main factor enhancing the bonding effect [17,18].

4.2.2. The Effect of Stay Time

Four groups of experiments were conducted for the stay time, with each group only changing the stay time of the pipetting needle on the base surface, while other conditions remain unchanged to study the effect of stay time on the transfer droplet volume. A pipetting needle with a diameter of 60 μm was selected, the initial distance was set to 0 μm, the stretching speed was set to 4 mm/s, and the stay time was 1 s, 2 s, 3 s, and 4 s. After the experiment, the bottom area of the transfer droplet was measured, and the micro-morphology of the transfer droplet was observed. The microscopic morphology of the transfer droplets formed under different stay times is shown in Figure 13. It can be seen from the figure that the bottom area of the transfer droplet increased with the increase in stay time. The microscopic morphology of the transfer droplet did not change due to the change in stay time.

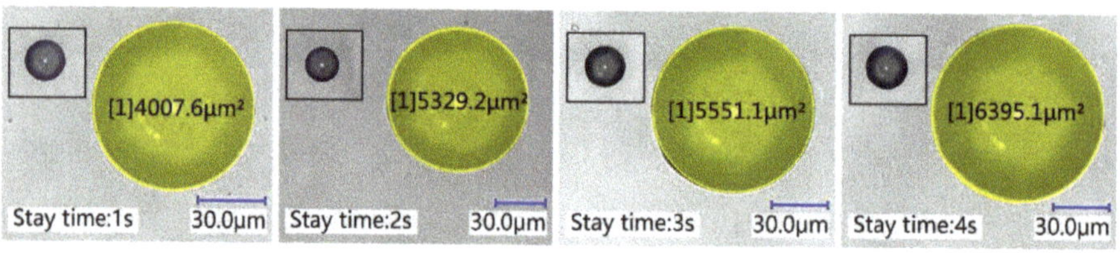

Figure 13. The transfer droplet obtained after changing the stay time.

Ten transfer droplets were randomly selected from the groups for bottom area measurement, and the measurement results of transfer droplet bottom area are shown in Figure 14a. It can be seen that changing the stay time of the pipetting needle on the base surface changed the bottom area of the transfer droplet, and the bottom area of transfer droplets obtained from the same group of experiments also fluctuated within a certain range. The average value of the transfer droplet diameter from each group was calculated, and the results are shown in Figure 14b. It can be seen more intuitively from the figure that there was a linear relationship between the transfer droplet diameter and the stay time. We found that a longer stay time led to a larger diameter of the transfer droplet. Upon only changing the stay time of the pipetting needle on the base surface, it was found that the volume of the initial droplet changed little. With the increase in stay time, the transfer droplet volume and the adhesive transfer rate increased. The specific values are shown in Figure 14c.

In order to ensure the quality of adhesive transfer, it is usually required that the adhesive remains relatively stationary for several seconds after contacting the base surface. In a relatively static state, the speed of the pipetting needle and the base surface acting on the liquid bridge tends to zero, and the transfer state of the adhesive belongs to a quasi-static state. Therefore, we can study the relationship between the stay time and the transfer rate by studying the transfer state of the adhesive in the quasi-static state, where surface tension plays an important role [16]. With the increase in residence time, the adhesive promoted the liquid bridge to gradually reach an equilibrium state under the action of mass force, resulting in an increase in the pressure of the liquid bridge on the base surface.

The surface tension decreased with increasing pressure [28]; thus, an increase in the contact area between the liquid bridge and the base surface occurred. The increase in contact area meant that there was a smaller receding contact angle between the liquid bridge and the base surface, whereby the attraction of the base surface to the adhesive was enhanced. With a stronger attraction of the base surface to the adhesive, more adhesive was transferred to the base surface, forming larger transfer droplets and increasing the transfer rate. This conclusion is consistent with the conclusion obtained in [28], proving the correctness of the experiment in this paper.

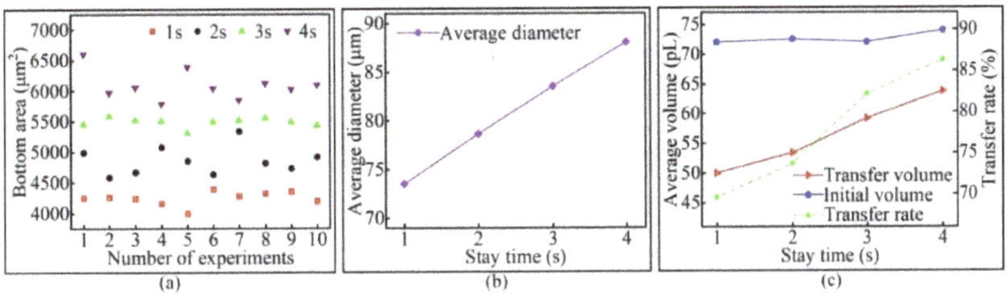

Figure 14. The relationship between the stay time and the transfer droplet volume: (**a**) the bottom area of the transfer droplet; (**b**) the average diameter of the transfer droplet; (**c**) volume comparison.

4.2.3. The Effect of the Initial Distance

Keeping the pipetting needle diameter, stay time, and stretching speed unchanged, different initial distances were set to study the effect of the initial distance on the volume of the transfer droplet. A pipetting needle with a diameter of 60 μm was selected, the stay time was set to 1 s, the stretching speed was set to 4 mm/s, and the initial distances were 0 μm, 6 μm, 12 μm, and 18 μm. The microscopic morphology of the transfer droplets formed under different initial distances is shown in Figure 15. It can be seen from the figure that transfer droplets of different sizes could be obtained by setting different initial distances.

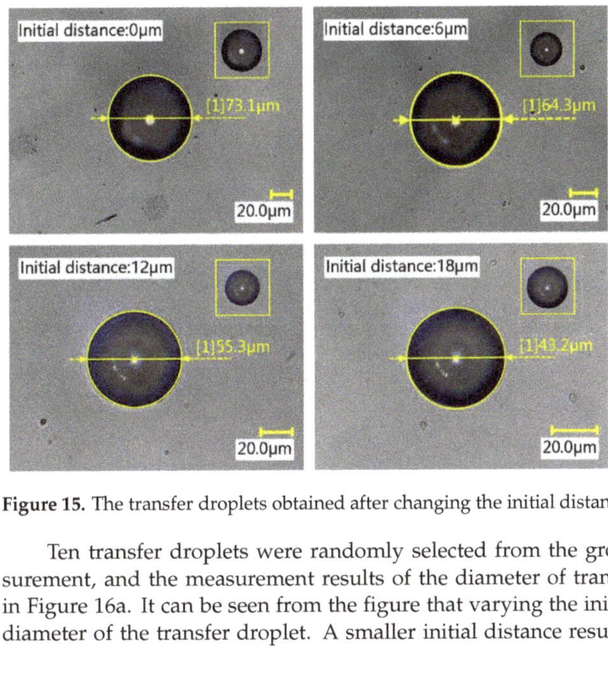

Figure 15. The transfer droplets obtained after changing the initial distance.

Ten transfer droplets were randomly selected from the groups for diameter measurement, and the measurement results of the diameter of transfer droplet are shown in Figure 16a. It can be seen from the figure that varying the initial distance altered the diameter of the transfer droplet. A smaller initial distance resulted in a larger transfer droplet average diameter. This result is shown in Figure 16b. Through further analysis, we found that changing the initial distance had little effect on the initial droplet volume, but an increase in the initial distance led to a decrease in the transfer droplet volume and adhesive transfer rate. This result is shown in Figure 16c.

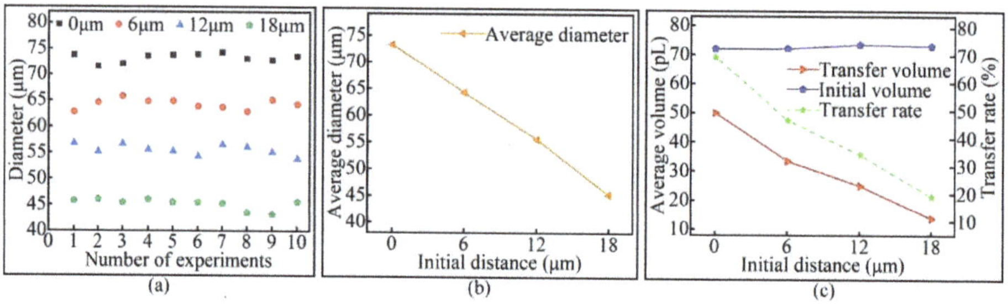

Figure 16. The relationship between the initial distance and the transfer droplet volume: (**a**) the diameter of the transfer droplet; (**b**) the average diameter of the transfer droplet; (**c**) volume comparison.

Without changing any configuration parameters, the initial droplet with almost the same volume could be obtained each time. When the initial droplet contacted the base surface and formed a liquid bridge, the volume of the liquid bridge became the initial droplet volume. As the initial distance increased, the liquid bridge became finer and longer, and the contact area between the adhesive and the base surface became smaller. Therefore, a larger initial distance resulted in a smaller diameter of the transfer droplet. The research results in [29] showed that the transfer rate is related to the viscous force. As the initial distance increased, the viscous force between the liquid bridge and the base surface gradually weakened, whereby the attraction of the base surface to the adhesive became smaller. Therefore, with the initial droplet volume almost constant, the initial distance increased, the transfer droplet volume decreased, and the adhesive transfer rate decreased. In this paper, the same conclusion as that of [29] was also obtained through experiments, indicating that the adhesive transfer rate under microscopic and macroscopic conditions had the same law and obeyed the requirements of conventional hydrodynamics.

4.2.4. The Effect of Stretching Speed

The stretching speed of the pipetting needle was varied, with other factors remaining unchanged, to analyze the relationship between the stretching speed and the volume of the transfer droplet. A pipetting needle with a diameter of 60 μm was selected, the stay time was set to 1 s, the initial distance was set to 0 μm, and the stretching speed was divided into seven groups (1–7 mm/s). The microscopic morphology of the transfer droplets formed under different stretching speeds is shown in Figure 17. It can be seen from the figure that changing the stretching speed had an effect on the size of the transfer droplets. However, the stretching speed had almost no effect on the morphology of transfer droplets, and a complete transfer droplet could be obtained at any stretching speed.

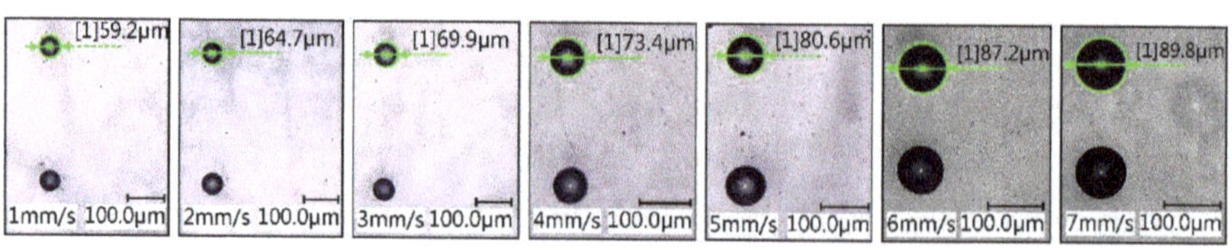

Figure 17. The transfer droplet obtained after changing the stretching speed.

Ten transfer droplets were randomly selected from the groups for diameter measurement, and the results are shown in Figure 18a. It can be seen from the figure that the diameter of the transfer droplet changed with the change in stretching speed. According to the relationship between the average diameter of the transfer droplet and the stretching speed shown in Figure 18b, it can be seen that a faster stretching speed led to a larger diameter of the transfer droplet. Changing the stretching speed had almost no effect on the volume of initial droplet; however, with the increase in stretching speed, the volume of the transfer droplet and adhesive transfer rate decreased. The specific values are shown in Figure 18c.

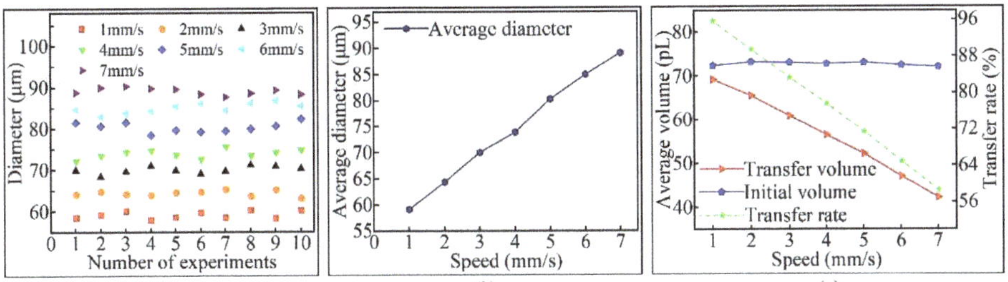

Figure 18. The relationship between the stretching speed and the transfer droplet volume: (**a**) the diameter of the transfer droplet; (**b**) the average diameter of the transfer droplet; (**c**) volume comparison.

The above research results are consistent with [30]. The phenomenon of the transfer droplet diameter increasing with the increase in stretching speed can be explained according to two aspects. Firstly, a smaller stretching speed resulted in a smaller velocity gradient in the vertical direction inside the adhesive. With an increase in stretching speed, the velocity gradient resulted in the two ends of the liquid bridge being very small while the middle was very large, leading to slow momentum propagation in the middle. Secondly, when different stretching speeds were used to stretch the liquid bridge to the same height, a faster stretching speed led to a shorter stretching time. Accordingly, at a faster stretching speed, the degree of inward sliding of the contact line of the base surface decreased, and the transfer droplet diameter increased.

The main reasons for the decrease in the adhesive transfer rate could be explained by the surface tension playing a leading role in the transfer process when the stretching speed was low. The main factor affecting surface tension is the receding contact angle. A smaller receding contact angle resulted in a greater attraction to the adhesive. Because the receding contact angle of the base surface was smaller than the pipetting needle, the base surface was more attractive to the adhesive, and more adhesive was transferred to the base surface. With an increase in stretching speed, the velocity gradient in the vertical direction of the liquid bridge increased, resulting in a slower momentum propagation of the adhesive and a lower adhesive transfer rate.

4.3. The Experiment of fL–pL Level Automatic Dispensing

4.3.1. Adhesive Automatic Distribution of fL–pL Level

Through the single-factor experiment, it was found that the initial droplet volume, stay time, initial distance, and stretching speed all affected the transfer droplet volume. By calculating the overall standard deviation, it was found that, relative to other levels, a diameter of the pipetting needle of 20 μm, a stay time of 3 s, an initial distance of 0 μm, and a stretching speed of 2 mm/s were optimal to obtain the transfer droplet with the best uniformity.

According to the design requirements, the movement trajectory was planned. Figure 19 shows a comparison diagram of the automatic dispensing effect. The preset distance between two adjacent vertices of the quadrangular star was 1600 μm, and the actual distances were 1607 μm, 1609 μm, 1610 μm, and 1609 μm. The difference rates were 0.4%, 0.6%, 0.6%, and 0.6%, respectively. The preset circle diameter was 2200 μm, and the preset distance between adjacent transfer droplets in the circle was 275 μm. The actual values were 2191 μm and 272 μm, with difference rates of 0.4% and 1.1%, respectively.

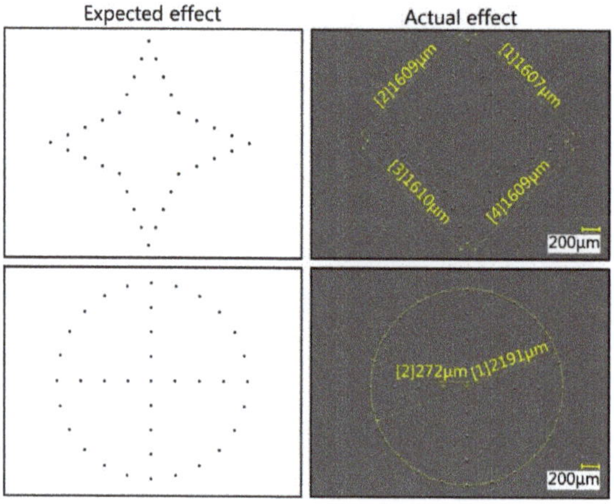

Figure 19. Comparison diagram of automatic dispensing effect.

In order to reflect the difference between the planned trajectory and the actual trajectory, the planned position of each transfer droplet in the quadrangular star was compared with the actual position. The compound effect of planned location and actual location is shown in Figure 20. It can be seen from the figure that the actual transfer droplet trajectory was basically consistent with the preset transfer droplet trajectory. After comparison, four representative transfer droplets were selected. Among them, transfer droplet 1 and transfer droplet 2 had good morphology and position accuracy, while the morphology of transfer droplet 3 was intermediate, and that of transfer droplet 4 was poor. After the transfer droplets were compounded, the diameter and position deviation of the transfer droplets were measured, and the deviation rate was calculated. After measurement, it was found that the minimum diameter of the transfer droplet was 23.5 μm, the maximum diameter was 24.5 μm, and the deviation between the planned position and the actual position was 0.6–2.6 μm. After calculation, the average diameter of the transfer droplets was obtained as 23.9 μm, the average deviation was 1.5 μm, the deviation rate was 2.5–10.8%, and the average deviation rate was 6.2%. The reason for the large position deviation of some transfer droplets could be that, during the dispensing process, the driving device had a slight vibration relative to the base surface, which caused the actual position of the transfer droplet to deviate.

It can be seen from Section 4.1 that the theoretical value of the volume of the transfer droplet was close to the measured value; thus, due to its simplicity, it was used to calculate the volume of the transfer droplet. After measurement, it was found that the minimum height of the transfer droplet was 3.1 μm, and the maximum height was 3.6 μm. The minimum contact angle of the transfer droplet was 34°, and the maximum contact angle was 40°. The minimum radius of the transfer droplet was 11.7 μm, and the maximum radius was 12.3 μm. Specific data are shown in Figure 21a. After calculation, the minimum volume of the initial droplet was obtained as 1188.6 fL, the maximum volume was 1456.2 fL, and the average volume was 1303.5 fL. The minimum volume of the transfer droplet was obtained as 573.3 fL, the maximum volume was 779.9 fL, and the average transfer volume was 659.6 fL. The minimum transfer rate of adhesive was obtained as 46.1%, the maximum transfer rate is 55.6%, and the average transfer rate is 50.6%. Specific data are shown in Figure 21b.

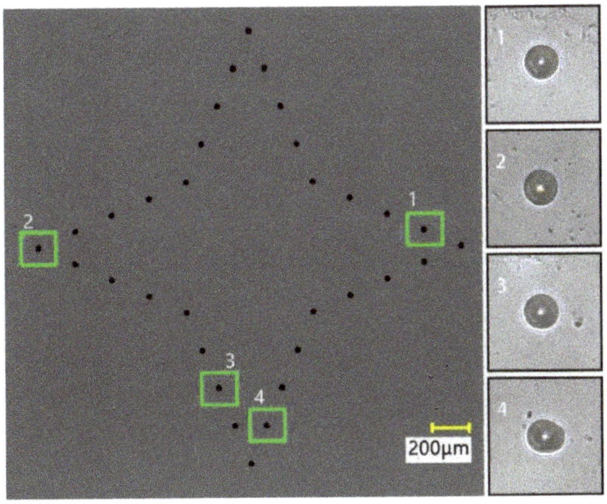

Figure 20. Compound effect comparison.

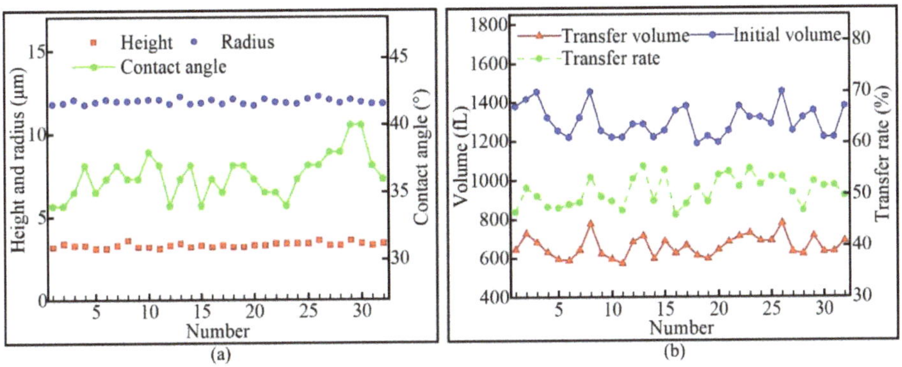

Figure 21. Quadrangular star parameter: (**a**) transfer droplet parameter; (**b**) adhesive transfer rate.

Through the fL-level adhesive automatic distribution experiment, it could be found that the actual position was basically consistent with the preset position, revealing an average deviation of 6.2% and a minimum volume of the transfer droplet of 573.3 fL. Therefore, the ultra-micro-dispensing system designed in this paper could realize the automatic dispensing of adhesive at an fL level. Compared with the existing adhesive transfer system, this system has the advantages of better stability and a smaller transfer volume.

4.3.2. The Application of Microporous Encapsulation

The internal structure of an MEMS is precise and complex. In the actual assembly process, ultra-micro-dispensing technology is often used to connect and encapsulate elements of the microelectromechanical system. Microporous encapsulation faces problems that need to be solved. Because the encapsulation experiment needs to be carried out in micron-level small holes, common dispensing methods cannot ensure an accurate position and appropriate adhesive filling volume. Therefore, it is necessary to design a new distribution mechanism to solve the problem of microporous encapsulation.

The automatic dispensing method designed in this paper can realize microporous encapsulation. During the adhesive filling process, the pipetting needle should not be in contact with the micropore; therefore, a pipetting needle with a diameter of 200 μm was selected for adhesive filling, the initial distance was set to 2 μm, and the rise amount was set to 1 μm. Under the action of viscous force, the adhesive was transferred to the micropore.

When the adhesive was observed to overflow the micropore, the microporous encapsulation was completed. The effect of microporous encapsulation is shown in Figure 22. It can be seen from the figure that, after the microporous encapsulation experiment, the microporous structure was complete, and the adhesive could be used to evenly fill the micropore. Therefore, the ultra-micro-automatic dispensing method based on surface tension designed in this paper could meet the actual encapsulation needs.

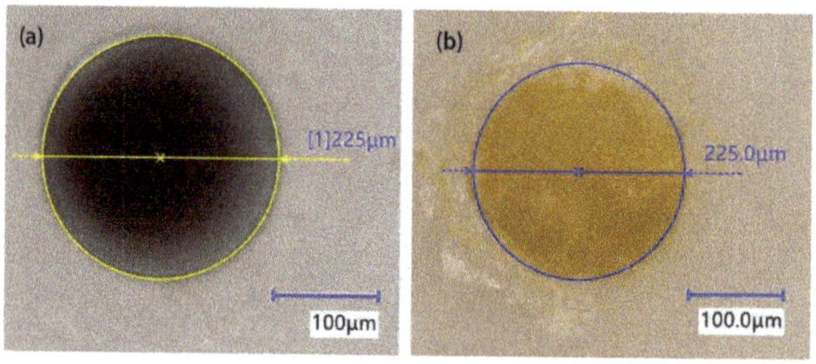

Figure 22. Microporous encapsulation effect: (**a**) before encapsulation; (**b**) after encapsulation.

5. Conclusions

According to the principle of liquid transfer printing, this paper studied the transfer mechanism of adhesive from a capillary tube to the base surface using a pipetting needle, and an ultra-micro-dispensing method was proposed. This method could realize the automatic distribution of adhesive with fL–pL resolution.

Firstly, the theoretical value of the transfer droplet volume was calculated, and then the actual value of the transfer droplet volume was measured using 3D scanning. The results showed that the measured volume was about 96.4% of the theoretical volume.

By studying the effects of initial droplet volume, stay time, initial distance, and stretching speed on the transfer droplet volume, it was found that the transfer droplet volume increased with the increase in the initial droplet volume, stay time, and stretching speed, whereas it decreased with the increase in initial distance. The adhesive transfer rate was also affected by these four factors, and the influence rule was consistent with the transfer droplet volume. The initial droplet volume increased with the increase in the pipetting needle diameter, whereas the stay time, initial distance, and stretching speed had little effect on the initial droplet volume.

By planning the movement trajectory, the automatic dispensing of transfer droplets with fL resolution in the plane was realized. After calculation and analysis, the dispensing results were as follows: a minimum radius of the transfer droplet of 11.7 μm, a minimum transfer volume of 573.3 fL, an average transfer volume of 659.6 fL, and an average transfer rate of 50.6%. The average deviation of the automatic dispensing position was 1.5 μm. In addition, the encapsulation experiment using a Φ225 μm × 70 μm micropore was successfully completed, further indicating that the ultra-micro-volume dispensing method proposed in this paper can be applied to the encapsulation of micro-sized parts.

Author Contributions: H.L. (corresponding author) contributed to conceptualization, project administration, funding acquisition, and methodology. X.C. contributed to investigation, data analysis, writing—original draft preparation, and software. S.W. helped with data processing and writing—review and editing. S.J. contributed to data processing and software application. Y.C. helped with control system debugging and data processing. F.L. helped with control system debugging and experiment planning. All authors have read and agreed to the published version of the manuscript.

Funding: This work was supported by the National Natural Science Foundation of China (Grant No. 52175428, 51775354), the Joint Fund of the Science and Technology Department of Liaoning Province and the State Key Laboratory of Robotics of China (Grant No. 2020-KF-12-07), the Liaoning Revitalization Talents Program (Grant No. XLYC2007072), the Central Government Guides Local Special Funds for Science and Technology Development (Grant No. 2020JH6/10500048), and the Transformation of Major Scientific and Technological Achievements in Shenyang (Grant No. 20-203-5-01).

Conflicts of Interest: The authors declare no conflict of interest.

References

1. Zhang, Q.; Xu, C.; Xu, C.Y.; Hisayuki, A. Approach and experiment of ultra-micro dispensing. *Opt. Precis. Eng.* **2013**, *21*, 2071–2078. [CrossRef]
2. Chen, C.P.; Dong, X.G.; Qin, W.; Fang, Z.F. Vision based control of micro-fluid dispensing for surface mount technology. *Adv. Mater. Res.* **2012**, *403*, 4380–4387. [CrossRef]
3. Shi, Y.L.; Fu-Dong, L.; Yang, X.; Zhang, Z.T.; Xu, D. pL class adhesive dispensing approach for micro bonding. *Opt. Precis. Eng.* **2012**, *20*, 2744–2750. [CrossRef]
4. Chen, X.B.; Kai, J. Modeling of positive-displacement fluid dispensing processes. *IEEE Trans. Electron. Packag. Manuf.* **2004**, *27*, 157–163. [CrossRef]
5. Zhang, Q.; Ye, S.L.; Luo, K.; Xu, C. Development and application on hole-sealing of pL ultra-micro dispenser. *Opt. Precis. Eng.* **2020**, *28*, 1132–1139.
6. Gu, S.D.; Yang, Z.G.; Jiang, H.; Liu, J.F.; Lu, S.; Liu, Q. Piezoelectric driven hydraulic amplification jetting system. *Opt. Precis. Eng.* **2015**, *23*, 1627–1634.
7. Fan, Z.H.; Rong, W.B.; Wang, L.F.; Sun, L.N. Control and experiment of micro-dispenser by piezoelectric drive. *Opt. Precis. Eng.* **2016**, *24*, 1042–1049.
8. He, X.; Elmer, J.W.; Debroy, T. Heat transfer and fluid flow in laser microwelding. *J. Appl. Phys.* **2005**, *97*, 084909. [CrossRef]
9. Zhu, X.; Liu, G.; Guo, Y.; Tian, Y. Study of PMMA thermal bonding. *Microsyst. Technol.* **2007**, *13*, 403–407. [CrossRef]
10. Cozma, A.; Puers, B. Characterization of the electrostatic bonding of silicon and pyrex glass. *J. Micromech. Microeng.* **1995**, *5*, 98–102. [CrossRef]
11. Yang, L.Y.; King, C.K.; Bernstein, J.B. Liquid dispensing encapsulation in semiconductor packaging. *Microelectron. Int.* **2003**, *20*, 29–35. [CrossRef]
12. Heidarpour, F.; Farahani, M.; Ghabezi, P. Experimental investigation of the effects of adhesive defects on the single lap joint strength. *Int. J. Adhes. Adhes.* **2018**, *80*, 128–132. [CrossRef]
13. Nguyen, Q.H.; Choi, M.K.; Choi, S.B. A new type of piezostack-driven jetting dispenser for semiconductor electronic packaging: Modeling and control. *Smart Mater. Struct.* **2008**, *17*, 015033. [CrossRef]
14. Chen, X.B.; Zhang, W.J.; Schoenau, G. Off-line control of time-pressure dispensing processes for electronics packaging. *IEEE Trans. Electron. Packag. Manuf.* **2003**, *26*, 286–293. [CrossRef]
15. Liao, X.; Braunschweig, A.B.; Zheng, Z.; Mirkin, C.A. Force- and time-dependent feature size and shape control in molecular printing via polymer-pen lithography. *Small* **2010**, *6*, 1082–1086. [CrossRef] [PubMed]
16. Yakhnin, E.D.; Chadov, A.V. Investigation of the transfer of a liquid from one solid surface to another. 2. Dynamic transfer. *Colloid J. USSR* **1983**, *45*, 1034–1039.
17. Cai, S.B.; Bhushan, B. Meniscus and viscous forces during normal separation of liquid-mediated contacts. *Nanotechnology* **2007**, *18*, 465704. [CrossRef]
18. Cai, S.B.; Bhushan, B. Meniscus and viscous forces during separation of hydrophilic and hydrophobic surfaces with liquid-mediated contacts. *Mater. Sci. Eng.* **2008**, *61*, 78–106. [CrossRef]
19. Zhang, X.; Padgett, R.S.; Basaran, O.A. Nonlinear deformation and breakup of stretching liquid bridges. *J. Fluid Mech.* **1996**, *329*, 207–245. [CrossRef]
20. Dodds, S.; Silveira, C.M.; Kumar, S. Stretching and slipping of liquid bridges near plates and cavities. *Phys. Fluids* **2009**, *21*, 092103.
21. Bai, S.-E.; Shim, J.-S.; Lee, C.-H.; Bai, C.-H.; Shin, K.-Y. Dynamic effect of surface contact angle on liquid transfer in a low speed printing process. *Jpn. J. Appl. Phys.* **2014**, *53*, 05HC05. [CrossRef]
22. Ahmed, D.H.; Sung, H.J.; Kim, D.S. Simulation of non-Newtonian ink transfer between two separating plates for gravure-offset printing. *Int. J. Heat Fluid Flow* **2011**, *32*, 298–307. [CrossRef]
23. Dodds, S.; Carvalho, M.; Kumar, S. Stretching liquid bridges with moving contact lines: The role of inertia. *Phys. Fluids* **2011**, *23*, 092101. [CrossRef]
24. Huynh, H.S.; Guan, J.P.; Vuong, T.; Ng, T.W. Comparisons of liquid and gaseous microdrops deposited on surfaces via a retreating tip. *Langmuir* **2013**, *29*, 11615–11622. [CrossRef] [PubMed]
25. Thompson, P.A.; Robbins, M.O. Simulations of contact-line motion: Slip and dynamic contact angle. *Phys. Rev. Lett.* **1989**, *63*, 766–769. [CrossRef] [PubMed]
26. Saha, A.A.; Mitra, S.K. Effect of dynamic contact angle in a volume of fluid (VOF) model for a microfluidic capillary flow. *J. Colloid Interface Sci.* **2009**, *339*, 461–480. [CrossRef]
27. Zhu, P.; Xu, Z.; Xu, X.; Wang, D.; Wang, X.; Yan, Y.; Wang, L. Squeezing dynamic mechanism of high-viscosity droplet and its application for adhesive dispensing in sub-nanoliter resolution. *Micromachines* **2019**, *10*, 728. [CrossRef]
28. Eugene, C.; Metin, S. Adhesion of biologically inspired oil-coated polymer micropillars. *J. Adhes. Sci. Technol.* **2008**, *22*, 569–589.

29. Zhu, P.; Xu, Z.; Wang, X.; Zheng, Y.; Xu, X.; Wang, L. Influence of initial distance between needle tip and substrate on contact dispensing of high-viscosity adhesive. *Int. J. Adhes. Adhes.* **2018**, *85*, 23–28. [CrossRef]
30. Chen, H.; Tang, T.; Amirfazli, A. Effects of surface wettability on fast liquid transfer. *Phys. Fluid* **2015**, *27*, 112102. [CrossRef]

micromachines

Article

A Microfluidic Approach for Probing Heterogeneity in Cytotoxic T-Cells by Cell Pairing in Hydrogel Droplets

Bart M. Tiemeijer [1,2], Lucie Descamps [1,2], Jesse Hulleman [1,2], Jelle J. F. Sleeboom [3] and Jurjen Tel [1,2,*]

1 Laboratory of Immunoengineering, Department Biomedical Engineering, Eindhoven University of Technology, 5600 MB Eindhoven, The Netherlands
2 Institute for Complex Molecular Systems, Eindhoven University of Technology, 5600 MB Eindhoven, The Netherlands
3 Microsystems, Department of Mechanical Engineering, Eindhoven University of Technology, 5600 MB Eindhoven, The Netherlands
* Correspondence: j.tel@tue.nl

Abstract: Cytotoxic T-cells (CTLs) exhibit strong effector functions to leverage antigen-specific anti-tumoral and anti-viral immunity. When naïve CTLs are activated by antigen-presenting cells (APCs) they display various levels of functional heterogeneity. To investigate this, we developed a single-cell droplet microfluidics platform that allows for deciphering single CTL activation profiles by multi-parameter analysis. We identified and correlated functional heterogeneity based on secretion profiles of IFNγ, TNFα, IL-2, and CD69 and CD25 surface marker expression levels. Furthermore, we strengthened our approach by incorporating low-melting agarose to encapsulate pairs of single CTLs and artificial APCs in hydrogel droplets, thereby preserving spatial information over cell pairs. This approach provides a robust tool for high-throughput and single-cell analysis of CTLs compatible with flow cytometry for subsequent analysis and sorting. The ability to score CTL quality, combined with various potential downstream analyses, could pave the way for the selection of potent CTLs for cell-based therapeutic strategies.

Keywords: single-cell; CD8 T-cell; cytokines; droplet microfluidics; microgel; heterogeneity

Citation: Tiemeijer, B.M.; Descamps, L.; Hulleman, J.; Sleeboom, J.J.F.; Tel, J. A Microfluidic Approach for Probing Heterogeneity in Cytotoxic T-Cells by Cell Pairing in Hydrogel Droplets. *Micromachines* **2022**, *13*, 1910. https://doi.org/10.3390/mi13111910

Academic Editor: Pingan Zhu

Received: 1 October 2022
Accepted: 3 November 2022
Published: 4 November 2022

1. Introduction

Cytotoxic T-cells (CTLs) are specialized cells that recognize and kill malignant or infected cells, making them the commander-in-chief and the soldiers of the immunological army. The efficiency of these CTLs to find and successfully kill their targets is dependent on the well-coordinated activation of naïve CTLs in the lymph nodes. Only upon encountering antigen-presenting cells (APCs) in the lymph node will they turn into effector CTLs, after which they will return to circulation and actively seek and destroy target cells at sites of inflammation in the tissue. Importantly, a few studies revealed that disease control depends on CTL quality rather than on quantity, where quality was defined by multiple functions [1–3]. In line with that, several sub-populations of CTLs were identified based on secreted cytokines and the presence of polarizing cytokines [4,5], suggesting that specific subsets of CTLs are relevant for immunity. Since interrogation of each individual cell is needed to truly identify CTLs of interest, the field has recently moved from conventional bulk experiments to single-cell approaches. Over the past two decades, single-cell technologies have emerged with the goal to dissect cellular heterogeneity and interrogate relevant sub-populations [6–10]. To efficiently and reproducibly probe the heterogeneity upon CTL activation, microfluidic tools were developed to allow precise manipulation and compartmentalization of single cells in small volumes (pico- to nano-liter) [11], thereby providing highly controlled environments acting like bioreactors to efficiently activate CTLs [12,13]. Although these efforts to pair CTLs either with target cells or APCs in microwell- and microtrap-based devices have proven to be effective, they are often restricted by a limited throughput since pairs are captured on-chip and the sample size is thus restricted by the dimensions of the device [12,14]. Droplet-based microfluidics has the major advantage that the sample size can be increased by longer running times thereby yielding higher throughputs [15–18], which is especially important when one is studying heterogeneity in immune cell responses and rare cell behavior is expected. Previous work aimed at pairing and studying CTL responses in droplets yielded interesting findings at a single-cell resolution [19,20]. However, these studies were often limited to the measurement of only a few parameters through microscopy imaging. As various factors, including both membrane marker expression and cytokine secretion, were previously used to score CTL quality, combining these will allow for true interrogation of the heterogenous CTL response [1–3]. To overcome both the challenge of limited throughput as well as allowing measurement in a multiparameter fashion, hydrogels can be incorporated during droplet production to obtain "microgels", in which cells can be paired. These will maintain the spatial coupling of single-cell pairs while allowing subsequent cell profiling using flow cytometry, facilitating a multiparameter measurement [21]. Hence, microgels could pave the way for novel and robust analytic tools to study single CTL and APC interactions [12,22,23].

Here, we generated a single-cell droplet microfluidics platform to probe CTL heterogeneity upon activation using soluble stimuli and artificial antigen-presenting cells (aAPCs). We incorporated ultra-low melting point agarose hydrogels to create microgels containing CTL/aAPC pairs for downstream analysis by flow cytometry. Activation of CTLs was measured in a multidimensional fashion, screening both expression of membrane markers as well as multiplexed cytokine secretion. We observed distinct CTL activation profiles induced by soluble stimuli compared to aAPCs. This approach allows for scoring the quality of CTL activation combined with various potential downstream applications after sorting, and can therefore greatly benefit future immune cell therapeutic applications.

2. Materials and Methods

2.1. Cell Isolation and Preparation

CD8+ T-cells were isolated from buffy coats obtained from healthy human donors (Sanquin bloodbank, Eindhoven, The Netherlands) after written informed consent per the Declaration of Helsinki and according to the institutional guidelines. Peripheral blood mononuclear cells were isolated using Lymphoprep (Stemcell Technologies, Vancouver, Canada) according to manufacturer's protocol, after which CD8+ cells were isolated using magnetic-activated cell sorting kit (Miltenyi Biotech, Bergisch Gladbach, Germany). Isolated CD8+ T-cells were resuspended in RPMI medium (Gibco, Life Technologies, Carlsbad, CA, USA) with 2% Human Serum (Sanquin Bloodbank) and 1% Penicillin–Streptomycin (Gibco, Life Technologies), hereafter referred to as culture medium. The cells were then coated with capture antibodies for IL-2, TNF-α, and IFN-γ (Miltenyi Biotech). After incubation and washing, the cells were resuspended in culture media for droplet encapsulation.

2.2. Microfluidic Device Fabrication

Droplet microfluidics devices were produced using soft lithography. Photomasks were ordered from CAD/Art Services, Inc. (Bandon, OR, USA). PDMS molds were produced by spin-coating wafers with SU-8 3000 photoresist (Microresist Technology, Berlin, Germany) according to manufacturer's protocol to obtain 30 µm of channel height. PDMS devices were fabricated by mixing SYLGARD® 184 PDMS with SYLGARD® 184 curing agent (both from Merck) at 10:1 *w*/*w* before pouring the mixture onto the PDMS molds and curing for 2 h at 65 °C. Using a 1 mm biopsy puncher, holes for the inlets and outlet were punched. The obtained PDMS devices were bonded to glass slides using a plasma asher (Emitech, K1050X, Montigny-le-Bretonneux, France). After bonding, the channels were treated with 5% perfluorooctyltriethoxysilane in HFE-7500 fluorinated oil (both from Fluorochem, Hadfield, United Kingdom) in order to make channel walls hydrofobic, incubated for 1 h at 65 °C, flushed again with HFE-7500, and incubated overnight at 65 °C for thermal bonding.

2.3. Production of Temperature Regulation Device

The designs for all heating devices were made in Siemens NX (Siemens AG, Munich, Germany) (designs are available as Electronic Supplementary Material). All devices were printed in clear resin (RS-F2-GPCL-04, Formlabs, Somerville, MA, USA) using a Formlabs Form 3 SLA printer. After printing, the uncured resin was removed from the channels by flushing them several times with clean isopropanol from a 20 mL syringe that was directly connected to the Luer-lock connections. The prints were then washed and cured per the manufacturer's instructions in a Form Wash (FH-WA-01, Formlabs) and Form Cure (FH-CU-01, Formlabs) station. The chip platform was finished by bonding a 0.5 mm PMMA plate to the bottom using super glue (Loctite, Düsseldorf, Germany) and a glass microscope slide to the top using Dowsil™732 silicon glue (Dow Corning, Midland, MI, USA). Luer-lock to barb connectors (Cole-Parmer, Vernon Hills, IL, USA) were used to connect the tubing. The three heating devices were connected to a water pump (7026898, RS PRO) in series and warm water was flushed through to maintain a temperature of 37 °C.

2.4. Droplet Production, Cell Encapsulation and CTL Stimulation

Droplet production was performed using a previously reported pipette tip method [24], and by attaching the droplet device to a neMESYS microfluidic pump (Cetoni, Korbußen, Germany). The first inlet was used for 2.5% Picosurf (Spherefluidics, Cambridge, United Kingdom) in HFE-7500 (30 µL/min), the second for CD8+ T-cells (5 µL/min), and the last for stimuli, or aAPC Dynabeads (Thermofisher) (5 µL/min). Cells were injected at a concentration of 4×10^6 cells/mL. When Dynabeads were included these were injected at a concentration of 1×10^7 particles/mL. Soluble stimuli were added at a concentration of 1 mg/mL for PMA and 10 mg/mL for ionomycin (both from Peprotech). When producing aqueous droplets cells, Dynabeads and stimuli were suspended in culture media, when producing microgels they were suspended in culture media containing 1% *w/v* of ultra-low melting point agarose (Merck, Kenilworth, NJ, USA). Droplets were collected in Eppendorf tubes and incubated at 37 °C and 5% CO_2 for 24 h.

2.5. Droplet Characterization

To determine the distribution of cells among droplets, contents were manually counted in brightfield microscopy-obtained images. Droplet size was determined using ImageJ software [25], where automated thresholding was used to create a greyscale image of the brightfield images after which the particle analysis function was used to automatically measure droplets. The diameter was calculated as the average between the major and minor axes of detected particles.

2.6. Cell/Microgel Retrieval and Flow Cytometric Measurement

After 24 h of incubation, droplets containing agarose were cooled to 4 °C for 30 min to obtain microgels. Both cells and microgels were retrieved from the emulsion by adding 20% of 1H,1H,2H,2H-perfluoro-1-octanol (PFO) in HFE-7500 onto the emulsion at a 1:1 *v/v* ratio. The obtained solution was washed and afterward stained using Zombie NIR viability kit (Biolegend). Next, they were stained with a cocktail of cytokine-detection antibodies for IFNγ, TNFα, and IL-2 (all from Miltenyi Biotech), along with antibodies to detect surface marker expression; CD8-Brilliant violet 605, CD69-Brilliant violet 650, and CD25-Brilliant violet 786 (all from Biolegend), according to manufacturer's protocols. During the staining of microgels, the incubation time was doubled to give antibodies more time to diffuse into the agarose. When washing microgels, the washing solution was kept on the microgels for 5 min before spinning down. Centrifugation of microgels was performed at 100 RCF for 10 min. After the staining procedures, fluorescent values of both cells and microgels were measured using FACSymphony (BD) and data were analyzed using Flowjo Software 10.7.0 (FLowJo LLC, Ashland, OR, USA).

2.7. Statistical Analysis

Data processing and statistical analysis were performed using PRISM 9 (Graphpad software). Data are shown as mean ± standard error of the mean (SEM) unless indicated differently. Statistical analysis was performed using repeated-measures one-way ANOVA with post-hoc Tukey's test after normality was proven using Shapiro–Wilk test. $p < 0.05$ was considered significant.

3. Results

3.1. Single CTL Activation in Droplets Reveals Highly Heterogeneous Responses

Stimulation of single cells with soluble stimuli has proven a great way to reveal heterogeneous behavior using droplet-based microfluidic platforms [26–28]. Droplets provide controlled environments containing cells along with stimuli and assay reagents, thereby preventing cells from influencing each other in a juxta- or paracrine fashion. We investigated primary human CTLs and encapsulated them in 70 picolitre-sized droplets along with Phorbol 12-myristate 13-acetate (PMA) and ionomycin as stimuli [13,29] (Figure 1A). Microscopy analysis showed that the cell encapsulation followed the predicted Poisson distribution (Supplementary Figure S1A) [30] and ensured that virtually all CTLs were encapsulated as single cells in monodisperse droplets. After single-cell culture, CTLs were retrieved from droplets by PFO-induced de-emulsification. Subsequently, CTL activation was assessed by the early activation markers CD69 and CD25. CD69 is an early inducible cell surface glycoprotein acquired during activation and functions as a signal-transmitting receptor [31]. CD25 is the alpha chain of the trimeric IL-2 receptor and is considered to be a prominent early-to-middle cellular activation marker [32]. Simultaneously, we investigated the secretion of the cytokines TNFα, IFNγ, and IL-2, captured via membrane-bound constructs. We observed three distinct phenotypes based on the expression of activation markers CD25 and CD69 (Figure 1C,D). Within these populations, a high degree of heterogeneity was found with respect to the secreted cytokines (Figure 1E). These encompassed most combinations possible, confirming the heterogeneous nature of the CTL activation. Interestingly, membrane marker upregulation appeared reversely correlated with cytokine secretion as the CD69-CD25- population appeared to have the lowest percentage of non-producing cells. Furthermore, the CD69-CD25- population showed an increased tendency to produce a combination of IFNγ and TNFα when compared to the other two populations. These results indicate the analytical potency of our droplet platform. CTL activation via PMA/Ionomycin is often used in the literature; however, it surpasses the natural way of CTL activation by TCR engagement and as such does not mimic a physiological setting. Thus, an activation model more resembling immune activation via direct cell contact would translate much better to any system of therapeutic value.

3.2. Spatial Pairing Data Are Lost in Aqueous Droplets upon CTL and aAPC Interactions

In order to achieve effective CTL activation, APCs need to engage with naïve CTLs in the lymph nodes. Activation involves three main signals [33,34]. The first being the TCR/CD3 complex binding to the peptide antigen presented by the major histocompatibility complex of APCs. Co-stimulatory molecules provide the second signal [35,36] by binding to the CD28 receptors located in close proximity to the TCR. Together, these signals lead to the release of cytokines, which further shape the development of the immune response [37]. To mimic the first two signals of this cell–cell contact-dependent activation in our platform, we aimed to co-encapsulate CTLs with aAPCS (Figure 2A,B and Supplementary Video S1), which contain anti-CD3 and anti-CD28 antibodies and are routinely used to activate T-cells both in bulk as well as at single-cell level [38,39]. By switching from soluble stimuli to aAPCs, the encapsulation efficiency is altered but the prevalence of all combinations of encapsulation still followed the predicted Poisson distribution (Figure 2B and Supplementary Figure S2). After in-droplet co-culture of single-cell pairs and verification of CTL/aAPC interaction based on cell morphology (Supplementary Video S2), we assessed CTL activation and observed heterogeneous expression of CD69 and CD25

(Figure 2C), and very little cytokine secretion in only a small percentage of CTLs (Figure 2D). Unlike in the PMA/ionomycin stimulation, here, we cannot ensure that individual cells are co-encapsulated with aAPCs. According to the Poisson distribution, the fraction of encapsulated CTLs with one or multiple aAPCs can be estimated at around 50% but only around 30% showed an increase in membrane marker expression. Therefore, the preservation of spatial coupling of a CTL and aAPC would be an elegant approach to maintain such information until analysis of activation.

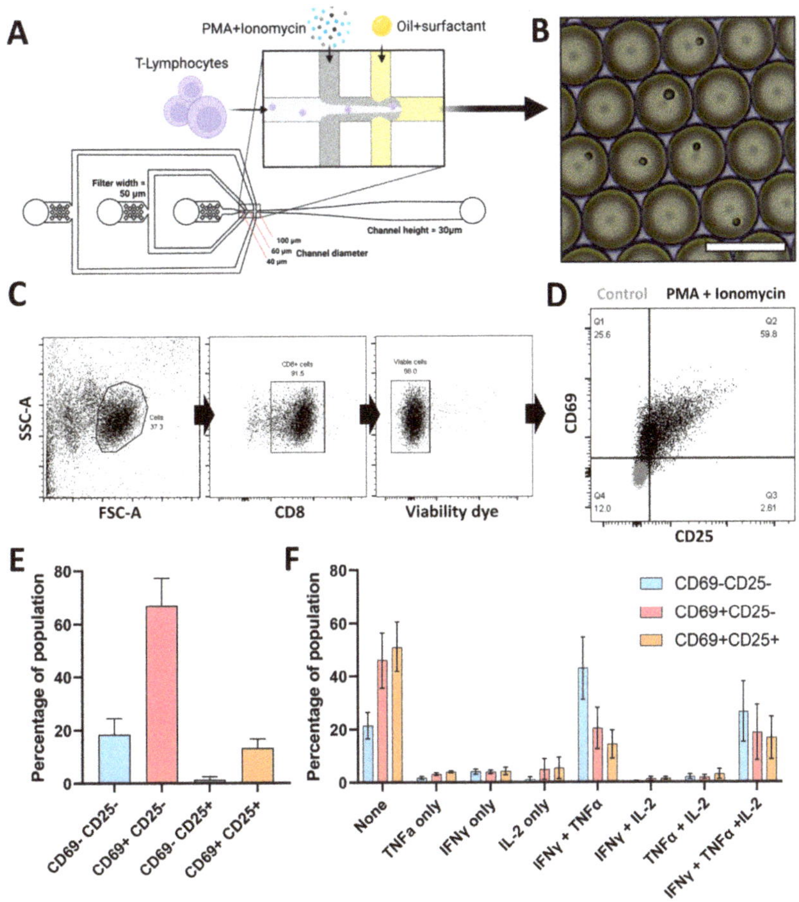

Figure 1. Single-cell CTL activation using soluble stimuli. (**A**) Layout of the microfluidic device used for droplet production including popped-out schematic of droplet formation, scale bar represents 500 μm. Layout is a direct copy of the photolithography mask design, and channel height is equal throughout. (**B**) Brightfield microscopy image of aqueous droplets containing single CTLs. Scale bar = 100 μm. (**C**) Gating strategy for flow cytometry data to select viable CD8-positive cells. (**D**) Marker expression of selected viable CD8-positive cells for CD69 and CD25 activation markers, either after stimulation with PMA + Ionomycin (black dots) or unstimulated (grey dots). Data display one representative donor. (**E**) Prevalence of populations of PMA + Ionomycin activated CTLs based on CD25 and CD69 expression. Data represent SEM of $n = 4$ biological replicates. (**F**) Bar graph displaying the frequency of different profiles of secretion as exhibited by the three prevalent populations of PMA + Ionomycin stimulated CTLs. Data represent average values of $n = 3$ biological replicates.

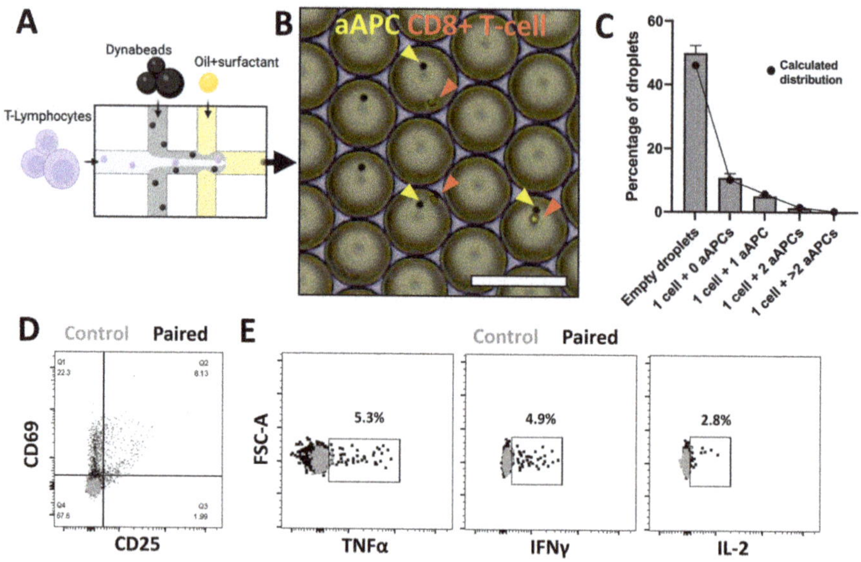

Figure 2. Single-cell CTL activation using aAPCs. (**A**) Schematic of droplet formation when CTLs are paired with Dynabeads. (**B**) Brightfield microscopy image of aqueous droplets containing CTLs (red) and aAPCs (yellow) in pairs. Scale bar = 100 µm. (**C**) Counted (grey bars) and calculated (black dots) Poisson distribution of droplet contents when co-encapsulating CTL and aAPCs. Error bars represent average ± SEM of 4 independent experiments in which at least 1000 droplets were counted. (**D**) Marker expression of CD69 and CD25 activation markers on CTLs retrieved from droplet co-culture with aAPC. (**E**) Cytokine secretion of CTLs retrieved from droplet co-culture with aAPCs (black dots) as compared to unstimulated control (grey dots).

3.3. Agarose Microgels as Bioreactors for Subsequent Flow Cytometry Measurement

To achieve spatial coupling of CTLs with aAPCs, hydrogel-based microgels were desired with properties that remain soluble during droplet production and culture but have a trigger-cross-linked ability. Ultra-low melting-point agarose proved an ideal candidate as it remains soluble under culture conditions at 37 °C and crosslinks below 18 °C prior to downstream analysis [40]. Advantages are pore size >~200 nm warranting diffusion of detection antibodies, and biocompatibility [27,41]. By using 3D-printed devices (Figure 3B), we ensured the encapsulation of CTLs in agarose solutions under temperature-controlled conditions to avoid clogging of the device by premature gelation, resulting in monodisperse droplet formation (Figure 3C). After cross-linking and de-emulsification, monodisperse agarose microgels containing cells were retrieved in PBS free of oil (Figure 3C). By comparing flow cytometric measurement of cells in microgels and empty microgels with respect to unencapsulated cells, the microgels appeared to primarily increase the scattering (SSC-A) of measured events (Supplementary Figure S3). Microgels containing CTLs were selected by intra-microgel staining for CD8 membrane protein (Figure 3D). We investigated whether the pore size of agarose hydrogels allowed diffusion of fluorophore-conjugated antibodies to ensure that CTL activation in microgels could be detected by our antibody panel. CTLs were therefore activated in bulk using PMA and ionomycin, partially stained in microgels, and partially stained in bulk. Altered expression, compared to unstimulated controls, indicated that intra-microgel and cell-specific staining of CTLs was achieved (Figure 3E). Additionally, microscopic images showed that staining was cell-specific and did not differ between encapsulated and unencapsulated cells (Supplementary Figure S4). Even for IL-2 detection, which uses the largest fluorophore–antibody conjugate [42], only a small difference could be observed in fluorescent intensity. Thus, microgel-encapsulated cells are compatible with fluorescent staining for flow cytometric analysis. Future additions of fluorophores can be reliably incorporated to extend the antibody panel for multidimensional analysis.

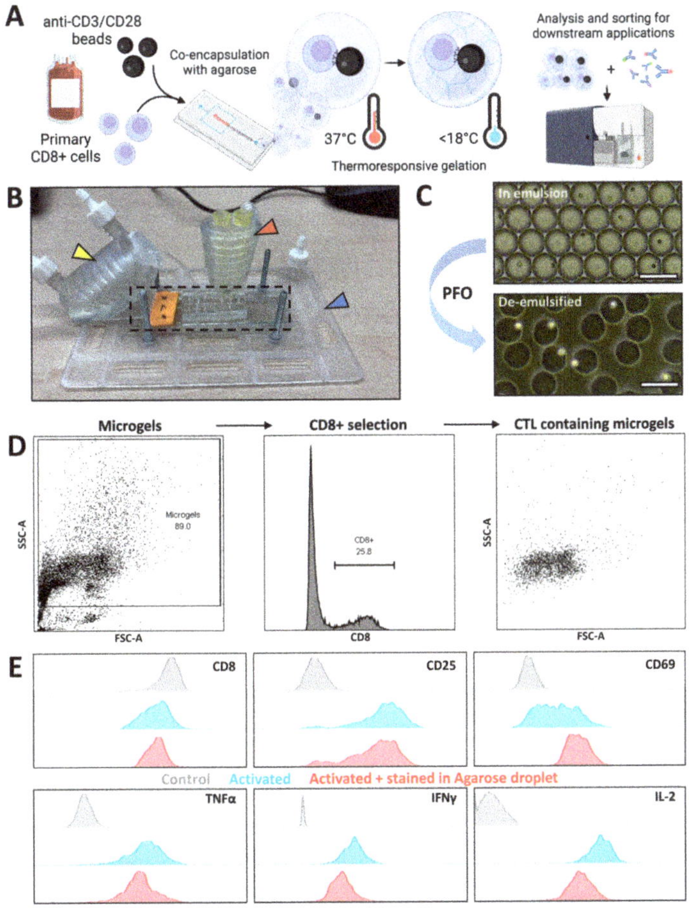

Figure 3. Encapsulation and flow cytometric measurement of CTL-containing agarose microgels. (**A**) Schematic of the workflow; cells are isolated and co-encapsulated with aAPCs in droplets along with agarose solution, droplets containing cells are cultured at 37 °C to allow interaction after which temperature is decreased below 18 °C to allow agarose to crosslink and create microgels containing cell pairs, microgels are retrieved and can be stained for membrane markers and measured or sorted using flow cytometry. (**B**) 3D printed temperature control devices. Device used to control temperature in pipette tips (red arrow) for injection of samples, microscope insert (blue arrow) containing microfluidic device (black dashed line), and heated holder for collection tube (yellow arrow). Hot water is flown through all devices in series to maintain 37 °C during droplet formation and after collection. (**C**) Brightfield microscopy images of water-in-oil emulsion before crosslinking and of de-emulsified microgels in PBS. Scale bars = 100 μm. (**D**) Gating strategy to retrieve cell-containing droplets; Microgels are gated based on the FSC/SSC of empty droplets, CD8+ events are selected. (**E**) Histograms depicting the effect of antibody staining in agarose hydrogels compared to staining in conventional cell solution. Comparing unstimulated control stained in solution (grey), PMA/Ionomycin bulk stimulated cells stained in solution (blue), PMA/Ionomycin bulk stimulated cells stained in microgels (red).

3.4. Phenotypic and Functional Analysis of CTL/aAPC Pairs in Microgels

Having established that detection of activation markers in microgels is possible, we moved forward to induce cell contact-mediated CTL activation. First, we monitored the location of cells and aAPCs to ensure that cell–bead interactions are not affected by agarose viscosity and still take place such as in aqueous droplets [43,44]. Brightfield images were captured during experiments directly after droplet production and again after 24 h of culture followed by agarose crosslinking (Figure 4A). Analysis of 532 droplets containing CTL/aAPC pairs demonstrated that single CTLs effectively interacted with aAPCs (Figure 4B), excluding that agarose viscosity affected cellular interactions. Additional temporal monitoring over the first 30 min after encapsulation was performed as well and displayed cells latching onto aAPCs within minutes, demonstrating that cell–cell interaction takes place, even in agarose droplets (Supplementary Video S3). Next, we selected all microgels which contained CTLs based on CD8 expression (Figure 4C) and subsequently checked pairing with aAPCs by autofluorescent properties of aAPCs. The position of pairs containing a single aAPC was verified based on overlaying the autofluorescence of a single aAPC (Supplementary Figure S5). The gated populations were compared and indeed indicated that 1:1 ratio single-cell contact-mediated activation occurred in a percentage of cells according to membrane markers (Figure 4D) as well as cytokines (Figure 4E). Furthermore, the observed percentages of activation approximated the percentages observed under aqueous conditions (Figure 2C,D). Especially when taking into account that in those experiments, ~50% of all measured cells actually encountered an aAPC. This demonstrated that agarose encapsulation does not affect mechanisms of activation and successfully serves its purpose of maintaining spatial information. Moreover, we could clearly distinguish microgels where CTLs were paired with one aAPC or multiple aAPCs and in this way probed the effect of multiple interactions. We demonstrated that multiple interactions resulted in an increased number of activated CTLs, primarily based on CD69 and CD25 marker expression (Figure 4E). Besides pairing of CTLs with beads, the platform can easily be adjusted to pair CTLs with another cell type. This would merely require the incorporation of an additional membrane marker staining, cell pairs can then be selected by gating for double-positive events (Supplementary Figure S6). Taken together, these results demonstrate that agarose encapsulation is a highly potent approach to investigating single-cell contact-mediated activation of CTLs in a high-throughput and multiparameter fashion.

3.5. Single-Cell Decoding of CTL Activation and Secretion Based on Different Stimuli

We investigated CTL activation using PMA/Ionomycin and aAPC-interaction as stimulation models at the single-cell level. This allowed us to establish a novel depth of comparing single-cell heterogeneity, based on both cell phenotype and function. Both models showed similar patterns in CD69 and CD25 expression, with differences in frequencies (Figures 1E and 4D). The most prevalent secreted cytokines within these populations were; (1) IFNγ in combination with TNFα; (2) IFNγ, TNFα and IL-2 in PMA/Ionomycin-stimulated CTLs (Figure 5A); (3) IFNγ in combination with TNFα and (4) TNFα only in aAPC-stimulated CTLs (Figure 5B). However, the correlation between membrane markers and cytokine production was vastly different between the two activation models. In PMA/Ionomycin-stimulated CTLs, membrane marker expression and cytokine production appeared to be negatively correlated since the double-negative phenotype (CD69-CD25-) showed the least amount of non-producing cells. On the contrary, membrane marker expression and cytokine production appear to be positively correlated in aAPC-stimulated CTLs, with the least amount of non-producing cells observed in the double positive phenotype (CD69+CD25+). Additionally, IL-2 secretion appears much more prevalent in the PMA/Ionomycin stimulated conditions, primarily when all three cytokines are secreted. These findings underline the difference between the two models, where PMA/ionomycin is a non-specific synthetic approach to obtain optimal cytokine secretion, and aAPCs are used to mimic biologically relevant mechanisms. Nevertheless, to the best of our knowledge, this difference has not been previously observed at this degree of single-cell resolution, which underlines the strengths of the demonstrated platform.

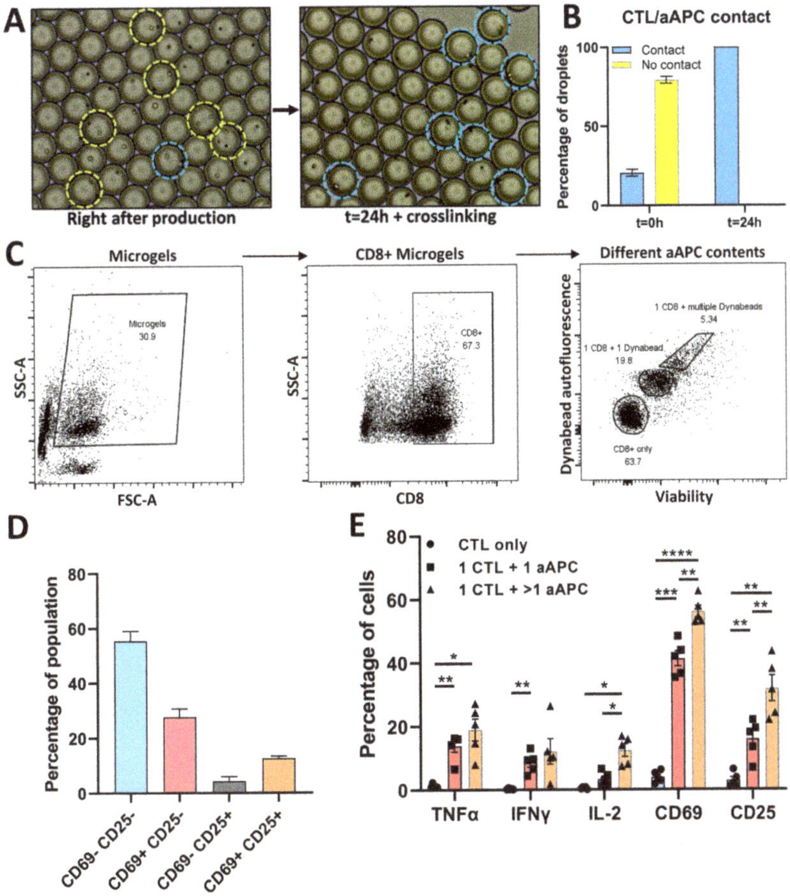

Figure 4. Encapsulation and flow cytometric measurement of CTL/aAPC pairs in microgels. (**A**) Brightfield microscopy images of droplets for CTL/aAPC pairing; Left image displays droplets directly after production at 37 °C where cells and particles are free floating in agarose solution droplets, right image displays droplets after 24 h of incubation at 37 °C and subsequent gelation at >18 °C where cells are fixed in crosslinked agarose. Yellow circles indicate CTL/aAPC combinations where no contact is observed, blue circles indicate combinations where contact is observed. (**B**) Results of manual quantification of $n = 46$ images containing $n = 532$ CTL/aAPC paired droplets from $n = 3$ independent experiments. error bars represent $n = 3$ experiments. (**C**) Gating strategy of CTL/aAPC droplets; all non-empty microgels are selected, all microgels containing at least a CD8+ cell are gated, based on Dynabead autofluorescence and viability dye intensity gates are drawn for single viable CTLs, 1:1 pairs of CTL/aAPCs, and microgels containing a CTL and multiple aAPCs. (**D**) Prevalence of populations of 1:1 aAPC activated CTLs based on CD25 and CD69 expression. Data represent SEM of $n = 5$ biological replicates. (**E**) Data summary of activation markers in microgels containing only CTLs (blue), microgels containing 1 CTL and 1 aAPC (red), and microgels containing 1 CTL with multiple aAPCs (orange). Data show $n = 5$ biological replicates with SEM. * $p < 0.05$, ** $p < 0.01$, *** $p < 0.001$, **** $p < 0.0001$, if no significance is indicated none was found.

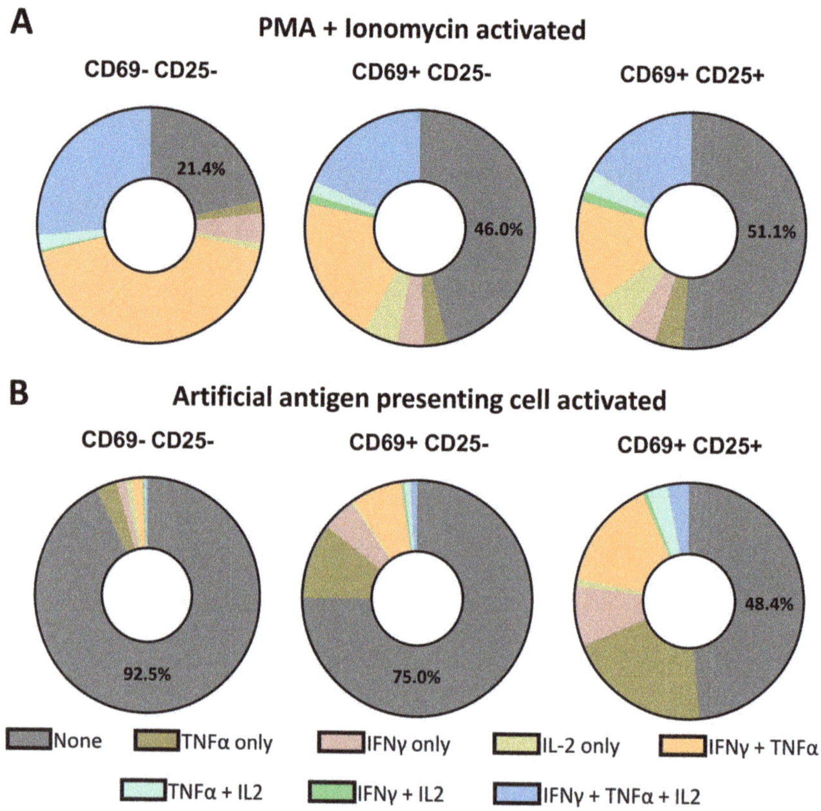

Figure 5. Comparing heterogeneous cell responses between PMA/Ionomycin and aAPC activation. (**A**) Pie graphs displaying the different profiles of secretion after PMA/Ionomycin stimulation of CTLs, $n = 3$ donors. (**B**) Pie graphs displaying the different profiles of secretion after aAPC stimulation of CTLs, $n = 5$ donors.

4. Discussion

We developed a droplet microfluidic-based single-cell activation platform for the screening of cytotoxic T-cell activation. In its standard form, the platform allowed us to activate CTLs at the single-cell level with synthetic soluble stimuli and detect heterogeneity based on multiparameter measurement of phenotype and function. In its more advanced form, we co-encapsulated CTLs and aAPCs in agarose microgels to study single-cell contact-mediated activation between single cells. In particular, microgel encapsulation allowed us to highlight the difference in CTL activation mechanisms upon stimulation with PMA/Ionomycin and aAPCs, whilst obtaining novel resolution over the resulting heterogeneous responses. Therefore, this platform can be used to dissect CTL multifunctionality based on the profiling of secreted cytokines, which is an important facet of strong adaptive immunity.

In comparison to previous research on droplets for cell pairing [20,44–48], our approach warrants unlimited throughput and is not limited by droplet capture in traps, wells, or observation chambers. Additionally, such methods often rely on monitoring via microscopy. Although this allows for temporal resolution, it limits the number of read-out parameters [49,50]. In applications where multiparameter functional measurements are possible, the ability to recover interrogated cells for downstream applications is often limited [51]. In our approach, by sacrificing temporal resolution, we are able to measure cell pairs using flow cytometry, which opens up the possibility for multiparameter (potentially up to 45 markers) measurement as well as sorting for downstream applications, which was previously reported with agarose microgels [52–56]. Combining single-cell analysis with fluorescence-activated droplet sorting was previously performed on-chip, where droplets are sorted either before [48,57] or after culture [19,47]. These works have great potential but are also limited by the number of measured parameters and the closed system they are performed in. For example, Gerard et al. demonstrated very potent sorting of specific IgG-secreting cells, but their sort is based on only a single parameter and adapting it for a different application will change the entire system [47]. Conversely, recovering cell pairs in microgels offers more flexibility towards fluorescent staining after culture, as well as the use of well-developed commercially available flow cytometers. Yanakieva et al. demonstrated this principle by sorting pairs of secretor cells and reporter cells in microgels in order to enrich yeast clones secreting biorelevant proteins [52]. However, such applications have previously only been demonstrated with yeast, bacteria or cell lines. Here, our microgel-based platform aims at monitoring the effect of physical cell–cell contact in primary immune cell activation. We achieved a higher number of screened parameters allowing for the study of both immune cell phenotype and functionality.

Nevertheless, a potential hurdle after the recovery of microgels is retrieving cells from encapsulation. For agarose microgels, this would require heating to above 70 degrees [40], or enzymatic digestion [58]. Potential candidates to avoid such harmful processes could be alginate [59–61] or thermo-reversible hydrogels [26,62]. In future adaptations, these could be utilized to extend our platform to cell retrieval. Furthermore, we demonstrated the compatibility of agarose microgels with downstream phenotypic and functional studies, but this approach could be even extended to genetics since agarose microgels have readily been shown to be suitable for PCR and sequencing purposes [63–65]. For example, single-cell sequencing could be performed to enable the study of TCR sequences which were shown to have high relevance and potential to design better vaccines or autoimmune therapies [66,67]. Reversely, the platform can be utilized to study the heterogeneity of APCs [10] as the panel of markers that is measured using flow cytometry can be readily switched to include different targets of interest. The potential of our platform is highlighted by the interesting observed differences in CTL activation approaches. This interesting difference might be explained by the different mechanisms exploited by the two activation models. On one hand, aAPCs target CD3 and CD28 receptors on the CTLs, partially mimicking how naïve cells are activated by APCs in the lymph node. On the other hand, PMA directly targets protein kinase C (PKC) and Ionomycin upregulates intracellular calcium, thus synergizing with PMA to activate PKC and completely bypassing membrane receptors [29,68]. PMA/Ionomycin is therefore much less physiologically relevant and might result in less intuitive and relevant results from a T-cell biology perspective.

Taken together, we believe that all these options illustrate the flexibility and potential of this droplet-based platform to investigate the activation of the adaptive immune system at the single-cell level.

Supplementary Materials: The following supporting information can be downloaded at: https://www.mdpi.com/article/10.3390/mi13111910/s1, Figure S1: "Droplet characterization", Figure S2: "CTL and aAPC distribution over monodisperse aqueous droplets", Figure S3: "Flow cytometry analysis of microgels", Figure S4: "Fluorescent staining of cells inside and outside of microgels stained for PE fluorescence", Figure S5: "Choosing gating strategy for 1:1 pair selection", Figure S6: "Monocyte and T-cell pairing". Video S1: "Droplet formation for CTL/aAPC pairs", Video S2: "CTL/aAPC pairs over time in aqueous droplets", Video S3: "CTL/aAPC pairs over time in agarose droplets". Design 1: "Eppendorf tube device", Design 2: "Pipette tip device", Design 3: "Platform device".

Author Contributions: B.M.T. performed conceptualization, methodology, investigation, formal analysis and writing—original draft preparation. L.D. performed writing—original draft preparation as well as writing—review and editing. J.H. performed investigation and formal analysis. J.J.F.S. worked on methodology and provided resources. J.T. performed supervision, writing—review and editing and funding acquisition. All authors have read and agreed to the published version of the manuscript.

Funding: This research was funded by the European Research Council (ERC) under the European Union's Horizon 2020 research and innovation program (Grant agreement No.802791).

Institutional Review Board Statement: Not applicable.

Informed Consent Statement: Not applicable.

Data Availability Statement: The raw data supporting the conclusions of this article will be made available by the authors upon request.

Acknowledgments: We acknowledge the generous support from the Eindhoven University of Technology.

Conflicts of Interest: The authors declare no conflict of interest.

References

1. Betts, M.R.; Nason, M.C.; West, S.M.; De Rosa, S.C.; Migueles, S.A.; Abraham, J.; Lederman, M.M.; Benito, J.M.; Goepfert, P.A.; Connors, M.; et al. HIV nonprogressors preferentially maintain highly functional HIV-specific CD8+ T cells. *Blood* **2006**, *107*, 4781–4789. [CrossRef] [PubMed]
2. Freel, S.A.; Lamoreaux, L.; Chattopadhyay, P.K.; Saunders, K.; Zarkowsky, D.; Overman, R.G.; Ochsenbauer, C.; Edmonds, T.G.; Kappes, J.C.; Cunningham, C.K.; et al. Phenotypic and Functional Profile of HIV-Inhibitory CD8 T Cells Elicited by Natural Infection and Heterologous Prime/Boost Vaccination. *J. Virol.* **2010**, *84*, 4998–5006. [CrossRef] [PubMed]
3. Zimmerli, S.C.; Harari, A.; Cellerai, C.; Vallelian, F.; Bart, P.-A.; Pantaleo, G. HIV-1-specific IFN-γ/IL-2-secreting CD8 T cells support CD4-independent proliferation of HIV-1-specific CD8 T cells. *Proc. Natl. Acad. Sci. USA* **2005**, *102*, 7239–7244. [CrossRef] [PubMed]
4. Paul, M.S.; Ohashi, P.S. The Roles of CD8+ T Cell Subsets in Antitumor Immunity. *Trends Cell Biol.* **2020**, *30*, 695–704. [CrossRef]
5. Woodland, D.L.; Dutton, R.W. Heterogeneity of CD4+ and CD8+ T cells. *Curr. Opin. Immunol.* **2003**, *15*, 336–342. [CrossRef]
6. Chattopadhyay, P.; Gierahn, T.M.; Roederer, M.; Love, J.C. Single-cell technologies for monitoring immune systems. *Nat. Immunol.* **2014**, *15*, 128–135. [CrossRef]
7. Shalek, A.K.; Satija, R.; Shuga, J.; Trombetta, J.J.; Gennert, D.; Lu, D.; Chen, P.; Gertner, R.S.; Gaublomme, J.T.; Yosef, N.; et al. Single-cell RNA-seq reveals dynamic paracrine control of cellular variation. *Nature* **2014**, *510*, 363–369. [CrossRef]
8. Satija, R.; Shalek, A.K. Heterogeneity in immune responses: From populations to single cells. *Trends Immunol.* **2014**, *35*, 219–229. [CrossRef]
9. Altschuler, S.J.; Wu, L.F. Cellular Heterogeneity: Do Differences Make a Difference? *Cell* **2010**, *141*, 559–563. [CrossRef]
10. Villani, A.C.; Satija, R.; Reynolds, G.; Sarkizova, S.; Shekhar, K.; Fletcher, J.; Griesbeck, M.; Butler, A.; Zheng, S.; Lazo, S.; et al. Single-cell RNA-seq reveals new types of human blood dendritic cells, monocytes, and progenitors. *Science* **2017**, *356*, eaah4573. [CrossRef]
11. Luo, X.; Chen, J.-Y.; Ataei, M.; Lee, A. Microfluidic Compartmentalization Platforms for Single Cell Analysis. *Biosensors* **2022**, *12*, 58. [CrossRef] [PubMed]
12. Zhou, Y.; Shao, N.; de Castro, R.B.; Zhang, P.; Ma, Y.; Liu, X.; Huang, F.; Wang, R.-F.; Qin, L. Evaluation of Single-Cell Cytokine Secretion and Cell-Cell Interactions with a Hierarchical Loading Microwell Chip. *Cell Rep.* **2020**, *31*, 107574. [CrossRef] [PubMed]
13. Chokkalingam, V.; Tel, J.; Wimmers, F.; Liu, X.; Semenov, S.; Thiele, J.; Figdor, C.G.; Huck, W.T.S. Probing cellular heterogeneity in cytokine-secreting immune cells using droplet-based microfluidics. *Lab Chip* **2013**, *13*, 4740–4744. [CrossRef] [PubMed]
14. Tang, X.; Liu, X.; Li, P.; Liu, F.; Kojima, M.; Huang, Q.; Arai, T. On-Chip Cell-Cell Interaction Monitoring at Single-Cell Level by Efficient Immobilization of Multiple Cells in Adjustable Quantities. *Anal. Chem.* **2020**, *92*, 11607–11616. [CrossRef] [PubMed]
15. Kemna, E.W.M.; Schoeman, R.M.; Wolbers, F.; Vermes, I.; Weitz, D.A.; Van Den Berg, A. High-yield cell ordering and deterministic cell-in-droplet encapsulation using Dean flow in a curved microchannel. *Lab Chip* **2012**, *12*, 2881–2887. [CrossRef]
16. Lagus, T.P.; Edd, J.F. High-throughput co-encapsulation of self-ordered cell trains: Cell pair interactions in microdroplets. *RSC Adv.* **2013**, *3*, 20512–20522. [CrossRef]
17. Sinha, N.; Subedi, N.; Tel, J. Integrating Immunology and Microfluidics for Single Immune Cell Analysis. *Front. Immunol.* **2018**, *9*, 2373. [CrossRef]
18. Shembekar, N.; Chaipan, C.; Utharala, R.; Merten, C.A. Droplet-based microfluidics in drug discovery, transcriptomics and high-throughput molecular genetics. *Lab Chip* **2016**, *16*, 1314–1331. [CrossRef]
19. Segaliny, A.I.; Li, G.; Kong, L.; Ren, C.; Chen, X.; Wang, J.K.; Baltimore, D.; Wu, G.; Zhao, W. Functional TCR T cell screening using single-cell droplet microfluidics. *Lab Chip* **2018**, *18*, 3733–3749. [CrossRef]
20. Sarkar, S.; Sabhachandani, P.; Stroopinsky, D.; Palmer, K.; Cohen, N.; Rosenblatt, J.; Avigan, D.; Konry, T. Dynamic analysis of immune and cancer cell interactions at single cell level in microfluidic droplets. *Biomicrofluidics* **2016**, *10*, 054115. [CrossRef]
21. Tiemeijer, B.M.; Tel, J. Hydrogels for Single-Cell Microgel Production: Recent Advances and Applications. *Front. Bioeng. Biotechnol.* **2022**, *10*, 891461. [CrossRef] [PubMed]

22. Rossi, J.; Paczkowski, P.; Shen, Y.-W.; Morse, K.; Flynn, B.; Kaiser, A.; Ng, C.; Gallatin, K.; Cain, T.; Fan, R.; et al. Preinfusion polyfunctional anti-CD19 chimeric antigen receptor T cells are associated with clinical outcomes in NHL. *Blood* **2018**, *132*, 804–814. [CrossRef] [PubMed]
23. Wimmers, F.; Aarntzen, E.H.J.G.; Duiveman-Deboer, T.; Figdor, C.G.; Jacobs, J.F.M.; Tel, J.; de Vries, I.J.M. Long-lasting multifunctional $CD8^{+}$T cell responses in end-stage melanoma patients can be induced by dendritic cell vaccination. *Oncoimmunology* **2015**, *5*, e1067745. [CrossRef] [PubMed]
24. Sinha, N.; Subedi, N.; Wimmers, F.; Soennichsen, M.; Tel, J. A Pipette-Tip Based Method for Seeding Cells to Droplet Microfluidic Platforms. *J. Vis. Exp.* **2019**, *144*, e57848. [CrossRef] [PubMed]
25. Rueden, C.T.; Schindelin, J.; Hiner, M.C.; Dezonia, B.E.; Walter, A.E.; Arena, E.T.; Eliceiri, K.W. ImageJ2: ImageJ for the next generation of scientific image data. *BMC Bioinform.* **2017**, *18*, 529. [CrossRef]
26. Tiemeijer, B.M.; Sweep, M.W.D.; Sleeboom, J.J.F.; Steps, K.J.; van Sprang, J.F.; De Almeida, P.; Hammink, R.; Kouwer, P.H.J.; Smits, A.I.P.M.; Tel, J. Probing Single-Cell Macrophage Polarization and Heterogeneity Using Thermo-Reversible Hydrogels in Droplet-Based Microfluidics. *Front. Bioeng. Biotechnol.* **2021**, *9*, 715408. [CrossRef]
27. Wimmers, F.; Subedi, N.; Van Buuringen, N.; Heister, D.; Vivié, J.; Beeren-Reinieren, I.; Woestenenk, R.; Dolstra, H.; Piruska, A.; Jacobs, J.F.M.; et al. Single-cell analysis reveals that stochasticity and paracrine signaling control interferon-alpha production by plasmacytoid dendritic cells. *Nat. Commun.* **2018**, *9*, 3317. [CrossRef]
28. van Eyndhoven, L.C.; Chouri, E.; Subedi, N.; Tel, J. Phenotypical Diversification of Early IFNα-Producing Human Plasmacytoid Dendritic Cells Using Droplet-Based Microfluidics. *Front. Immunol.* **2021**, *12*, 1592. [CrossRef]
29. Ai, W.; Li, H.; Song, N.; Li, L.; Chen, H. Optimal Method to Stimulate Cytokine Production and Its Use in Immunotoxicity Assessment. *Int. J. Environ. Res. Public Health* **2013**, *10*, 3834–3842. [CrossRef]
30. Collins, D.J.; Neild, A.; Demello, A.; Liu, A.-Q.; Ai, Y. The Poisson distribution and beyond: Methods for microfluidic droplet production and single cell encapsulation. *Lab Chip* **2015**, *15*, 3439–3459. [CrossRef]
31. Cambiaggi, C.; Scupoli, M.T.; Cestari, T.; Gerosa, F.; Carra, G.; Tridente, G.; Accolla, R.S. Constitutive Expression of CD69 in Interspecies T-Cell Hybrids and Locus Assignment to Human Chromosome 12. *Immunogenetics* **1992**, *36*, 117–120. [CrossRef] [PubMed]
32. Kmieciak, M.; Gowda, M.; Graham, L.; Godder, K.; Bear, H.D.; Marincola, F.M.; Manjili, M.H. Human T cells express CD25 and Foxp3 upon activation and exhibit effector/memory phenotypes without any regulatory/suppressor function. *J. Transl. Med.* **2009**, *7*, 89. [CrossRef] [PubMed]
33. Banchereau, J.; Steinman, R.M. Dendritic cells and the control of immunity. *Nature* **1998**, *392*, 245–252. [CrossRef] [PubMed]
34. Bakdash, G.; Sittig, S.P.; van Dijk, T.; Figdor, C.G.; de Vries, I.J.M. The nature of activatory and tolerogenic dendritic cell-derived signal II. *Front. Immunol.* **2013**, *4*, 53. [CrossRef]
35. Viola, A.; Lanzavecchia, A. T Cell Activation Determined by T Cell Receptor Number and Tunable Thresholds. *Science* **1996**, *273*, 104–106. [CrossRef] [PubMed]
36. Acuto, O.; Michel, F.M. CD28-mediated co-stimulation: A quantitative support for TCR signalling. *Nat. Rev. Immunol.* **2003**, *3*, 939–951. [CrossRef]
37. Kaliński, P.; Hilkens, C.; Wierenga, E.A.; Kapsenberg, M.L. T-cell priming by type-1and type-2 polarized dendritic cells: The concept of a third signal. *Immunol. Today* **1999**, *20*, 561–567. [CrossRef]
38. Dura, B.; Dougan, S.K.; Barisa, M.; Hoehl, M.M.; Lo, C.T.; Ploegh, H.L.; Voldman, J. Profiling lymphocyte interactions at the single-cell level by microfluidic cell pairing. *Nat. Commun.* **2015**, *6*, 5940. [CrossRef]
39. Li, Y.; Kurlander, R.J. Comparison of anti-CD3 and anti-CD28-coated beads with soluble anti-CD3 for expanding human T cells: Differing impact on CD8 T cell phenotype and responsiveness to restimulation. *J. Transl. Med.* **2010**, *8*, 104. [CrossRef]
40. Xiong, J.-Y.; Narayanan, J.; Liu, X.-Y.; Chong, T.K.; Chen, S.B.; Chung, T.-S. Topology Evolution and Gelation Mechanism of Agarose Gel. *J. Phys. Chem. B* **2005**, *109*, 5638–5643. [CrossRef]
41. Narayanan, J.; Xiong, J.-Y.; Liu, X.-Y. Determination of agarose gel pore size: Absorbance measurements vis a vis other techniques. *J. Physics: Conf. Ser.* **2006**, *28*, 83–86. [CrossRef]
42. Ritter, S.; Hiller, R.G.; Wrench, P.M.; Welte, W.; Diederichs, K. Crystal Structure of a Phycourobilin-Containing Phycoerythrin at 1.90-Å Resolution. *J. Struct. Biol.* **1999**, *126*, 86–97. [CrossRef] [PubMed]
43. Antona, S.; Platzman, I.; Spatz, J.P. Droplet-Based Cytotoxicity Assay: Implementation of Time-Efficient Screening of Antitumor Activity of Natural Killer Cells. *ACS Omega* **2020**, *5*, 24674–24683. [CrossRef] [PubMed]
44. Subedi, N.; Van Eyndhoven, L.C.; Hokke, A.M.; Houben, L.; Van Turnhout, M.C.; Bouten, C.V.C.; Eyer, K.; Tel, J. An automated real-time microfluidic platform to probe single NK cell heterogeneity and cytotoxicity on-chip. *Sci. Rep.* **2021**, *11*, 17084. [CrossRef] [PubMed]
45. Antona, S.; Abele, T.; Jahnke, K.; Dreher, Y.; Göpfrich, K.; Platzman, I.; Spatz, J.P. Droplet-Based Combinatorial Assay for Cell Cytotoxicity and Cytokine Release Evaluation. *Adv. Funct. Mater.* **2020**, *30*, 2003479. [CrossRef]
46. Hondroulis, E.; Movila, A.; Sabhachandani, P.; Sarkar, S.; Cohen, N.; Kawai, T.; Konry, T. A Droplet-Merging Platform for Comparative Functional Analysis of M1 and M2 Macrophages in Response to *E. coli*-Induced Stimuli. *Biotechnol. Bioeng.* **2017**, *114*, 705–709. [CrossRef]

47. Gérard, A.; Woolfe, A.; Mottet, G.; Reichen, M.; Castrillon, C.; Menrath, V.; Ellouze, S.; Poitou, A.; Doineau, R.; Briseno-Roa, L.; et al. High-throughput single-cell activity-based screening and sequencing of antibodies using droplet microfluidics. *Nat. Biotechnol.* **2020**, *38*, 715–721. [CrossRef]
48. Madrigal, J.L.; Schoepp, N.G.; Xu, L.; Powell, C.S.; Delley, C.L.; Siltanen, C.A.; Danao, J.; Srinivasan, M.; Cole, R.H.; Abate, A.R. Characterizing cell interactions at scale with made-to-order droplet ensembles (MODEs). *Proc. Natl. Acad. Sci. USA* **2022**, *119*, e2110867119. [CrossRef]
49. Sarkar, S. T Cell Dynamic Activation and Functional Analysis in Nanoliter Droplet Microarray. *J. Clin. Cell Immunol.* **2015**, *6*, 334. [CrossRef]
50. Bounab, Y.; Eyer, K.; Dixneuf, S.; Rybczynska, M.; Chauvel, C.; Mistretta, M.; Tran, T.; Aymerich, N.; Chenon, G.; Llitjos, J.-F.; et al. Dynamic single-cell phenotyping of immune cells using the microfluidic platform DropMap. *Nat. Protoc.* **2020**, *15*, 2920–2955. [CrossRef]
51. Xue, Q.; Bettini, E.; Paczkowski, P.; Qiong, X.; Kaiser, A.; McConnell, T.; Kodrasi, O.; Quigley, M.F.; Heath, J.; Fan, R.; et al. Single-cell multiplexed cytokine profiling of CD19 CAR-T cells reveals a diverse landscape of polyfunctional antigen-specific response. *J. Immunother. Cancer* **2017**, *5*, 85. [CrossRef] [PubMed]
52. Yanakieva, D.; Elter, A.; Bratsch, J.; Friedrich, K.; Becker, S.; Kolmar, H. FACS-Based Functional Protein Screening via Microfluidic Co-encapsulation of Yeast Secretor and Mammalian Reporter Cells. *Sci. Rep.* **2020**, *10*, 10182. [CrossRef] [PubMed]
53. Fang, Y.; Chu, T.H.; Ackerman, M.E.; Griswold, K.E. Going native: Direct high throughput screening of secreted full-length IgG antibodies against cell membrane proteins. *mAbs* **2017**, *9*, 1253–1261. [CrossRef] [PubMed]
54. Zhang, H.; Jenkins, G.; Zou, Y.; Zhu, Z.; Yang, C.J. Massively Parallel Single-Molecule and Single-Cell Emulsion Reverse Transcription Polymerase Chain Reaction Using Agarose Droplet Microfluidics. *Anal. Chem.* **2012**, *84*, 3599–3606. [CrossRef]
55. Li, M.; Liu, H.; Zhuang, S.; Goda, K. Droplet flow cytometry for single-cell analysis. *RSC Adv.* **2021**, *11*, 20944–20960. [CrossRef]
56. Bai, Y.; Weibull, E.; Joensson, H.N.; Andersson-Svahn, H. Interfacing picoliter droplet microfluidics with addressable microliter compartments using fluorescence activated cell sorting. *Sens. Actuators B Chem.* **2014**, *194*, 249–254. [CrossRef]
57. White, A.M.; Zhang, Y.; Shamul, J.G.; Xu, J.; Kwizera, E.A.; Jiang, B.; He, X. Deep Learning-Enabled Label-Free On-Chip Detection and Selective Extraction of Cell Aggregate-Laden Hydrogel Microcapsules. *Small* **2021**, *17*, 2100491. [CrossRef]
58. Fu, X.T.; Kim, S.M. Agarase: Review of Major Sources, Categories, Purification Method, Enzyme Characteristics and Applications. *Mar. Drugs* **2010**, *8*, 200–218. [CrossRef]
59. Ahmed, H.; Stokke, B.T. Fabrication of monodisperse alginate microgel beads by microfluidic picoinjection: A chelate free approach. *Lab Chip* **2021**, *21*, 2232–2243. [CrossRef]
60. Mao, A.S.; Shin, J.-W.; Utech, S.; Wang, H.; Uzun, O.; Li, W.; Cooper, M.; Hu, E.; Zhang, L.; Weitz, D.A.; et al. Deterministic encapsulation of single cells in thin tunable microgels for niche modelling and therapeutic delivery. *Nat. Mater.* **2016**, *16*, 236–243. [CrossRef]
61. Håti, A.G.; Bassett, D.C.; Ribe, J.M.; Sikorski, P.; Weitz, D.A.; Stokke, B.T. Versatile, cell and chip friendly method to gel alginate in microfluidic devices. *Lab Chip* **2016**, *16*, 3718–3727. [CrossRef] [PubMed]
62. Dolega, M.E.; Abeille, F.; Picollet-D'Hahan, N.; Gidrol, X. Controlled 3D culture in Matrigel microbeads to analyze clonal acinar development. *Biomaterials* **2015**, *52*, 347–357. [CrossRef] [PubMed]
63. Leng, X.; Zhang, W.; Wang, C.; Cui, L.; Yang, C.J. Agarose droplet microfluidics for highly parallel and efficient single molecule emulsion PCR. *Lab Chip* **2010**, *10*, 2841–2843. [CrossRef] [PubMed]
64. Novak, R.; Zeng, Y.; Shuga, J.; Venugopalan, G.; Fletcher, D.A.; Smith, M.T.; Mathies, R.A. Single-Cell Multiplex Gene Detection and Sequencing with Microfluidically Generated Agarose Emulsions. *Angew. Chem. Int. Ed.* **2010**, *50*, 390–395. [CrossRef]
65. Zhu, Z.; Zhang, W.; Leng, X.; Zhang, M.; Guan, Z.; Lu, J.; Yang, C.J. Highly sensitive and quantitative detection of rare pathogens through agarose droplet microfluidic emulsion PCR at the single-cell level. *Lab Chip* **2012**, *12*, 3907–3913. [CrossRef]
66. Huang, H.; Wang, C.; Rubelt, F.; Scriba, T.J.; Davis, M.M. Analyzing the Mycobacterium tuberculosis immune response by T-cell receptor clustering with GLIPH2 and genome-wide antigen screening. *Nat. Biotechnol.* **2020**, *38*, 1194–1202. [CrossRef]
67. Chiou, S.-H.; Tseng, D.; Reuben, A.; Mallajosyula, V.; Molina, I.S.; Conley, S.; Wilhelmy, J.; McSween, A.M.; Yang, X.; Nishimiya, D.; et al. Global analysis of shared T cell specificities in human non-small cell lung cancer enables HLA inference and antigen discovery. *Immunity* **2021**, *54*, 586–602.e8. [CrossRef]
68. Chatila, T.; Silverman, L.; Miller, R.; Geha, R. Mechanisms of T cell activation by the calcium ionophore ionomycin. *J. Immunol.* **1989**, *143*, 1283–1289.

micromachines

Article

Strategy for Fast Decision on Material System Suitability for Continuous Crystallization Inside a Slug Flow Crystallizer

Anne Cathrine Kufner [1,†], Adrian Krummnow [2,3,†], Andreas Danzer [2] and Kerstin Wohlgemuth [1,*]

1 Department of Biochemical and Chemical Engineering, Laboratory of Plant and Process Design, TU Dortmund University, D-44227 Dortmund, Germany
2 Department of Biochemical and Chemical Engineering, Laboratory of Thermodynamics, TU Dortmund University, D-44227 Dortmund, Germany
3 AbbVie Deutschland GmbH & Co. KG, Global Pharmaceutical R&D, Knollstraße, D-67061 Ludwigshafen am Rhein, Germany
* Correspondence: kerstin.wohlgemuth@tu-dortmund.de; Tel.: +49-(0)-231-755-3020
† These authors contributed equally to this work.

Abstract: There is an increasing focus on two-phase flow in micro- or mini-structured apparatuses for various manufacturing and measurement instrumentation applications, including the field of crystallization as a separation technique. The slug flow pattern offers salient features for producing high-quality products, since narrow residence time distribution of liquid and solid phases, intensified mixing and heat exchange, and an enhanced particle suspension are achieved despite laminar flow conditions. Due to its unique features, the slug flow crystallizer (SFC) represents a promising concept for small-scale continuous crystallization achieving high-quality active pharmaceutical ingredients (API). Therefore, a time-efficient strategy is presented in this study to enable crystallization of a desired solid product in the SFC as quickly as possible and without much experimental effort. This strategy includes pre-selection of the solvent/solvent mixture using heuristics, verifying the slug flow stability in the apparatus by considering the static contact angle and dynamic flow behavior, and modeling the temperature-dependent solubility in the supposed material system using perturbed-chain statistical associating fluid theory (PC-SAFT). This strategy was successfully verified for the amino acids L-alanine and L-arginine and the API paracetamol for binary and ternary systems and, thus, represents a general approach for using different material systems in the SFC.

Keywords: continuous crystallization; microfluidics; slug flow; contact angle; solid–liquid interaction; solubility modeling

Citation: Kufner, A.C.; Krummnow, A.; Danzer, A.; Wohlgemuth, K. Strategy for Fast Decision on Material System Suitability for Continuous Crystallization Inside a Slug Flow Crystallizer. *Micromachines* **2022**, *13*, 1795. https://doi.org/10.3390/mi13101795

Academic Editor: Pingan Zhu

Received: 30 September 2022
Accepted: 19 October 2022
Published: 21 October 2022

1. Introduction

In recent years, there has been increased research in the area of two-phase flow combined with micro- or minifluidics for a variety of application areas, such as measurement devices in life science and chemistry, as medical devices, microreactors, and heat exchangers, to name a few [1]. A further field of interest is crystallization as isolation technology for the small-scale production of active pharmaceutical ingredients (APIs). A typical production quantity for API production lies in the range of 250–1000 kg a^{-1}. Compared to other separation techniques, crystallization processes offer some key benefits, such as adjustable particular product properties and high product purity [2]. Therefore, one or even more crystallization steps are used in more than 90% of API production pathways [3]. The main specifications of the final product are a uniform particle size and shape to ensure constant bioavailability and dosage uniformity [4,5]. Operating modes for crystallization are batch, semi-batch, and continuous. The first mentioned are the most common ones in pharmaceutical crystallization due to the simplicity of apparatuses [4]. However, batch operation has some characteristic drawbacks, such as variability in product quality between batches, encrustation, and high capital costs [5,6]. Therefore, continuous crystallization is advantageous due to high process reproducibility and uniform product quality while operating in a steady state. Furthermore, the higher process efficiency in terms of used substrates, and the use of the same equipment for research and design, as well as for industrial production by the extend of operating time, make the application highly attractive.

A distinction is made between two types of continuous crystallizers for the small-scale production range: mixed-suspension mixed-product removal (MSMPR) crystallizers and tubular plug-flow crystallizers (PFC) [7–9]. This work exploits the advantages of a special PFC, the slug flow crystallizer (SFC). The SFC is characterized by a gentle particle treatment with respect to particle suspension and a narrow residence time distribution (RTD) of the liquid and solid phases—basic prerequisites for obtaining a narrow particle size distribution (PSD) and high purity in a reproducible manner during crystallization. In the SFC, managing two immiscible fluids creates a slug flow pattern. Due to easier handling in further downstream and to avoid cross-contamination, gas–liquid segmentation is used in this study. Because of the wall friction, Taylor vortices are induced inside the liquid segments (slugs), which provide increased mixing of the liquid and suspension of particles. Accordingly, crystallization phenomena such as secondary nucleation or agglomeration are reduced.

However, the characteristics and advantages mentioned here only apply if a stable and uniform slug flow is achieved, which depends on many parameters: First, the choice of inner diameter for the tubing is decisive in order to force dominant surface effects and allow the simplified formation of slugs over the entire diameter [10,11]. Therefore, several criteria are postulated for the transition from macro- to mini- and microchannel, but there is no clear definition. Often, the Eötvös number *Eö* ($Eö < 3.368$ [12], $Eö < 0.88$ [13], $Eö < (2 \cdot \pi)^2$ [14]) or the hydraulic diameter d_h [15–18] are used for confinement into the microscale range.

Second, the consideration of tubing material and material system (solute and solvent) combination is crucial. The property that receives special focus for the configuration of the slug shape is the three-phase contact angle Θ. It provides information on which of the two phases (gas or liquid) the wall-wetting phase is and whether convex or concave slugs are formed. Obtaining convex slugs is desirable in the case of crystallization since the RTD of the liquid phase corresponds to the RTD of the solid phase (particles) present in the liquid. This was demonstrated in our previous publication for crystallizing L-alanine from aqueous solution by using fluorinated ethylene propylene (FEP) as tubing material [19]. For aqueous systems, the utilization of a hydrophobic tubing material, such as FEP, leads to the avoidance of a wall film and the formation of convex slugs in the gas–liquid flow [20,21], whereas hydrophilic tubing material favors the undesirable wall crystallization [20]. In the literature, besides FEP [22–24], mostly silicone [20,21,25–29] or polyvinyl chloride (PVC) [20] are used as the tubing material for continuous crystallization applications. A wall film leads to broad RTDs if a material system is combined with a tubing material, where concave slugs are formed (rounded gas bubbles). Since gas bubbles roll over particles at the bottom, particles can be exchanged between neighboring slugs. Particles smaller than or equal to the wall film thickness are especially affected [29]. Accordingly, concave slugs lead to deviations between the RTDs of liquid and particles and negate the advantage of absent axial dispersion. Higher flow velocities increase this effect [29].

Consequently, it is necessary to consider gas–liquid interfacial tension $\sigma_{G/L}$, but also the solid–liquid ($\sigma_{S/L}$) and solid–gas ($\sigma_{S/G}$) interfacial tension. This relationship is described by Θ, which Young [30] defined (Equation (1)).

$$\sigma_{G/L} \cdot \cos(\Theta) = \sigma_{S/G} - \sigma_{S/L} \quad (1)$$

In the following, Θ serves as a parameter for the suitability of a material system for use in the crystallization process. Based on the literature on two-phase flow and the dependence of Θ on the tubing material and material system used [31–33], the wettability is divided into the intervals of highly wetting ($\Theta < 50°$), marginally wetting ($50 < \Theta < 90°$), and poorly wetting ($\Theta > 90°$). A distinction is also made between wet and dry flow patterns [31,34]. The dry flow pattern describes the absence of the wall film and is desirable for crystallization operation, but it also has the disadvantage that the pressure loss is higher due to the higher contact of the liquid with the wall in the apparatus. The dry pattern is not linked to a contact angle range, but its occurrence can be estimated using the capillary number *Ca* (Equation (2)), which implies that for $Ca < 10^{-3}$ (under flow boiling situations) [34–36] or, respectively, $Ca < 10^{-2}$ [37], dry plug flow occurs. *Ca* is determined by the ratio of flow velocity u and dynamic viscosity η with respect to the gas–liquid interfacial tension $\sigma_{G/L}$.

$$Ca = \frac{u \cdot \eta}{\sigma_{G/L}} \tag{2}$$

In general, particularly in further literature dealing with slug flow crystallization, Θ is given based on static contact angle (Θ_{stat}) measurements. However, a dynamic equilibrium is established in the apparatus so that the dynamic contact angle (Θ_{dyn}) may deviate from Θ_{stat} under certain conditions. For Θ_{dyn}, a distinction is made between a receding (at the front of a liquid slug) and an advanced (at the back of the liquid slug) contact angle. If these differ, this is referred to as contact angle hysteresis (CAH). The higher the hysteresis, the higher the probability of an unstable slug flow. This can be forced by surface disturbances or hydrodynamics, for example. Therefore, not only the Θ_{stat} is decisive for obtaining a stable slug flow and for the shapes of slugs, but also the CAH generated in the apparatus itself. Theoretically, by measuring the Θ of the slugs moving in the microchannel, it may be possible to correlate Θ_{dyn} as a function of the fluid velocities, as well as the material properties, and use this model to select the favored combination of the material system and tubing material in the SFC. Due to the difficulties in measuring Θ_{dyn} on such a small scale, as well as the complexity of the multiphase flow, other methods are preferred. Therefore, Θ_{stat} measurements are used in most cases to approximate the dynamic behavior. For instance, in a rectangular tubing for the range of capillary number *Ca* from 10^{-6} to 10^{-4}, Skartsis et al. [38] have shown that the dynamic contact angle can be well approximated by the static one. However, this conclusion cannot be directly transferred to all tubing geometries, material systems, and operating conditions. Therefore, Θ_{stat} and Θ_{dyn} should be checked qualitatively for a new material system.

In combination with crystallization, the choice of a new material system poses a number of additional challenges for operation in the SFC, besides the requirements for slug flow stability, since a high product quality in terms of mean particle size and width of PSD is aimed for, but also a high yield should be maintained. Therefore, besides demands for process safety, which include toxicity, explosion-proof environment, and more, the component's solubility in the solvent and its temperature dependency are crucial. If the solute solubility in the solvent is very low, the amount of solvent required is very high, and the mass of solids per volume of solvent is minimal. Furthermore, for the application in cooling crystallization, the temperature dependency of solubility is of decisive importance. In addition to other criteria, such as the lowest possible toxicity and chemical stability of the solute in the considered application range, the solvent's viscosity should be low for good mass and heat transfer.

In conclusion, selecting a new and suitable material system for crystallization in the SFC is challenging and linked to many constraints to maintain slug flow stability and consequently obtain high product quality at the end of the apparatus. Therefore, this work aims to present a systematic approach to decide material system suitability as fast as possible and with low experimental effort in order to enable the continuous operation of the desired product using cooling crystallization inside the SFC. This structured procedure includes the selection of a suitable solvent for the desired solid via a screening of different solvents and tubing materials and the evaluation of the suitability for the SFC reviewing Θ_{stat}. Furthermore, the suitability of the selected solvent/tubing material combination is validated by examining the dynamic behavior in the apparatus with regard to flow stability. As the last step, the temperature-dependent solute's solubility in the solvent is modeled and predicted to evaluate the possible yield for crystallization processes. With the help of this strategy, it is possible to ensure the crystallization of a new material system in the SFC within four steps for binary and ternary systems.

(1) Pre-selection of solvents for the crystallization of the desired solid compound
(2) Static contact angle measurements
(3) Proof of slug flow stability inside the apparatus
(4) Solubility modeling

2. Substances Used

Several common solvents were considered to select an appropriate solvent for crystallization inside the SFC by Θ_{stat} measurements. This involves the use of ultrapure, deionized, and bacteria-free filtered water (Milli-Q®, $\sigma_{G/L}(298.15\ K) = 72.04\ mN\ m^{-1}$ [39]) with a total organic carbon content of maximal 3 ppb, purified by a Milli-Q® Advantage A10 apparatus of *Merck KGaA*. As one product component, L-alanine (Ala) purchased by *Evonik Industries AG* with a purity of 99.7% was selected as it has similar particulate properties as high-priced APIs. The saturated aqueous solution was set according to the measurements and regressed data of Wohlgemuth et al. [40] (Equation (3)).

$$c^*\left(g_{Ala}g_{solution}{}^{-1}\right) = 0.11238 \cdot exp\left(9.0849 \cdot 10^{-3} \cdot \vartheta^*(^\circ C)\right) \tag{3}$$

As a further solid compound, L-arginine (Arg, purity > 99%, *Merck KGaA*) was chosen in order to prove the concept transferability with another amino acid. The following solubility equation was used (Equation (4)) to prepare a saturated aqueous solution and is based on gravimetric measurements conducted in this work.

$$c^*\left(g_{Arg}g_{solution}{}^{-1}\right) = 0.089 \cdot exp\left(2.57 \cdot 10^{-2} \cdot \vartheta^*(^\circ C)\right) \tag{4}$$

Since the water solubility of APIs is usually low, leading to limited bioavailability [41], paracetamol (APAP, acetaminophen according to USP, > 99%, *Merck KGaA*) was used as a third solid compound to demonstrate the application field of APIs. The regression curve of the saturated aqueous solution was calculated and used according to the measured data from Grant et al. [42] given in Equation (5).

$$c^*\left(g_{APAP}g_{solution}{}^{-1}\right) = 0.0067 \cdot exp\left(3.18 \cdot 10^{-2} \cdot \vartheta^*(^\circ C)\right) \tag{5}$$

Further solvents studied were ethanol absolute (99.9%, *VWR*, $\sigma_{G/L}(298.15\ K) = 21.72\ mN\ m^{-1}$ [39]), 2-propanol (99.9%, *VWR*, $\sigma_{G/L}(298.15\ K) = 21.74\ mN\ m^{-1}$ [43]), acetone ($\geq$ 99.5%, *Roth*, $\sigma_{G/L}(298.15\ K) = 22.57\ mN\ m^{-1}$ [44]) and n-hexane (95%, *VWR*, $\sigma_{G/L}(298.15\ K) = 17.78\ mN\ m^{-1}$ [45]). These (along with water) five solvents are considered first because they are common solvents used for other separation technologies such as extraction processes and are often upstream of crystallization processes [46].

Requirements for tubing material selection are thermal and chemical resistance to a broad spectrum of solvents. For analytical reasons, the transparency of the tubing would be advantageous but not mandatory. Besides FEP, which was successfully used in our previous publications [19,47,48], other materials such as aluminum, glass, polystyrene, and silicone were also tested via the Θ_{stat} measurements for their suitability as a tubing material for the SFC.

Synthetic air (Grade 5.0, *Messer Griesheim*) was utilized as second fluid phase to generate slug flow inside the tubing.

3. Modeling of Solubilities Using PC-SAFT Equation of State

According to the presented strategy, the modeling of the solubility within the system under consideration is carried out after selecting the solvent based on Θ_{stat} measurements.

The mole fraction solubility x_i^{L} of component i in a solvent (or solvent mixture) was determined by considering an equilibrium between a pure solid phase and a liquid phase according to Prausnitz (Equation (6)) [49]:

$$x_i^{\mathrm{L}} = \frac{1}{\gamma_i^{\mathrm{L}}} exp\left[\frac{\Delta h_i^{\mathrm{SL}}}{RT_i^{\mathrm{SL}}}\left(1 - \frac{T_i^{\mathrm{SL}}}{T}\right) - \frac{1}{RT}\int_{T_i^{\mathrm{SL}}}^{T} \Delta c_{\mathrm{p},i}^{\mathrm{SL}}(T)\, dT + \frac{1}{R}\int_{T_i^{\mathrm{SL}}}^{T} \frac{\Delta c_{\mathrm{p},i}^{\mathrm{SL}}(T)}{T}\, dT\right] \tag{6}$$

R is the universal gas constant, T is the system temperature, T_i^{SL} is the melting temperature of component i, and Δh_i^{SL} is the melting enthalpy at the melting temperature of component i. For amino acids, the difference between the component's liquid and solid heat capacity $\Delta c_{\mathrm{p},i}^{\mathrm{SL}}$ was recently assumed to be linear dependent on temperature (Equation (7)) [50]:

$$\Delta c_{\mathrm{p},i}^{\mathrm{SL}}(T) = \Delta a_{c_{\mathrm{p},i}^{\mathrm{SL}}}\, T + \Delta b_{c_{\mathrm{p},i}^{\mathrm{SL}}} \tag{7}$$

The combination of a slope $\Delta a_{c_{\mathrm{p},i}^{\mathrm{SL}}}$ and an intercept $\Delta b_{c_{\mathrm{p},i}^{\mathrm{SL}}}$ was used for Ala and Arg in this work. For active pharmaceutical ingredients, such as APAP, the difference between the component's liquid and solid heat capacity is often assumed to be insensitive to temperature [51]. Thus, APAP was modeled with the heat capacity difference at the melting temperature in this work. All melting properties were taken from the literature and are listed in Table 1.

Table 1. Pure-component melting properties of Ala, APAP, and Arg at 0.1 MPa.

Component	T_i^{SL}/K	Δh_i^{SL}/kJ mol^{-1}	$\Delta a_{c_{\mathrm{p},i}^{\mathrm{SL}}}$/J mol^{-1} K^{-2}	$\Delta b_{c_{\mathrm{p},i}^{\mathrm{SL}}}$/J mol^{-1} K^{-1}
Ala	608.0 [52]	25.99 [50]	−0.057 [50]	39.923 [50]
Arg	558.0 [50]	32.00 [50]	−0.364 [50]	237.991 [50]
APAP	443.6 [53]	27.10 [53]	0 [1]	99.800 [54]

[1] Assumption of $\Delta c_{\mathrm{p},i}^{\mathrm{SL}}(T) = \Delta c_{\mathrm{p},i}^{\mathrm{SL}}(T_i^{\mathrm{SL}})$ in this work.

The activity coefficient γ_i^{L} of a component i accounts for deviation from ideal-mixture behavior in the liquid phase and is related to the partial derivatives of the residual Helmholtz energy with respect to the mole fraction. Within the perturbed-chain statistical associating fluid theory (PC-SAFT) equation of state, the residual Helmholtz energy a^{res} is expressed by the sum of a hard-chain (a^{hc}), dispersion (a^{disp}), and association (a^{assoc}) Helmholtz energy contribution (Equation (8)) [55]:

$$a^{\mathrm{res}} = a^{\mathrm{hc}} + a^{\mathrm{disp}} + a^{\mathrm{assoc}} \tag{8}$$

These contributions account for repulsion, van der Waals attractions, and hydrogen bonds. Calculation requires the segment number m_i^{seg}, the segment diameter σ_i, the dispersion energy parameter $u_i k_{\mathrm{B}}^{-1}$, the association energy parameter $\varepsilon^{A_iB_i} k_{\mathrm{B}}^{-1}$, the association volume $\kappa^{A_iB_i}$, and the number of association sites N_i^{assoc} of every component i. k_{B} is the Boltzmann constant. The pure-component parameters used in this work were available in the literature and are summarized in Table 2.

Table 2. PC-SAFT pure-component parameters of Ala, Arg, APAP, water, and ethanol.

Component	M_i/g mol^{-1}	$m_i^{\mathrm{seg}} M_i^{-1}$/mol g^{-1}	σ_i/Å	$u_i k_{\mathrm{B}}^{-1}$/K	$\varepsilon^{A_iB_i} k_{\mathrm{B}}^{-1}$/K	$\kappa^{A_iB_i}$	N_i^{assoc}
Ala [56]	89.090	0.0613	2.5222	287.590	3176.600	0.0819	1/1
Arg [56]	174.210	0.0569	2.6572	349.710	2555.450	0.0393	3/1
APAP [57]	151.160	0.0498	3.5080	398.284	1994.200	0.0100	2/2
Water [58]	18.015	0.0669	σ_{water} *	353.950	2425.700	0.0451	1/1
Ethanol [55]	46.069	0.0517	3.1770	198.237	2653.384	0.0320	1/1

* $\sigma_{\mathrm{water}} = 2.7927 + 10.11\, exp(-0.01755\, T/\mathrm{K}) - 1.417\, exp(-0.01146\, T/\mathrm{K})$.

To calculate the segment diameter and the dispersion energy in mixtures of components i and j the combining rules as suggested by Berthelot [59] and Lorentz [60] were used:

$$\sigma_{ij} = \frac{1}{2}(\sigma_i + \sigma_j) \tag{9}$$

$$u_{ij} = \sqrt{u_i u_j}(1 - k_{ij}) \tag{10}$$

The binary interaction parameter k_{ij} was introduced for correction of deviations from the geometric mean of the dispersion energies of the pure components and is usually fitted to experimental data of binary mixtures. A linear temperature dependency was assumed in this work:

$$k_{ij} = k_{ij,\mathrm{m}} T + k_{ij,\mathrm{b}} \tag{11}$$

The slope $k_{ij,\mathrm{m}}$ and intercept $k_{ij,\mathrm{b}}$ for Ala/water, Arg/water, APAP/water, and water/ethanol were taken from the literature and are given in Table 3. The interaction parameters for Ala/ethanol, APAP/ethanol, and Arg/ethanol were fitted to solubility data from An et al. [61], this work, and Jiménez and Martínez [62], respectively, and are also available in Table 3.

Table 3. PC-SAFT interaction parameters for mixtures of Ala, Arg, APAP, water, and ethanol.

Mixture	$k_{ij,\mathrm{m}}$/K^{-1}	$k_{ij,\mathrm{b}}$
Ala/water [56]	2.910×10^{-4}	−0.147962
Ala/ethanol [1]	1.140×10^{-3}	−0.3513
Arg/water [56]	0	−0.0145
Arg/ethanol [2]	2.075×10^{-4}	−0.134529
APAP/water [63]	1.770×10^{-4}	−0.051
APAP/ethanol [3]	1.250×10^{-4}	−0.08764
water/ethanol [64]	6.860×10^{-4}	−0.2662

[1] Fitted to solubility data of An et al. [61] in this work. [2] Fitted to solubility data from this work. [3] Fitted to solubility data of Jiménez and Martínez [62] in this work.

The association energy and association volume in mixtures of components i and j were determined by applying the combining rules of Wolbach and Sandler [65]:

$$\varepsilon^{A_iB_j} = \frac{1}{2}\left(\varepsilon^{A_iB_i} + \varepsilon^{A_jB_j}\right) \tag{12}$$

$$\kappa^{A_iB_j} = \sqrt{\kappa^{A_iB_i}\kappa^{A_jB_j}}\left(\frac{\sqrt{\sigma_i\sigma_j}}{\frac{1}{2}(\sigma_i + \sigma_j)}\right)^3 \tag{13}$$

4. Strategy for Solvent Selection

In order to limit the solvent selection for a specific solid product, in this case Ala, Arg, and APAP, it is important to ensure stable slug flow inside the crystallizer. Furthermore, since the objective is to obtain a high product quality in terms of purity and maintain a narrow PSD, for the latter it is necessary to prevent a wall film (i.e., realize dry plug flow) and, thus, facilitate convex slugs. Therefore, the static contact angle is used in order to select appropriate solvents for crystallization in this apparatus. Afterwards, the dynamic behavior and slug flow stability are proved and correlated to the capillary number *Ca*. Subsequently, the solubility modeling of the selected material system is carried out, which completes the requirements for successful crystallization in the apparatus.

4.1. Static Contact Angle Measurements

The Θ_{stat} between solid wall material, liquid, and gaseous phase is crucial for these applications. Therefore, Θ_{stat} measurements were carried out with a drop shape analyzer (DSA30, *Krüss*) equipped with a DS4210 dosing unit and the software ADVANCED to provide initial estimates of the suitability of solvents for SFC application. The measurements are based on the sessile drop method, where a droplet is placed onto the solid surface to be examined by a cannula (d_i = 0.75 mm). In each case, ten measurements of Θ at the right and left edge of the droplet are measured for three drops (3–20 µL) of a solvent/mixture, and the average value is calculated.

Results of Static Contact Angle Measurements

The results of Θ_{stat} measurements of solvents on different tubing materials are presented in Table 4.

Table 4. The measured Θ_{stat} for different tubing materials and solvent combinations.

	Θ_{stat}/° Glass	Θ_{stat}/° Aluminum	Θ_{stat}/° Polystyrene	Θ_{stat}/° Silicone	Θ_{stat}/° FEP
n-Hexane	<20	<20	<20	<20	<20
Isopropanol	<20	<20	<20	<20	36.65 ± 1.37
Acetone	<20	<20	<20	<20	47.33 ± 3.48
Ethanol	<20	<20	<20	23.10 ± 1.34	45.51 ± 3.68
Water	20.89 ± 1.09	66.44 ± 1.41	81.98 ± 0.43	97.99 ± 0.38	100.66 ± 1.37

Values < 20° indicate that a measurement of a droplet via the software was not possible, caused by the low Θ_{stat}. Therefore, the combination of solvent, solid material, and air was assumed to be completely wetted and unsuitable for the use inside the SFC. Using a non-polar solvent (n-hexane) leads to complete wetting of all tested wall materials; hence, n-hexane is not a suitable solvent for SFC. Comparing the surface tensions of ethanol, acetone, and isopropanol (see Section 2), they are in a similar range, and, therefore, the Θ_{stat} are also in a similar range, despite their different solvent nature of polar protic and polar aprotic. However, since Θ_{stat} of isopropanol, acetone, and ethanol are <50° with all materials tested, they are also classified as highly wetting and, thus, not suitable for SFC crystallization. It is becoming clear that water forms higher Θ_{stat} with all the materials tested than the other solvents due to its high $\sigma_{G/L}$. It also can be seen that the most hydrophobic wall material tested, FEP, forms higher Θ_{stat} compared to the other materials. Therefore, the following focuses on FEP as a wall material.

Figure 1 shows the Θ_{stat} of solvent/FEP combinations, which are listed in Table 4. According to the literature, a poorly wetting system (Θ > 90°) is preferred. Consequently, combining an aqueous system and FEP as tubing material is the most suitable setup for crystallization purposes inside SFC for the tested combinations since Θ_{stat} > 100°. This indicates that convex slugs are formed, and the presence of a wall film is negligible.

Conversely, solvents that form a Θ_{stat} < 90° with FEP as the wall material do not appear suitable for use in the SFC. This correlation based on Θ_{stat} measurements is further verified in the next section by reviewing the transport of slugs of all tested solvents in an FEP tubing in order to evaluate the corresponding dynamic behavior.

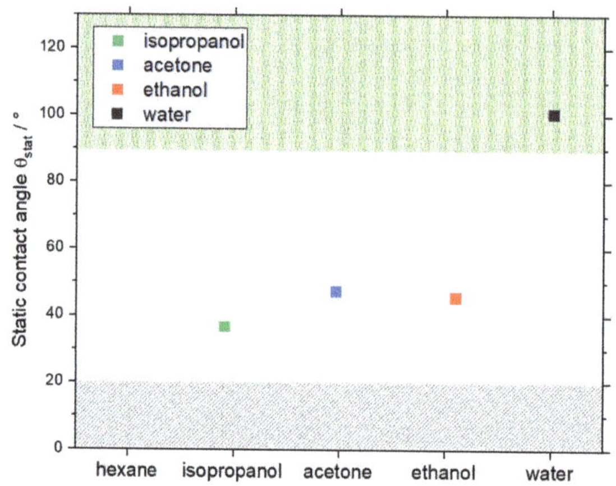

Figure 1. Θ_{stat} for the respective solvents is shown. The grey marked area indicates the region in which Θ_{stat} measurements were not possible ($\Theta_{stat} < 20°$) with the method described before in Section 4.1. The green area ($\Theta_{stat} \geq 90°$) marks the Θ_{stat} at which a non-wetting behavior is expected, and a stable slug flow might be generated.

4.2. *Proof of Stable Slug Flow Inside SFC*

In the following, in order to connect the results from Θ_{stat} measurements and the solubility modeling with the operation inside the SFC and proof of the suitability of the solvent for crystallization purposes, the setup as schematically shown in Figure 2 was applied. The setup can be divided into slug formation and slug flow zone, including image analysis. A solvent, solvent mixture, or saturated solution, which applicability has to be checked, is given inside a feed vessel. For the case of using a saturated solution, those were prepared by setting up a slightly supersaturated solution according to Equation (3), Equation (4), or Equation (5), followed by stirring for 48 h and filtering. The liquid is pumped via a peristaltic pump (*Ismatec Reglo Digital MS-4/12*, d_i = 2.29 mm Pharmed) to the slug formation zone. This consists of a T-junction (polypropylene, PP) in which the feed is fed in from one side and synthetic air supplied via pipeline pressure from a gas cylinder from the other, forming alternating gas and liquid segments of equal size, respectively. The gaseous volume flow rate was controlled by a high-resolution needle valve (NV-001-HR, *Bronkhorst*) and a flow meter (El-Flow-Select, *Bronkhorst*). The choice of a T-junction is critical for the slug length distribution and its reproducibility throughout the slug flow zone since a squeezing mechanism is evoked as slug formation mechanism, which has already been demonstrated in our previous publication [47]. The inner diameter of the T-junction was chosen to d_i = 3 mm similar to the tubing's inner diameter ($d_{i,tubing}$ = 3.18 mm) to minimize the slug length variability [66]. After the slug flow was built, the slugs were transported through the FEP tubing (L_{tubing} = 7.5 m, $d_{out,tubing}$ = 4.76 mm) in the slug flow zone covered with a polyvinyl chloride (PVC) cooling jacket (d_i = 15 mm), which is filled with deionized water. As no cooling crystallization experiment was performed in this study, no cooling profile was adjusted, and the experiments were conducted at ambient temperature ($\vartheta_{amb} \approx 22$ °C).

At the end of SFC tubing, a camera (*Samsung NX 300*, 18–55 mm lens) is placed in order to evaluate the slug shape and slug length distribution by image analysis. Therefore, the process tubing is placed in a glass box (14 cm × 6 cm × 6 cm) filled with degassed water at ambient temperature. The back of the box is darkened with black cardboard and a LED lamp (*LED Panel Light* from *LED Universum*) is placed above the box for indirect illumination. This procedure minimizes reflections and light influences from the environment and creates a high contrast for the evaluation of the resulting videos. The image evaluation is conducted by an in-house MATLAB script, which has already been described in a previous publication [47]. The slug length and CAH are evaluated quantitatively and the slug shape qualitatively.

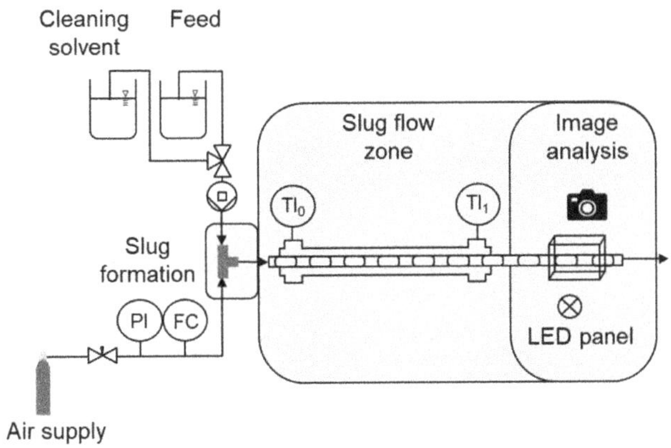

Figure 2. Schematic setup for the validation of solvent suitability for slug flow crystallization.

Results of Proof of Slug Flow Stability

Images of slugs during operation with different solvents at the end of the apparatus consisting of an FEP tubing are shown in Figure 3.

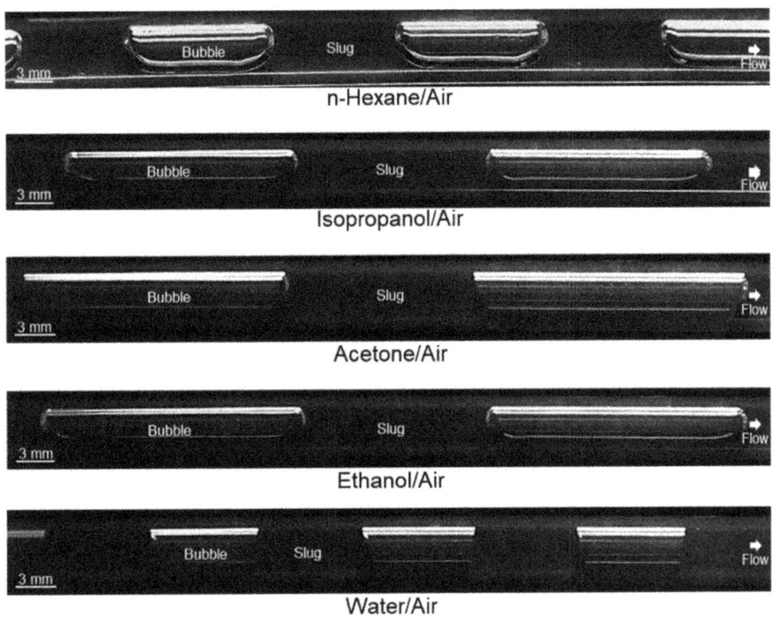

Figure 3. Images of slugs at the end of the apparatus (L = 7.5 m) during operation with different solvents in an FEP tubing. The liquid and gas flow rates were set to Q = 10 mL min^{-1} each. The experiments were conducted at ambient temperature ($\vartheta_{amb} \approx 22$ °C).

In all cases, the gas bubble extends over the entire diameter of the tubing, and stable slug flow is formed independently of the solvent used and its Θ_{stat} described in Table 4. Via image analysis, it was quantitatively demonstrated that for each solvent, slugs of equal size are formed, but differences in slug length occur depending on the solvent applied (Table 5). During experiments, it was observed that this is due to the changing slug mechanism in the slug formation zone. While in the case of n-hexane, isopropanol, acetone, and ethanol, the dripping mechanism was observed to form the slugs, in the case of water, the squeezing mechanism produces the slugs.

Table 5. Median slug length L_{50} and slug length distribution $L_{90\text{-}10}$ at the end of the apparatus (L_{tubing} = 7.5 m) for the operation with different solvents. The experiments were conducted at ambient temperature ($\vartheta_{amb} \approx 22$ °C).

	L_{50}/mm	$L_{90\text{-}10}$/mm
n-Hexane	10.70	2.23
Isopropanol	24.30	1.27
Acetone	16.37	2.45
Ethanol	20.58	0.85
Water	10.16	0.99

The squeezing mechanism is desirable for the reproducibility of the slug length according to the literature [47,67,68] and the desired and set operating conditions of short slugs. In general, the slug length is relevant because it influences pressure drop, mixing, suspension of particles, and heat transfer [47,66,69]. The slug length is influenced by mixer geometry and dimension, mixer material/three-phase contact angle, phase ratio/liquid hold-up, velocity of the phases, properties of the phases, supplying of the phases, and the bubble detachment mechanism [47,70,71]. All these influences have to be considered when designing a new apparatus for a new material system.

However, not only slug length, but also slug shape is crucial for crystallization purposes. According to Figure 3, a significant difference in slug shape can be seen for the different solvents tested. While the gas bubbles are rounded in the case of n-hexane, ethanol, and isopropanol, clear edges can be seen in the case of water. For acetone, a transitional form is visible, showing less curvature of the bubble cap compared to the bubbles for n-hexane, ethanol, and isopropanol slugs. The rounded shape of the gas bubbles can lead to the fact that in the presence of crystals in the liquid phase, the crystals can migrate between neighboring slugs and, thus, significantly broaden the RTD of the solid phase. Based on the curvature of the gas bubbles, it can be seen that for the n-hexane, ethanol, and isopropanol slugs, a wall film is present, whereas the contact of the gas bubble and the wall is increased in the case of acetone and water. Therefore, the slug flow for acetone and water is visually assigned to the dry pattern, which is preferred for crystallization. This assignment is confirmed by considering the *Ca* number for the used solvents (Figure 4). The values for hexane, water, and acetone are higher than the values for ethanol and isopropanol ($Ca_{EtOH} = 2.3 \cdot 10^{-3}$; $Ca_{IPA} = 4.8 \cdot 10^{-3}$; $Ca_{Ac} = 5.9 \cdot 10^{-4}$; $Ca_{wat} = 5.8 \cdot 10^{-4}$; $Ca_{hexane} = 7.3 \cdot 10^{-4}$). Since all calculated *Ca* numbers are $<10^{-2}$, the higher limit postulated in the literature [37] is not sufficient to predict the dry pattern in our case. The limit of $Ca < 10^{-3}$ [34] is applicable to our purposes in the meantime as hexane has already been excluded for suitability due to its non-polar properties.

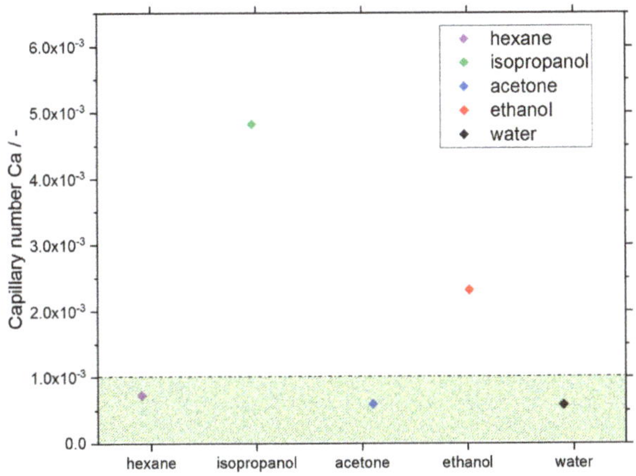

Figure 4. Calculated *Ca* numbers for the tested solvents in the SFC. The calculation was performed for the operating parameters based on the slug flow stability experiments at liquid and gas flow rates of Q = 10 mL min^{-1} each and a ambient temperature of $\vartheta_{amb} \approx 22$ °C. The green area marks the dry pattern slug flow range according to the limit of $Ca < 10^{-3}$.

Furthermore, it is recognizable that no constant Θ_{dyn} is formed in the apparatus itself, as in the case of Θ_{stat}, but different dynamic contact angles Θ_{dyn} are present depending on whether at the bottom or top of the slug, left or right. The sampled Θ_{dyn} from the experiment shown in Figure 3 indicates that the receding contact angle is smaller than the advanced contact angle for rounded bubbles, confirming the literature. In addition, the literature states that as *Ca* increases, the CAH range becomes broader [34]. The measured ranges of receding and advanced contact angles for the cases considered are shown in Table 6.

Table 6. Measured receding and advanced dynamic contact angle ranges for the different solvent used in the experiments. CAH_{max} is built by the lowest value for receding and highest advanced dynamic contact angle value for the respective solvent.

Solvent	Θ_{dyn}/° Receding	Θ_{dyn}/° Advanced	CAH_{max}/°
n-Hexane	20–34	25–36	16
Isopropanol	36–63	47–64	28
Acetone	50–60	65–73	23
Ethanol	26–41	40–56	30
Water	82–92	84–92	10

Although the difference in receding and advanced Θ_{dyn} observed in the experiments differ significantly, no instability of the slug flow has been observed, so the critical CAH has not been exceeded. If the absolute values of the dynamic contact angles are considered, the static contact angles are near the range of the dynamic contact angles for water, ethanol, isopropanol, and hexane slugs. However, this does not apply to the acetone slugs. In this case, the static contact angle is lower than the dynamic ones. Due to the lower curvature of the gas bubble and correspondingly higher Θ_{dyn}, it is to be expected that the particles in the slugs can be better transported along. This shows that, in addition to the wetting properties, the hydrodynamics are also decisive and must be taken into account. Furthermore, solvents' volatility should be checked to avoid the expansion of the gas bubble and the change in slug form during the operation. Summarizing, the combination of estimating the *Ca* number and using the Θ_{stat} appears to be sufficient to have a first indication of the slug shape. A dry pattern is expected for $Ca < 10^{-3}$ and $\Theta_{stat} \geq 90°$.

4.3. Proving the Operability of Slug Flow Crystallizer with the Solutes

As described in the introduction, some material systems have already been used for SFC crystallization, mainly amino acids or proteins. A previous publication has already demonstrated the evidence of convex slugs for a saturated Ala/water solution in the crystallization process [19]. The measured Θ_{stat} = 98.45° ± 1.69° and calculated $Ca_{AlaSolution} = 6.3 \cdot 10^{-4}$ for the same operating conditions fulfills the above-mentioned criteria.

To verify the hypothesis to other solid compounds, another amino acid, Arg, which to our best knowledge has not yet been used in the literature for SFC, and APAP are chosen. The solvent selection has already been made in Section 4.1, so for the solutes Arg and APAP water seems ideal as a solvent in an FEP tubing with regard to the slug flow stability and slug shape. The solute influences the contact angle depending on the hydrophobicity of the amino acid side chain and the structure of the molecule [72,73]. The measured Θ_{stat} of the saturated aqueous solutions of both solutes on the FEP tubing material result in $\Theta_{stat,Arg}$ = 93.26 ± 1.22° and $\Theta_{stat,APAP}$ = 90.61 ± 1.62°. Therefore, both solutes have higher impact on the Θ_{stat} than Ala but are still >90°. The *Ca* numbers were calculated to $Ca_{ArgSolution} = 8.7 \cdot 10^{-4}$ for a saturated Arg/water solution and $Ca_{APAPSolution} = 6.3 \cdot 10^{-4}$ for the saturated APAP/water solution. Consequently, both solute/solvent combinations seem to be suitable for crystallization inside the SFC by fulfilling the criteria so that convex slugs should be formed. Evidence of slug shape and stable slug flow in the apparatus can be seen in Figure 5.

Figure 5. Images of saturated Arg/water (**top**) and APAP/water (**bottom**) slugs at the end of the apparatus (L_{tubing} = 7.5 m) during operation inside an FEP tubing of SFC. The liquid and gas flow rates were set to Q = 10 mL min^{-1} each. The experiments were conducted at ambient temperature ($\vartheta_{amb} \approx$ 22 °C).

In both cases, stable slug flow was observed over the entire tubing length. The dry pattern, which was already anticipated by the *Ca* numbers, can be confirmed visually for both solutes. Furthermore, the slug shape is convex according to the heuristics, so no wall film is present, and a narrow RTD of liquid and solid phases could be achieved. For evaluation of slug flow stability and slug length reproducibility for the saturated Arg/water solution, the characteristic values of slug length distribution were calculated to L_{50} = 10.04 mm and $L_{90\text{-}10}$ = 0.59 mm. For the saturated aqueous APAP solution, slug lengths of L_{50} = 9.82 mm and $L_{90\text{-}10}$ = 2.24 mm were achieved. Based on the Θ_{stat} measurements and the verification of the dynamic behavior inside the SFC, the combination of the two solutes Arg and APAP, respectively, with water is suitable for crystallization in the apparatus and fulfills the requirements with respect to slug flow stability and slug shape for obtaining a high product quality. For further evaluation of the material systems with respect to the suitability for cooling crystallization, temperature-dependent solute solubilities are modeled for the three presented solutes as a next step.

4.4. Solubilities for Binary Systems

Figure 6 shows the modeled temperature-dependent solubilities of Ala, Arg, and APAP in water (a) and ethanol (b) at 0.1 MPa using PC-SAFT compared to experimental data. The solubility lines were modeled with the parameters from Tables 1–3. Modeling results and experimental data (from the literature, as well as from gravimetric measurements performed in this work) are in very good agreement. PC-SAFT reveals a high accuracy in predicting solubilities over a wide temperature range.

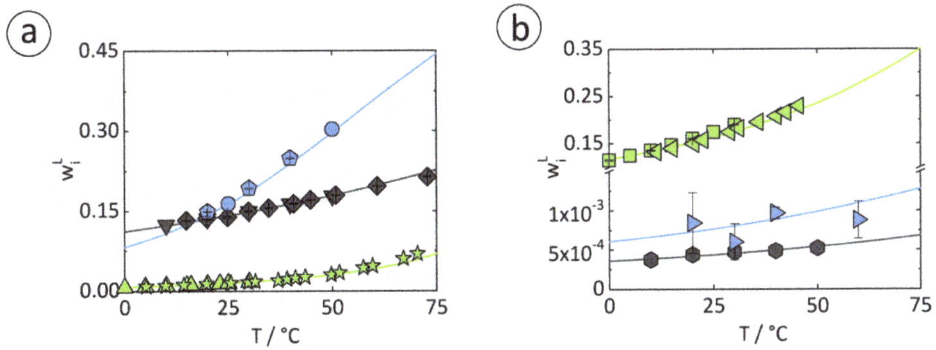

Figure 6. Solubilities of Ala (gray), Arg (blue), and APAP (green) in water (**a**) and ethanol (**b**) at 0.1 MPa: Down-pointing triangles, diamonds, circles, up-pointing triangles, and stars depict measured solubilities in water from An et al. [61], Grosse Daldrup et al. [74], Amend and Helgeseon [75], Granberg et al. [76], and Grant et al. [42]. Hexagons, squares, and left-pointing triangles denote solubility measurements in ethanol from An et al. [61], Granberg et al. [53], and Matsuda et al. [77]. Pentagons and right-pointing triangles are measurements in water and in ethanol performed in this work, respectively. The solid lines are modeled solubility lines using PC-SAFT.

Above the solubility lines are the two-phase regions in which crystallization occurs if an initially homogeneous mixture is cooled starting below the solubility line across the solubility line. Arg solubilities in water are high, whereas the solubility line strongly depends on temperature. Thus, the Arg/water system is well suitable for cooling crystallization as it enables high yields. This is not the case for the other modeled systems. Ala solubilities in water are high, but the temperature dependence of the solubility line is much less pronounced, so yields are too low. In the APAP/water system, the solubilities are comparatively lower than in the other systems. Moreover, the temperature dependence is again too low to reach sufficient yields when applying cooling crystallization.

To increase the yield in the case of APAP/water for crystallization in the SFC, the application in the ternary system is possible. In order to solubilize, ethanol seems to be useful as an increase in yield due to the higher solubility of APAP in ethanol compared to water. Furthermore, ethanol is suitable as a wash liquid in the downstream processing of Ala [78] and can also serve as an antisolvent in the crystallization process of Ala and Arg in aqueous solutions.

5. Consideration of Ternary Systems for Slug Flow Crystallizer Application

In order to extend the possible field of application, the strategy shown above is applied to ethanol/water mixtures. Therefore, solubility modeling of Ala, Arg, and APAP in a ternary system is first performed to determine the influence of different compositions of ethanol/water on solute solubility. Afterwards, it is identified which fractions of ethanol in water are usable for a possible crystallization in the SFC based on Θ_{stat} measurements and *Ca* number calculation.

Figure 7 depicts exemplified the solubilities of Ala from 10 °C to 20 °C, 30 °C, 40 °C, and 50 °C in the ternary phase diagram Ala/water/ethanol at 0.1 MPa. The solubility-reducing effect of ethanol is precisely predicted over the entire temperature range due to the excellent agreement between modeled and experimental solubilities. Modeling was performed based on the parameters from Tables 1–3, which were used for the binary systems in Figure 6. No additional parameters were fitted. Thus, PC-SAFT shows a high capability in predicting solubilities in solvent mixtures at different temperatures. This is valid in the same way for the ternary systems of Arg/water/ethanol, and APAP/water/ethanol as shown in Figures S1 and S2 in the supplementary information.

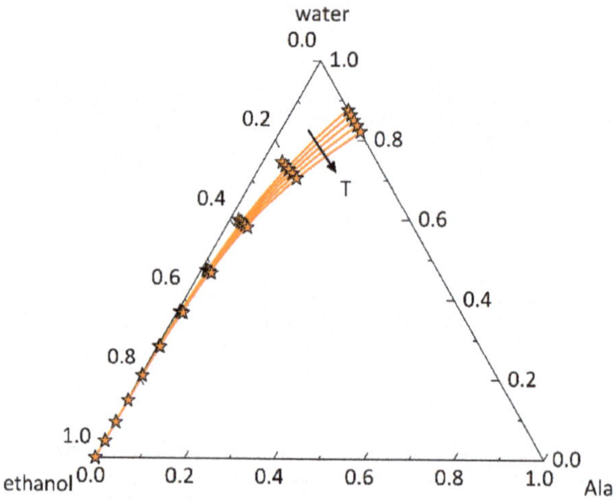

Figure 7. Ternary phase diagram of Ala/water/ethanol at 0.1 MPa with compositions given in mass fractions: Solubility lines were predicted in this work using PC-SAFT, and symbols denote solubility measurements from An et al. [61]. The arrow indicates the direction of increasing temperature from 10 °C to 20 °C, 30 °C, 40 °C, and 50 °C.

In order to obtain the indication about the slug flow stability in the SFC, Table 7 shows the Θ_{stat} measured for different ethanol/water compositions on FEP, as well as the calculated *Ca* numbers for the respective operating parameters.

From Table 7, it is clear that Θ_{stat} decreases with increasing ethanol fraction. Based on the general definition of non-wetting property for $\Theta > 90°$, the mixtures with w_{EtOH} up to 10 wt.-% are suitable to evoke stable slugs for crystallization inside the SFC. The *Ca* number increases with higher velocity and rising ethanol content in the mixture.

Figure 8a shows the calculated *Ca* numbers plotted against the measured Θ_{stat}. The green region marks the previous limits from the literature ($\Theta > 90°$ and $Ca < 10^{-3}$) for a dry pattern. The white regions are the transient regions where data points fulfill only one of the criteria, and, therefore, the slug flow stability at these operating points or for this composition should be experimentally verified. Within the gray region, no stable slug flow should be observable. In our case, only the data points for pure water at volume flow rates of 20 mL min^{-1} and 30 mL min^{-1} are in the green area. None of the mixtures with ethanol content seems to be suitable since they are located in the gray region.

Table 7. Results of Θ_{stat} measurements, observed slug forming mechanism, CAH_{max}, and *Ca* calculation for different ethanol/water mixtures and velocities in the apparatus. The densities, viscosities, and surface tensions for the mixtures were taken from [79] and used for the calculation of *Ca* number at ϑ = 20 °C. The Θ_{stat} measurements were conducted at ambient temperature ($\vartheta_{amb} \approx$ 22 °C) and FEP was used as tubing material.

w_{EtOH}/ wt.-%	Θ_{stat}/° FEP	Volume Flow Rate/ mL min^{-1}	*Ca*/ -	Flow Pattern Based on Literature Limit ($Ca < 10^{-3}$)	Flow Pattern Based on Experiments	Slug Form Mechanism	CAH_{max}/ -
0	100.66 ± 1.37	20	$5.83 \cdot 10^{-4}$	Dry	Dry	Squeezing	10
		30	$8.74 \cdot 10^{-4}$	Dry	Dry	Squeezing	9
		40	$1.17 \cdot 10^{-3}$	Wet	Dry	Squeezing	9
		60	$1.75 \cdot 10^{-3}$	Wet	Dry	Squeezing	10
10	89.64 ± 4.10	20	$1.18 \cdot 10^{-3}$	Wet	Dry	Squeezing	7
		30	$1.77 \cdot 10^{-3}$	Wet	Dry	Squeezing	11
		40	$2.36 \cdot 10^{-3}$	Wet	Dry	Squeezing	8
		60	$3.55 \cdot 10^{-3}$	Wet	Dry	Transition	11
20	83.39 ± 1.96	20	$2.09 \cdot 10^{-3}$	Wet	Dry	Transition	12
		30	$3.14 \cdot 10^{-3}$	Wet	Dry	Transition	12
		40	$4.19 \cdot 10^{-3}$	Wet	Dry	Transition	9
		60	$6.28 \cdot 10^{-3}$	Wet	Dry	Dripping	16
30	78.65 ± 1.84	20	$3.07 \cdot 10^{-3}$	Wet	Dry	Transition	13
		30	$4.61 \cdot 10^{-3}$	Wet	Dry	Transition	10
		40	$6.15 \cdot 10^{-3}$	Wet	Dry	Dripping	12
		60	$9.22 \cdot 10^{-3}$	Wet	Wet	Dripping	19
50	44.20 ± 2.74	20	$4.25 \cdot 10^{-3}$	Wet	Dry	Dripping	13
		30	$6.37 \cdot 10^{-3}$	Wet	Wet	Dripping	21
		40	$8.49 \cdot 10^{-3}$	Wet	Wet	Dripping	19
		60	$1.27 \cdot 10^{-2}$	Wet	Wet	Dripping	17

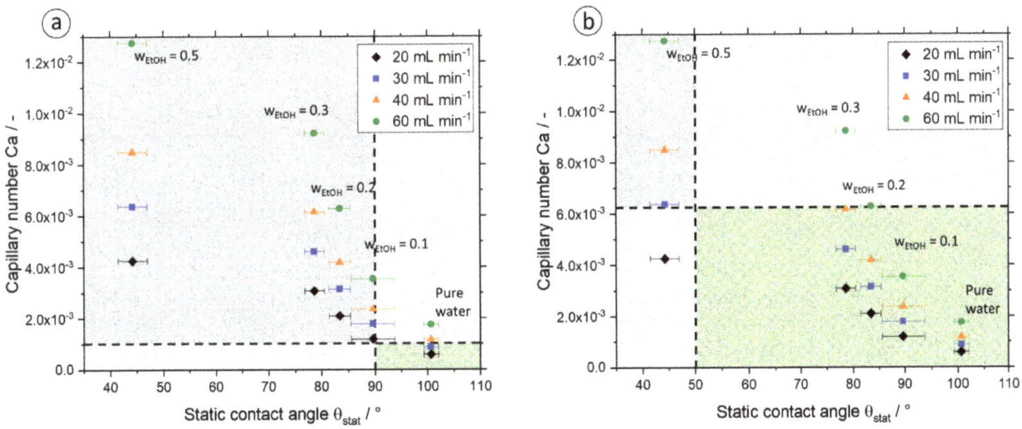

Figure 8. The *Ca* number is plotted against the Θ_{stat} for different EtOH/water compositions and volume flow rates. Delineations for the dry pattern are shown via the black dashed lines based on the literature (**a**) and based on the observations in this work (**b**). The green area marks the dry pattern, the white area the transition, and the gray area the wet region. The latter is unsuitable for crystallization in the SFC.

This contradicts the observations from our experiments for stable slug flow. Figure 9 shows exemplarily the slug shape formed in the apparatus at a total volumetric flow rate of Q_{tot} = 20 mL min^{-1} for all tested ethanol/water mixtures from w_{EtOH} = 0–50 wt.-% (All other pictures are given in the supplementary information Figures S3–S7). According to our results, there is a broader green region where stable slug flow is possible and dry flow pattern is formed, also with solvent mixtures, as this is the case for all mixtures in Figure 9.

w_{EtOH} = 0 %

w_{EtOH} = 10 %

w_{EtOH} = 20 %

w_{EtOH} = 30 %

w_{EtOH} = 50 %

Figure 9. Depiction of the slug flow obtained in the experiments for evaluating the slug shape for different compositions of ethanol/water mixtures at a total volumetric flow rate of Q_{tot} = 20 mL min^{-1} at ambient temperature ($\vartheta_{amb} \approx 22$ °C).

Thus, according to the literature, the suitable region is too small and does not fit to the experimental results. From this, the limits for a stable slug flow for a system and the operating parameters under consideration were modified and given in Figure 8b. The limit of partly wetting region ($\Theta_{stat} > 50°$) can be set as a new limit for the contact angle and $Ca < 6.3 \cdot 10^{-3}$ is set according to the observations. Thus, the range of application is significantly increased, as the majority of the tested mixtures also lie in the stable slug flow area (green area). Operating points or compositions that do not meet any of the criteria should be avoided for crystallization in the SFC.

However, other experimental observations have emerged that should definitely be taken into account for performing reproducible and reliable crystallization in the apparatus. For higher ethanol content, the front interface appears to be more rounded than for lower ethanol content or no ethanol content (experimental observation), confirming the dependence of the meniscus curvature on the *Ca* number [80].

Furthermore, differences in operation with respect to slug formation could be identified since inside the T-junction a transition state from squeezing to dripping mechanism at the compositions w_{EtOH} = 20 wt.-% and w_{EtOH} = 30 wt.-%, and at w_{EtOH} = 50 wt.-% the undesired dripping mechanism was observed. Consequently, the lower interfacial tension of the mixture with rising ethanol content and, correspondingly, the lower contact angle between wall material (T-junction, polypropylene (PP), $\Theta_{PP/wat/air}$ = 102°), liquid, and gas lead to a change in the slug formation mechanism. These observations have also been noticed during the handling of higher total volumetric flow rates in the apparatus. The observed slug formation mechanisms and contact angle hysteresis for different compositions and flow velocities are summarized in Table 7. It can be seen that with increasing flow velocity, the transition from squeezing mechanism to dripping mechanism takes place with decreasing ethanol fraction. Accordingly, the operating range in which the SFC is to operate is decisive for the selection of the maximum permissible ethanol content in the mixture. Consequently, at a volume flow of Q_{tot} = 20 mL min^{-1}, an ethanol content of up to w_{EtOH} = 10 wt.-% is permissible in order to enable continuous crystallization operation with high product quality. At a volume flow rate of Q_{tot} = 60 mL min^{-1}, on the other hand, an ethanol content should be dispensed with so that a constant slug length and, correspondingly, the same crystallization conditions for each crystal and the condition $RTD_L = RTD_S$ can be fulfilled. With regard to the static contact angles in Table 7, this means that contact angles from Θ_{stat} = 80° (w_{EtOH} = 0.3) are also theoretically usable in the apparatus, but here, the restriction applies whereby the flow velocity should not be excessively high, since in this case, the slug formation mechanism does not meet the specifications.

6. Conclusions

A quick decision on the suitability of a new solvent system for obtaining a uniform and reproducible product yield with particles of desired size, a narrow width of particle size distribution, and a high purity inside the slug flow crystallizer is possible with the help of an efficient strategy. In this context, particular attention must be paid to slug flow stability and residence time distribution along the tubing, but also to material system-specific criteria for crystallization. In this study, it was shown for several solid components (two amino acids, one API) that selecting a suitable solvent is simplified by using static contact angle measurements and evaluating a dimensionless parameter. Based on the experimental results in this study, the conventional criteria from the literature for the static contact angle (non-wetting behavior for $\Theta_{stat} > 90°$) or the classification of the flow pattern via the capillary number ($Ca < 10^{-3}$) are not sufficient due to the various relationships in the complex two-phase flow (velocity influence, three-phase interactions, slug formation mechanism, solute influence in the system under consideration, and more). However, by combining the static contact angle and the capillary number, the range of a dry pattern can be reliably estimated. For the system under consideration, three areas can be defined: the dry pattern area, which is limited by $\Theta_{stat} > 50°$ and $Ca < 6.3 \cdot 10^{-3}$; the transition range, in which only one criterion is reached, and a check for suitability for slug flow stability should be carried out depending on the compositions or the operating conditions; and the range for which no criterion is reached and crystallization in this composition should be avoided in combination with the operating parameters, since a wet pattern results. By using the ranges defined here for Θ_{stat} and *Ca* number, the possible range of applications for the SFC has been extended.

In addition to these two indicating parameters (Θ_{stat} and *Ca*), the volatility of the solvent and the solute solubility, and its effect on the contact angle, should also be monitored and backed up by viewing the dynamic contact angle behavior inside the apparatus. Further consideration of the solute's solubility in the solvent via criteria and modeling of the temperature-dependent solubility ensures the prerequisites for successful crystallization in the SFC and reduces the experimental and time effort. However, this is only valid if contamination, and disturbances in the material, for example, can be prevented and uniform operating conditions are maintained. This strategy has been demonstrated for both the binary and ternary systems and, thus, represents a general approach for using different material systems in the apparatus.

Supplementary Materials: The following supporting information can be downloaded at https://www.mdpi.com/article/10.3390/mi13101795/s1, Figure S1: Ternary phase diagram of Arg/water/ethanol at 0.1 MPa; Figure S2: Ternary phase diagram of APAP/water/ethanol at 0.1 MPa; Figure S3: Images of water slugs at different flow velocities inside the SFC; Figure S4: Images of ethanol/water slugs (w_{EtOH} = 10%) at different flow velocities inside the SFC; Figure S5: Images of ethanol/water slugs (w_{EtOH} = 20%) at different flow velocities inside the SFC; Figure S6: Images of ethanol/water slugs (w_{EtOH} = 30%) at different flow velocities inside the SFC; Figure S7: Images of ethanol/water slugs (w_{EtOH} = 50%) at different flow velocities inside the SFC.

Author Contributions: Conceptualization, A.C.K. and K.W.; methodology, A.C.K. and K.W.; software, A.K.; investigation, A.C.K. and A.K.; writing—original draft preparation, A.C.K. and A.K. (equal authorship); writing—review and editing, A.D. and K.W.; visualization, A.C.K. and A.K.; supervision, A.D. and K.W. All authors have read and agreed to the published version of the manuscript.

Funding: This research received no external funding.

Data Availability Statement: All data are contained within the article or the supplementary information.

Acknowledgments: The authors express their special thanks to Daniela Ermeling, Tobias Pape, Marén Schwandt, and Aaron Hiese for the technical support during the experiments.

Conflicts of Interest: The authors declare no conflict of interest. The funders had no role in the design of the study; in the collection, analyses, or interpretation of data; in the writing of the manuscript; or in the decision to publish the results.

Abbreviations

A	Association site
assoc	Association
Ac	Acetone
Ala	L-alanine
amb	Ambient
APAP	Acetaminophen
API	Active pharmaceutical ingredient
Arg	L-arginine
B	Association site
b	Intercept
disp	Dispersion
dyn	Dynamic contact angle
EtOH	Ethanol
FEP	Fluorinated ethylene propylene
hc	Hard chain
i	Component i
IPA	Isopropanol
j	Component j
L	Liquid phase
m	Slope
MSMPR	Mixed-suspension mixed-product removal
PC-SAFT	Perturbed-chain statistical associating fluid theory
PFC	Plug-flow crystallizer
PP	Polypropylene
PSD	Particle size distribution
PVC	Polyvinyl chloride
res	Residual
RTD	Residence time distribution
RTD_L	Residence time distribution of the liquid phase
RTD_S	Residence time distribution of the solid phase
S	Solid phase
SFC	Slug flow crystallizer
stat	Static contact angle
wat	Water

Latin Symbols

a	Helmholtz energy/J mol^{-1}
c	Concentration/g g^{-1}
c^*	Saturation concentration/g g^{-1}
Ca	Capillary number/-
c_p	Heat capacity/J mol^{-1}
d_h	Hydraulic diameter/mm
d_i	Inner diameter/mm
d_out	Outer diameter/mm
$E\ddot{o}$	Eötvös number/-
h	Enthalpy/J mol^{-1}
k	Interaction parameter/-
k_B	Boltzmann constant/J K^{-1}
L_{50}	Median slug length
$L_{90\text{-}10}$	Width of slug length
L_tubing	Length of tubing/m
M	Molar mass/g mol^{-1}
m^seg	Segment number/-
N	Number of sites/-
Q_air	Gas volume flow rate/mL min^{-1}
Q_liq	Liquid volume flow rate/mL min^{-1}
Q_tot	Total volume flow rate/mL min^{-1}
R	Universal gas constant/J mol^{-1} K^{-1}
T	Temperature/K
u	Flow velocity/m s^{-1}
u	Dispersion energy/J mol^{-1}
w	Mass fraction/wt.-%
x	Mole fraction/mol mol^{-1}

Greek Symbols

η	Dynamic viscosity/Pa s
Θ	Three-phase contact angle/°
$\Delta a_{c_{p,i}^{\mathrm{SL}}}$	Slope of linear temperature-dependent heat capacity/J mol^{-1} K^{-2}
$\Delta b_{c_{p,i}^{\mathrm{SL}}}$	Intercept of linear temperature-dependent heat capacity/J mol^{-1} K^{-1}
$\sigma_{\mathrm{G/L}}$	Gas–liquid interfacial tension/N m^{-1}
$\sigma_{\mathrm{S/G}}$	Solid–gas interfacial tension/N m^{-1}
$\sigma_{\mathrm{S/L}}$	Solid–liquid interfacial tension/N m^{-1}
ϑ	Temperature/°C
ϑ^*	Saturation temperature/°C
γ	Activity coefficient/-
ε	Association energy/J mol^{-1}
κ	Association volume/-
σ	Segment diameter/Å

References

1. Etminan, A.; Muzychka, Y.S.; Pope, K. Liquid film thickness of two-phase slug flows in capillary microchannels: A review paper. *Can. J. Chem. Eng.* **2021**, *100*, 325–348. [CrossRef]
2. Kleinebudde, P.; Khinast, J.; Rantanen, J. *Continuous Manufacturing of Pharmaceuticals*; John Wiley & Sons: Hoboken, NJ, USA, 2017.
3. Chen, J.; Sarma, B.; Evans, J.M.B.; Myerson, A.S. Pharmaceutical Crystallization. *Cryst. Growth Des.* **2011**, *11*, 887–895. [CrossRef]
4. Jiang, M.; Braatz, R.D. Designs of continuous-flow pharmaceutical crystallizers: Developments and practice. *CrystEngComm* **2019**, *21*, 3534–3551. [CrossRef]
5. Ma, Y.; Wu, S.; Macaringue, E.G.J.; Zhang, T.; Gong, J.; Wang, J. Recent Progress in Continuous Crystallization of Pharmaceutical Products: Precise Preparation and Control. *Org. Process Res. Dev.* **2020**, *24*, 1785–1801. [CrossRef]
6. Hofmann, G. *Kristallisation in der Industriellen Praxis*; Wiley-VCH: Weinheim, Germany, 2004.

7. Orehek, J.; Teslić, D.; Likozar, B. Continuous Crystallization Processes in Pharmaceutical Manufacturing: A Review. *Org. Process Res. Dev.* **2020**, *25*, 16–42. [CrossRef]
8. Cote, A.; Erdemir, D.; Girard, K.P.; Green, D.A.; Lovette, M.A.; Sirota, E.; Nere, N.K. Perspectives on the Current State, Challenges, and Opportunities in Pharmaceutical Crystallization Process Development. *Cryst. Growth Des.* **2020**, *20*, 7568–7581. [CrossRef]
9. Wood, B.; Girard, K.; Polster, C.S.; Croker, D. Progress to Date in the Design and Operation of Continuous Crystallization Processes for Pharmaceutical Applications. *Org. Process Res. Dev.* **2019**, *23*, 122–144. [CrossRef]
10. Cubaud, T.; Ulmanella, U.; Ho, C.-M. Two-phase flow in microchannels with surface modifications. *Fluid Dyn. Res.* **2006**, *38*, 772–786. [CrossRef]
11. Etminan, A.; Muzychka, Y.; Pope, K. A Review on the Hydrodynamics of Taylor Flow in Microchannels: Experimental and Computational Studies. *Processes* **2021**, *9*, 870. [CrossRef]
12. Bretherton, F.P. The motion of long bubbles in tubes. *J. Fluid Mech.* **1961**, *10*, 166. [CrossRef]
13. Suo, M.; Griffith, P. Two-Phase Flow in Capillary Tubes. *J. Basic Eng.* **1964**, *86*, 576–582. [CrossRef]
14. Brauner, N.; Maron, D.M. Stability analysis of stratfied liquid-liquid flow. *Int. J. Multiph. Flow* **1992**, *18*, 103–121. [CrossRef]
15. Kandlikar, S.G.; Grande, W.J. Evolution of Microchannel Flow Passages—Thermohydraulic Performance and Fabrication Technology. *Heat Transf. Eng.* **2003**, *24*, 3–17. [CrossRef]
16. Triplett, K.; Ghiaasiaan, S.; Abdel-Khalik, S.; Sadowski, D. Gas–liquid two-phase flow in microchannels Part I: Two-phase flow patterns. *Int. J. Multiph. Flow* **1999**, *25*, 377–394. [CrossRef]
17. Chen, L.; Tian, Y.; Karayiannis, T. The effect of tube diameter on vertical two-phase flow regimes in small tubes. *Int. J. Heat Mass Transf.* **2006**, *49*, 4220–4230. [CrossRef]
18. Shah, M.M. A general correlation for heat transfer during film condensation inside pipes. *Int. J. Heat Mass Transfer* **1979**, *22*, 547–556. [CrossRef]
19. Termühlen, M.; Etmanski, M.M.; Kryschewski, I.; Kufner, A.C.; Schembecker, G.; Wohlgemuth, K. Continuous slug flow crystallization: Impact of design and operating parameters on product quality. *Chem. Eng. Res. Des.* **2021**, *170*, 290–303. [CrossRef]
20. Su, M.; Gao, Y. Air–Liquid Segmented Continuous Crystallization Process Optimization of the Flow Field, Growth Rate, and Size Distribution of Crystals. *Ind. Eng. Chem. Res.* **2018**, *57*, 3781–3791. [CrossRef]
21. Jiang, M.; Braatz, R.D. Low-Cost Noninvasive Real-Time Imaging for Tubular Continuous-Flow Crystallization. *Chem. Eng. Technol.* **2018**, *41*, 143–148. [CrossRef]
22. Robertson, K.; Flandrin, P.-B.; Klapwijk, A.R.; Wilson, C. Design and Evaluation of a Mesoscale Segmented Flow Reactor (KRAIC). *Cryst. Growth Des.* **2016**, *16*, 4759–4764. [CrossRef]
23. Scott, C.D.; Labes, R.; Depardieu, M.; Battilocchio, C.; Davidson, M.G.; Ley, S.V.; Wilson, C.C.; Robertson, K. Integrated plug flow synthesis and crystallisation of pyrazinamide. *React. Chem. Eng.* **2018**, *3*, 631–634. [CrossRef]
24. Levenstein, M.A.; Wayment, L.E.; Scott, C.D.; Lunt, R.A.; Flandrin, P.-B.; Day, S.J.; Tang, C.C.; Wilson, C.C.; Meldrum, F.C.; Kapur, N.; et al. Dynamic Crystallization Pathways of Polymorphic Pharmaceuticals Revealed in Segmented Flow with Inline Powder X-ray Diffraction. *Anal. Chem.* **2020**, *92*, 7754–7761. [CrossRef] [PubMed]
25. Pu, S.; Hadinoto, K. Comparative evaluations of bulk seeded protein crystallization in batch versus continuous slug flow crystallizers. *Chem. Eng. Res. Des.* **2021**, *171*, 139–149. [CrossRef]
26. Pu, S.; Hadinoto, K. Improving the reproducibility of size distribution of protein crystals produced in continuous slug flow crystallizer operated at short residence time. *Chem. Eng. Sci.* **2021**, *230*, 116181. [CrossRef]
27. Eder, R.J.P.; Schrank, S.; Besenhard, M.O.; Roblegg, E.; Gruber-Woelfler, H.; Khinast, J.G. Continuous Sonocrystallization of Acetylsalicylic Acid (ASA): Control of Crystal Size. *Cryst. Growth Des.* **2012**, *12*, 4733–4738. [CrossRef]
28. Neugebauer, P.; Khinast, J.G. Continuous Crystallization of Proteins in a Tubular Plug-Flow Crystallizer. *Cryst. Growth Des.* **2015**, *15*, 1089–1095. [CrossRef]
29. Besenhard, M.O.; Neugebauer, P.; Scheibelhofer, O.; Khinast, J.G. Crystal Engineering in Continuous Plug-Flow Crystallizers. *Cryst. Growth Des.* **2017**, *17*, 6432–6444. [CrossRef] [PubMed]
30. Young, T. An essay on the cohesion of fluids. *Philos. Trans. R. Soc. Lond.* **1805**, *95*, 65–87. [CrossRef]
31. Lee, C.Y.; Lee, S.Y. Pressure drop of two-phase plug flow in round mini-channels: Influence of surface wettability. *Exp. Therm. Fluid Sci.* **2008**, *32*, 1716–1722. [CrossRef]
32. Barajas, A.; Panton, R. The effects of contact angle on two-phase flow in capillary tubes. *Int. J. Multiph. Flow* **1993**, *19*, 337–346. [CrossRef]
33. Santos, R.M.; Kawaji, M. DEVELOPMENTS ON WETTING EFFECTS IN MICROFLUIDIC SLUG FLOW. *Chem. Eng. Commun.* **2012**, *199*, 1626–1641. [CrossRef]
34. Srinivasan, V.; Khandekar, S. Thermo-hydrodynamic transport phenomena in partially wetting liquid plugs moving inside micro-channels. *Sādhanā* **2017**, *42*, 607–624. [CrossRef]
35. Serizawa, A.; Feng, Z.; Kawara, Z. Two-phase flow in microchannels. *Exp. Therm. Fluid Sci.* **2002**, *26*, 703–714. [CrossRef]
36. Cubaud, T.; Ho, C.-M. Transport of bubbles in square microchannels. *Phys. Fluids* **2004**, *16*, 4575–4585. [CrossRef]
37. Thulasidas, T.; Abraham, M.; Cerro, R. Flow patterns in liquid slugs during bubble-train flow inside capillaries. *Chem. Eng. Sci.* **1997**, *52*, 2947–2962. [CrossRef]
38. Skartsis, L.; Khomami, B.; Kardos, J.L. The effect of capillary pressure on the impregnation of fibrous media. *SAMPE J.* **1992**, *28*, 19–24.

39. Maximino, R.B. Surface tension and density of binary mixtures of monoalcohols, water and acetonitrile: Equation of correlation of the surface tension. *Phys. Chem. Liq.* **2009**, *47*, 475–486. [CrossRef]
40. Wohlgemuth, K.; Schembecker, G. Modeling induced nucleation processes during batch cooling crystallization: A sequential parameter determination procedure. *Comput. Chem. Eng.* **2013**, *52*, 216–229. [CrossRef]
41. Vetter, T.; Burcham, C.; Doherty, M.F. Regions of attainable particle sizes in continuous and batch crystallization processes. *Chem. Eng. Sci.* **2014**, *106*, 167–180. [CrossRef]
42. Grant, D.; Mehdizadeh, M.; Chow, A.-L.; Fairbrother, J. Non-linear van't Hoff solubility-temperature plots and their pharmaceutical interpretation. *Int. J. Pharm.* **1984**, *18*, 25–38. [CrossRef]
43. Vazquez, G.; Alvarez, E.; Navaza, J.M. Surface Tension of Alcohol Water + Water from 20 to 50 °C. *J. Chem. Eng. Data* **1995**, *40*, 611–614. [CrossRef]
44. Howard, K.S.; McAllister, R.A. Surface tension of acetone-water solutions up to their normal boiling points. *AIChE J.* **1957**, *3*, 325–329. [CrossRef]
45. Klein, T.; Yan, S.; Cui, J.; Magee, J.W.; Kroenlein, K.; Rausch, M.H.; Koller, T.M.; Fröba, A.P. Liquid Viscosity and Surface Tension of *n*-Hexane, *n*-Octane, *n*-Decane, and *n*-Hexadecane up to 573 K by Surface Light Scattering. *J. Chem. Eng. Data* **2020**, *64*, 4116–4131. [CrossRef] [PubMed]
46. Schreiber, M.; Brunert, M.; Schembecker, G. Extraction on a Robotic Platform—Autonomous Solvent Selection under Economic Evaluation Criteria. *Chem. Eng. Technol.* **2021**, *44*, 1578–1584. [CrossRef]
47. Termühlen, M.; Strakeljahn, B.; Schembecker, G.; Wohlgemuth, K. Characterization of slug formation towards the performance of air-liquid segmented flow. *Chem. Eng. Sci.* **2019**, *207*, 1288–1298. [CrossRef]
48. Termühlen, M.; Strakeljahn, B.; Schembecker, G.; Wohlgemuth, K. Quantification and evaluation of operating parameters' effect on suspension behavior for slug flow crystallization. *Chem. Eng. Sci.* **2021**, *243*, 116771. [CrossRef]
49. Prausnitz, J.M. *Molecular Thermodynamics of Fluid-Phase Equilibria*; Prentice-Hall: Englewood Cliffs, NJ, USA, 1969.
50. Do, H.T.; Chua, Y.Z.; Kumar, A.; Pabsch, D.; Hallermann, M.; Zaitsau, D.; Schick, C.; Held, C. Melting properties of amino acids and their solubility in water. *RSC Adv.* **2020**, *10*, 44205–44215. [CrossRef]
51. Grant, D.J.; Grant, D.J.W.; Higuchi, T. *Solubility Behavior of Organic Compounds*; Wiley: New York, NY, USA, 1990.
52. Chua, Y.Z.; Do, H.T.; Schick, C.; Zaitsau, D.; Held, C. New experimental melting properties as access for predicting amino-acid solubility. *RSC Adv.* **2018**, *8*, 6365–6372. [CrossRef]
53. Granberg, R.A.; Rasmuson, C. Solubility of Paracetamol in Pure Solvents. *J. Chem. Eng. Data* **1999**, *44*, 1391–1395. [CrossRef]
54. Neau, S.H.; Bhandarkar, S.V.; Hellmuth, E.W. Differential Molar Heat Capacities to Test Ideal Solubility Estimations. *Pharm. Res.* **1997**, *14*, 601–605. [CrossRef]
55. Gross, J.; Sadowski, G. Application of the Perturbed-Chain SAFT Equation of State to Associating Systems. *Ind. Eng. Chem. Res.* **2002**, *41*, 5510–5515. [CrossRef]
56. Held, C.; Cameretti, L.F.; Sadowski, G. Measuring and Modeling Activity Coefficients in Aqueous Amino-Acid Solutions. *Ind. Eng. Chem. Res.* **2011**, *50*, 131–141. [CrossRef]
57. Ruether, F.; Sadowski, G. Modeling the Solubility of Pharmaceuticals in Pure Solvents and Solvent Mixtures for Drug Process Design. *J. Pharm. Sci.* **2009**, *98*, 4205–4215. [CrossRef] [PubMed]
58. Cameretti, L.F.; Sadowski, G. Modeling of aqueous amino acid and polypeptide solutions with PC-SAFT. *Chem. Eng. Process. Process Intensif.* **2008**, *47*, 1018–1025. [CrossRef]
59. Berthelot, D. Sur le mélange des gaz, Comptes rendus hebdomadaires des séances de l'Académie des. *Sciences* **1898**, *126*, 1703–1855.
60. Lorentz, H.A. Ueber die Anwendung des Satzes vom Virial in der kinetischen Theorie der Gase. *Ann. Phys.* **1881**, *248*, 127–136. [CrossRef]
61. An, M.; Qiu, J.; Yi, D.; Liu, H.; Hu, S.; Han, J.; Huang, H.; He, H.; Liu, C.; Zhao, Z.; et al. Measurement and Correlation for Solubility of l-Alanine in Pure and Binary Solvents at Temperatures from 283.15 to 323.15 K. *J. Chem. Eng. Data* **2020**, *65*, 549–560. [CrossRef]
62. Jiménez, J.A.; Martínez, F. Thermodynamic magnitudes of mixing and solvation of acetaminophen in ethanol + water cosolvent mixtures. *Rev. Acad. Colomb. Scienc.* **2006**, *30*, 87–99.
63. Paus, R.; Ji, Y. Modeling and predicting the influence of variable factors on dissolution of crystalline pharmaceuticals. *Chem. Eng. Sci.* **2016**, *145*, 10–20. [CrossRef]
64. Held, C.; Sadowski, G. *Manual 'ePC-SAFT 1.0'*; TU Dortmund University: Dortmund, Germany, 2017.
65. Wolbach, J.P.; Sandler, S.I. Using Molecular Orbital Calculations to Describe the Phase Behavior of Cross-associating Mixtures. *Ind. Eng. Chem. Res.* **1998**, *37*, 2917–2928. [CrossRef]
66. Kashid, M.N.; Agar, D.W. Hydrodynamics of liquid–liquid slug flow capillary microreactor: Flow regimes, slug size and pressure drop. *Chem. Eng. J.* **2007**, *131*, 1–13. [CrossRef]
67. Fu, T.; Ma, Y. Bubble formation and breakup dynamics in microfluidic devices: A review. *Chem. Eng. Sci.* **2015**, *135*, 343–372. [CrossRef]
68. Fu, T.; Ma, Y.; Funfschilling, D.; Zhu, C.; Li, H.Z. Squeezing-to-dripping transition for bubble formation in a microfluidic T-junction. *Chem. Eng. Sci.* **2010**, *65*, 3739–3748. [CrossRef]

69. Muzychka, Y.S.; Walsh, E.; Walsh, P. Heat Transfer Enhancement Using Laminar Gas-Liquid Segmented Plug Flows. *J. Heat Transf.* **2011**, *133*, 041902. [CrossRef]
70. Shao, N.; Gavriilidis, A.; Angeli, P. Effect of Inlet Conditions on Taylor Bubble Length in Microchannels. *Heat Transf. Eng.* **2011**, *32*, 1117–1125. [CrossRef]
71. Qian, D.; Lawal, A. Numerical study on gas and liquid slugs for Taylor flow in a T-junction microchannel. *Chem. Eng. Sci.* **2006**, *61*, 7609–7625. [CrossRef]
72. Zhu, C.; Gao, Y.; Li, H.; Meng, S.; Li, L.; Francisco, J.S.; Zeng, X.C. Characterizing hydrophobicity of amino acid side chains in a protein environment via measuring contact angle of a water nanodroplet on planar peptide network. *Proc. Natl. Acad. Sci. USA* **2016**, *113*, 12946–12951. [CrossRef]
73. Atta, D.Y.; Negash, B.M.; Yekeen, N.; Habte, A.D.; Malik, A.B.A. Influence of natural L-amino acids on the interfacial tension of an oil-water system and rock wettability alterations. *J. Pet. Sci. Eng.* **2020**, *199*, 108241. [CrossRef]
74. Daldrup, J.-B.G.; Held, C.; Ruether, F.; Schembecker, G.; Sadowski, G. Measurement and Modeling Solubility of Aqueous Multisolute Amino-Acid Solutions. *Ind. Eng. Chem. Res.* **2010**, *49*, 1395–1401. [CrossRef]
75. Amend, J.P.; Helgeson, H.C. Solubilities of the common L-α-amino acids as a function of temperature and solution pH. *Pure Appl. Chem.* **1997**, *69*, 935–942. [CrossRef]
76. Granberg, R.A.; Rasmuson, C. Solubility of Paracetamol in Binary and Ternary Mixtures of Water + Acetone + Toluene. *J. Chem. Eng. Data* **2000**, *45*, 478–483. [CrossRef]
77. Matsuda, H.; Mori, K.; Tomioka, M.; Kariyasu, N.; Fukami, T.; Kurihara, K.; Tochigi, K.; Tomono, K. Determination and prediction of solubilities of active pharmaceutical ingredients in selected organic solvents. *Fluid Phase Equilibria* **2015**, *406*, 116–123. [CrossRef]
78. Terdenge, L.-M.; Wohlgemuth, K. Impact of agglomeration on crystalline product quality within the crystallization process chain. *Cryst. Res. Technol.* **2016**, *51*, 513–523. [CrossRef]
79. Khattab, I.S.; Bandarkar, F.; Fakhree, M.A.A.; Jouyban, A. Density, viscosity, and surface tension of water+ethanol mixtures from 293 to 323K. *Korean J. Chem. Eng.* **2012**, *29*, 812–817. [CrossRef]
80. Giavedoni, M.D.; Saita, F.A. The rear meniscus of a long bubble steadily displacing a Newtonian liquid in a capillary tube. *Phys. Fluids* **1999**, *11*, 786–794. [CrossRef]

micromachines

MDPI

Article

Ultrasoft and Ultrastretchable Wearable Strain Sensors with Anisotropic Conductivity Enabled by Liquid Metal Fillers

Minjae Choe [1], Dongho Sin [1], Priyanuj Bhuyan [1], Sangmin Lee [1], Hongchan Jeon [2,*] and Sungjune Park [1,*]

1 Department of Polymer-Nano Science and Technology, Department of Nano Convergence Engineering, Jeonbuk National University, Jeonju 54896, Republic of Korea
2 Sustainable Materials Research Team, Research & Development Division, Hyundai Motor Group, Uiwang 16082, Republic of Korea
* Correspondence: hongchan@hyundai.com (H.J.); s.park@jbnu.ac.kr (S.P.)

Abstract: Herein, ultrasoft and ultrastretchable wearable strain sensors enabled by liquid metal fillers in an elastic polymer are described. The wearable strain sensors that can change the effective resistance upon strains are prepared by mixing silicone elastomer with liquid metal (EGaIn, Eutectic gallium-indium alloy) fillers. While the silicone is mixed with the liquid metal by shear mixing, the liquid metal is rendered into small droplets stabilized by an oxide, resulting in a non-conductive liquid metal elastomer. To attain electrical conductivity, localized mechanical pressure is applied using a stylus onto the thermally cured elastomer, resulting in the formation of a handwritten conductive trace by rupturing the oxide layer of the liquid metal droplets and subsequent percolation. Although this approach has been introduced previously, the liquid metal dispersed elastomers developed here are compelling because of their ultra-stretchable (elongation at break of 4000%) and ultrasoft (Young's modulus of <0.1 MPa) mechanical properties. The handwritten conductive trace in the elastomers can maintain metallic conductivity when strained; however, remarkably, we observed that the electrical conductivity is anisotropic upon parallel and perpendicular strains to the conductive trace. This anisotropic conductivity of the liquid metal elastomer film can manipulate the locomotion of a robot by routing the power signals between the battery and the driving motor of a robot upon parallel and perpendicular strains to the hand-written circuit. In addition, the liquid metal dispersed elastomers have a high degree of deformation and adhesion; thus, they are suitable for use as a wearable sensor for monitoring various body motions.

Keywords: liquid metal elastomers; soft and stretchable electronics; wearable strain sensors; soft robotics

Citation: Choe, M.; Sin, D.; Bhuyan, P.; Lee, S.; Jeon, H.; Park, S. Ultrasoft and Ultrastretchable Wearable Strain Sensors with Anisotropic Conductivity Enabled by Liquid Metal Fillers. *Micromachines* **2023**, *14*, 17. https://doi.org/10.3390/mi14010017

Academic Editor: Pingan Zhu

Received: 8 September 2022
Revised: 28 November 2022
Accepted: 16 December 2022
Published: 21 December 2022

1. Introduction

Soft and stretchable electronics have gained a great attention due to their potential applications in wearable sensors [1,2], soft robotics [3,4], and electronics skins [5,6]. The ability to change electric behavior of the soft and stretchable devices in response to various external stimuli can be utilized for designing the wearable sensors with targeted sensing mechanism such as piezoelectric, piezocapacitive, and piezoresistive. Among them, the piezoresistive sensors that can change the electrical resistance upon deformation of the devices have been widely utilized for measuring external strain, pressure, and forces straight forwardly [7,8].

Stretchable and soft piezoresistive sensors that can undergo a high degree of deformation can be utilized for wearable strain sensors, which are appealing for application in biomedical devices [9,10] and soft robotics [11,12] for monitoring human motion and utilizing human–machine interfaces. They can be created using elastic polymers embedded with conductive fillers and exhibit changes in electrical resistance in response to external strains [13,14]. These wearable sensors typically require mechanically competent circuits to electrically interconnect and mechanically support electronic components [15,16]. To fabricate soft and stretchable circuits, electrodes such as metal particles [17,18], metal wires [19,20], and carbonaceous fillers [21,22] have been included in elastic components; however, these conductors may compromise the mechanical softness of the elastic components owing to the rigidity of the materials.

Liquid metals (gallium and gallium alloys) are considered compelling electrodes for creating soft and stretchable circuits because they can preserve electrical conductivity while strained [23,24]. The liquid metals can form a thin (~3 nm) surface oxide layer in air [25,26]; thus, the metal can adhere to various substrates and can be patterned into desired shapes by injection [27,28], vacuum-assisted capillary filling [29,30], printing [31,32], and forced wetting [33,34]. An alternative method to form the liquid metal circuits is to utilize the liquid metal particles rendered by ultrasonication [35–37]. Liquid metal can break into micro- and nanosized particles in elastomers by ultrasonication and are subsequently stabilized by the oxide. Although the liquid metal particles with tunable sizes enabled by various loading volumes can benefit for use in dielectric elastomers [38,39], electronic ink [36,40,41], and catalysts [42,43], the liquid metal particle-included elastomer composites are electrically nonconductive because of the oxide layer on the particles [44,45]. Thus, various methods have been explored to restore electrical conductivity by rupturing the oxide layer of the particles and coalescing into conductive pathways. Laser- and thermal sintering of the liquid metal particles can result in conductive metallic circuits [46,47] and applying dielectrophoresis can also directly guide the liquid metal particles to align and to form the circuits [48,49]. Although these approaches can be used to pattern the metallic circuits, localizing mechanical pressure onto the liquid metal elastomer to rupture the oxide layer on the liquid metal particles is the probably most convenient method for creating circuits that can be utilized for stretchable and soft electronics because these approaches are fairly facile and can enable to create electric circuit non-lithographically, thus, it is economically favorable [50–55].

In this study, we demonstrate stretchable and soft wearable strain sensors enabled by liquid metal fillers in an elastic polymer. Once the oxide layer of the liquid metal droplets is ruptured by localizing mechanical pressure on the elastic polymer, a handwritten conductive trace is formed. Although this concept has been introduced by several groups [50,53,56], we clarify the novel aspects of the present study by demonstrating anisotropic conductivity of the liquid metal elastomer film upon parallel and perpendicular strains to the handwritten conductive trace. The handwritten circuit can typically provide an electrical pathway in the plane direction; however, we observed the varied electrical behavior of the liquid metal circuits under parallel and perpendicular strain along the direction of the conductive traces and we termed this 'anisotropic conductivity.' This behavior can be realized by utilizing a silicone elastomer (ExSil 100) that exhibits ultrahigh elongation at break (4000%) and ultrasoft properties (Young's modulus of <0.1 MPa) owing to non-crosslinked and highly entangled polymeric chains. We utilized this principle to manipulate the locomotion of a robot by routing the power signals between the battery and the driving motor of a robot upon parallel and perpendicular strains to the handwritten circuit. The elastomer used in this study is ultrasoft and readily deformable in response to external strains. Thus, we utilized the liquid metal elastomer as a stretchable and soft wearable strain sensor to monitor various human motions by attaching it to curvilinear body surfaces. These ultrastretchable and soft wearable sensors with anisotropic conductivity enabled by liquid metal fillers would be useful for application in soft robotics, artificial skins, and stretchable and soft electronics.

2. Materials and Methods

2.1. Fabrication of Liquid Metal Elastomer Films

The liquid metal elastomer composites were prepared by shear mixing silicone elastomers (Sylgard 184 from Dow Corning and ExSil 100 from Gelest, Morrisville, PA, USA) and liquid metal (eutectic gallium-indium alloy, Indium Corporation, Clinton, New York, NY, USA) with various wt% using a planetary mixer at 800 rpm for 10 min (Thinky mixer

ARE-310, Sotokanda, Chiyoda-ku, Tokyo, Japan). Once the silicone elastomer and liquid metal was completely mixed, the liquid metal elastomer films were prepared by thermal curing at 60 °C for 4 h, and localized mechanical pressure (normal force of 0.1 MPa, which was measured by dividing the maximum weight multiple and the gravitational acceleration by the area) was applied onto the liquid metal elastomer films using a stylus with a diameter of 8 mm and subsequently scratched along the surface of the elastomer to create conductive traces. Copper tapes were attached to both ends of the conductive traces to characterize their electrical behavior.

2.2. Characterization

The mechanical properties of the liquid metal elastomer films were characterized using an extensometer (Quasar 2.5 Single column, Galdabini, Cardano al Campo, Italy) with a constant load application of 1 kN at an extension rate of 10 mm/min until the samples failed. All electrical characterizations were performed using a benchtop multimeter (Keysight 34461a, Keysight Technologies, Santa Rosa, CA, USA). By connecting a benchtop multimeter and extension rate of 10 mm/min by extensometer, data was collected to evaluate how resistance changed as a function of strain. The morphology of liquid metal elastomers according to the liquid metal content was observed using an optical microscopy (Olympus CX23, Olympus Corporation, Tokyo, Japan).

3. Results and Discussion

We used two silicone elastomers (Sylgard 184 and ExSil 100) to fabricate liquid metal elastomer films with conductive traces created by hand writing, as shown in Figure 1a,b. The liquid metal elastomer films were prepared by shear mixing the silicone elastomers with liquid metal (LM) fillers with different loading amounts. Once the liquid metal elastomer film was formed by shear mixing, followed by thermal curing, the LM droplets stabilized by the oxides were dispersed in the elastomeric matrix. The diameter of the LM droplets can be manipulated by varying the mixing conditions, such as the concentration of the LM and rotating speed during mixing [44,57]. The liquid metal elastomer film did not initially exhibit electrical conductivity owing to the presence of a thin oxide layer on the LM droplet surface, however, localizing mechanical pressure onto the liquid metal elastomer films resulted in rupturing of the oxide layer and percolating the liquid metal to form a conductive trace (Figure 1c). Previously, the high rigidity-induced high energy dissipation coefficient of the polymeric matrix with LM droplets can help to generate mechanically induced conductive traces by applying localized forces onto the polymer [58]. Although the elastomer used in this study is soft (Young's modulus of 3.9 and 0.1 MPa for Sylgard 184 and ExSil 100, respectively) [59], we note that the liquid metal elastomer film can also be directly and conductively traced by multiple applications of localized mechanical pressure.

The liquid metal elastomer films with various ratios of the LM fillers were used for characterizing the mechanical properties. Figure 2a,b shows Young's modulus of the elastic films with the LM according to the content of the LM fillers. As the liquid metal content increases in the range of 0 to 80 wt%, the Young's modulus of the Sylgard 184 films with the liquid metals (Sylgard 184-LM) and the ExSil 100 film with the liquid metals (ExSil 100-LM) decreases from 1.6 MPa to 0.5 MPa and from 0.3 MPa to 0.1 MPa, respectively, due to fluidic nature of the LM fillers [39,51,52,55]. The presence of fluid droplets with a high surface tension (around 360 mN m^{-1}) [60–63] can affect the Young's modulus of elastomer composites as predicted by Equation (1).

$$E_c = E\left\{\frac{1+\frac{5\gamma}{2ER}}{\frac{5\gamma}{2ER}(1-\varphi)+\left(1+\frac{5\varphi}{3}\right)}\right\} \quad (1)$$

where, E_c, E, γ, φ, and R represent the Young's modulus of composite, Young modulus of encasing elastomer, surface tension of the LM droplet (360 mN m^{-1}), volume percent (φ) of the LM droplet and the radius of the LM droplet. The experimental results are almost identical to theoretical values, i.e., difference is less than 0.6 MPa. As shown in Figure 2c,d, elongation at break of the elastomer films with the LM fillers also decrease inversely as a function of the contents of the LM fillers presumably due to large area of surface oxide of the LM droplets.

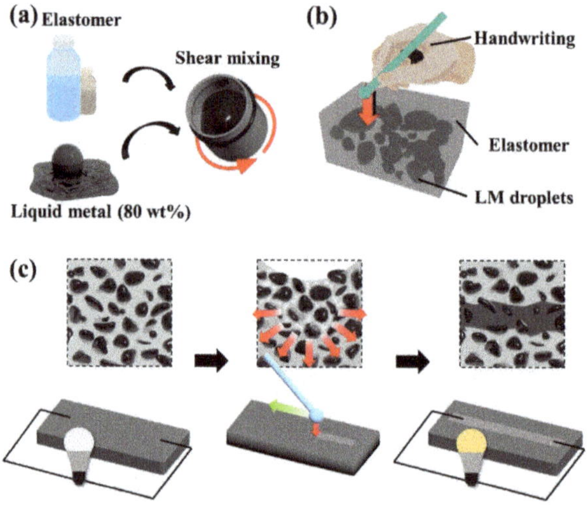

Figure 1. Schematics showing the fabrication process of the liquid metal elastomer film with a hand-written circuit. (**a**) Shear mixing a silicone elastomer with the liquid metal, (**b**) thermally cured silicone elastomer with the liquid metal droplets dispersed, and (**c**) conductive trace by sintering the liquid metal droplets via applying localized mechanical pressure onto the liquid metal elastomer film.

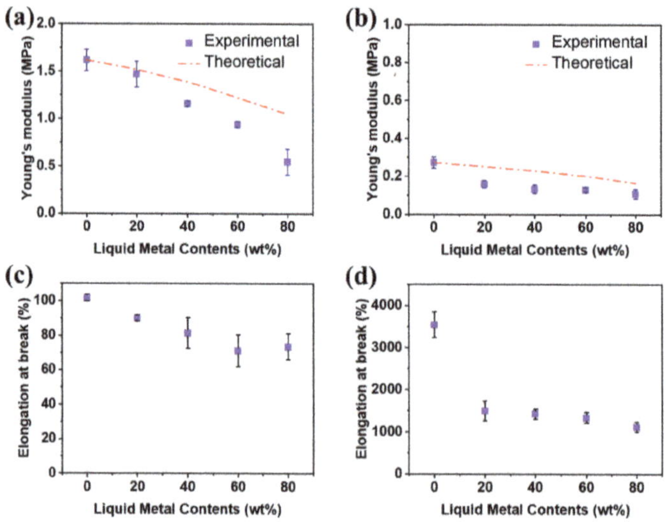

Figure 2. Mechanical properties of the liquid metal elastomer films made of Sylgard 184 (**a**,**c**) and ExSil 100 (**b**,**d**). (**a**,**b**) Young's modulus of the liquid metal elastomer films as a function of the contents of liquid metal fillers. (**c**,**d**) Plot of elongation at break of the liquid metal elastomer films as a function of the contents of the liquid metal fillers.

As previously studied, Young's modulus of the elastic composites can be changed by various diameters of the fillers [38,39,64,65]. As shown in the optical microscopy images of the elastomer composites with LM fillers (Figure 3), the diameter of the LM droplets dispersed in the elastomers decreases inversely as a function of the contents of the LM fillers due to the viscosity of the composite increased as a function of LM fillers content [12,38,44]. As the viscosity of the composite increases, more force is applied to maintain the constant rotational speed (800 rpm), resulting in the smaller radius LM droplets.

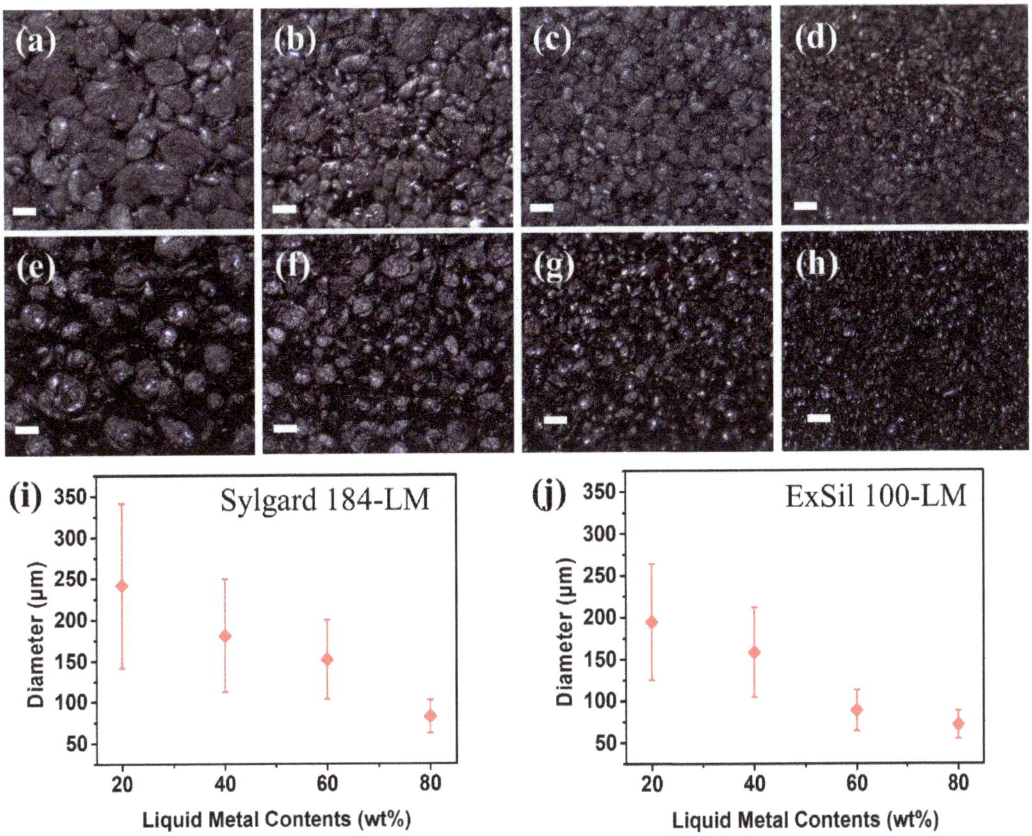

Figure 3. Optical microscopy images showing (**a–d**) the Sylgard 184 and (**e–h**) the ExSil 100 with the various contents of the liquid metal fillers. (**a**) 20 wt%, (**b**) 40 wt%, (**c**) 60 wt%, and (**d**) 80 wt% of the liquid metal fillers. The scale bar represents 200 μm. (**i,j**) The diameters of the liquid metal droplets dispersed in (**i**) Sylgard 184 and (**j**) ExSil 100 according to various contents of the liquid metal fillers.

As shown in Figure 4a–h, the liquid metal elastomer films (ExSil 100-LM and Sylgard 184-LM) with the handwritten circuits shows the electrical conductivity maintained upon various deformations, such as stretching, folding, and twisting as demonstrated by LED activation (Video S1, Supplementary Materials). The elastomer-LM films exhibits the effective resistance increased as a function of strain in parallel direction along the conductive trace (Figure 4c,i) as theoretically expected by the well-known equation $R = \rho(L/A)$, where R denotes the effective resistance and ρ is the bulk resistivity of the liquid metal. As the film is stretched, the length of the trace (L) increases, whereas the cross-sectional area (A) narrows, resulting in increased resistance (R). However, it is observed that there is a little change in the effective resistance of the handwritten liquid metal circuit upon twisting because the length and area of the conductive trace is almost constant (Figure 4d,j).

We also scratched the conductive trace region of the liquid metal elastomer film and punched it to make a hole with 3 mm in diameter to demonstrate the electrical conductivity of the handwritten circuit after mechanical damages without leakage owing to the liquid metal electrode stabilized by the oxides (Figure 4e,f).

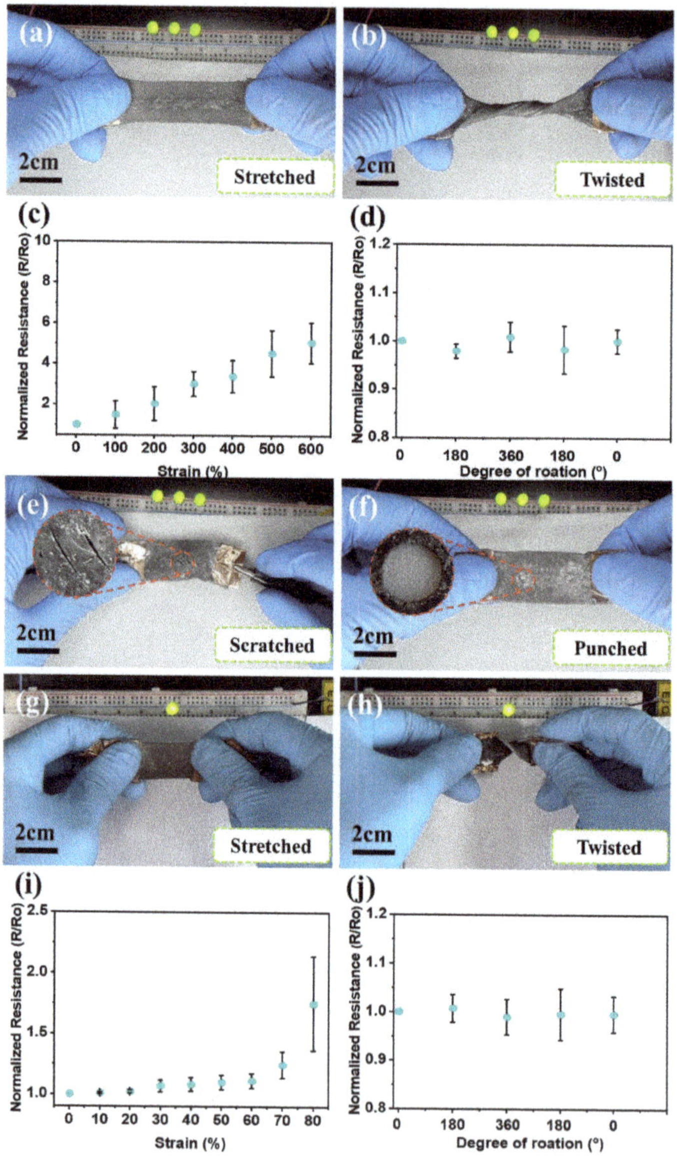

Figure 4. Photographs of (**a–f**) the ExSil 100-LM films and (**g–j**) the Sylgard 184-LM films showing electrical conductivity while strained. The ExSil 100-LM films were subjected to (**a**) stretching, (**b**) twisting and folding, (**e**) scratching, and (**f**) punching. (**c**,**d**) The effective normalized resistance of the ExSil 100-LM films upon (**c**) stretching and (**d**) twisting. The Sylgard 184-LM films were also subjected to (**g**) stretching and (**h**) twisting. (**i**,**j**) The effective normalized resistance of the Sylgard 184-LM films upon (**i**) stretching and (**j**) twisting.

We also characterized electrical stability of the liquid metal elastomer films as shown in Figure 5. The normalized resistance increased by 3% over 50 cycles of tensile testing at 100% strain. Although the fluctuation of the effective resistance increased because of the less entangled polymeric network upon consecutive tensile strains, the normalized effective resistance value increased slightly, thereby demonstrating the electrical stability of the handwritten conductive trace.

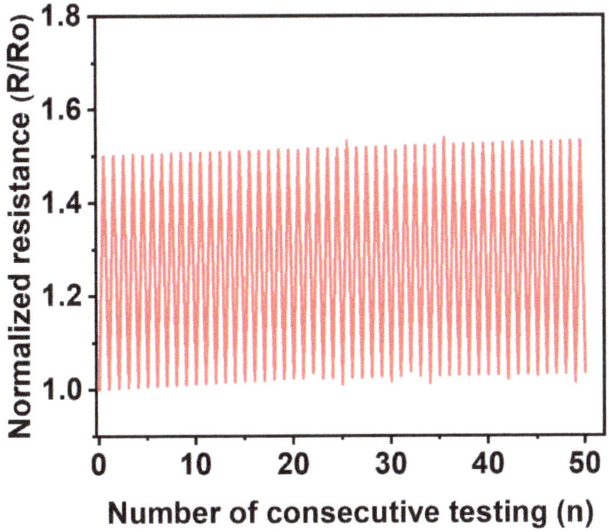

Figure 5. Normalized effective resistance of the liquid metal elastomer film with handwritten circuit during 50 consecutive cycles of tensile testing at strains of 100%.

We characterized the electrical behavior of the conductive trace when the elastic films were strained (Figure 6). We used ExSil 100-LM films due to their extremely high elongation at break induced by non-crosslinked and highly entangled polymeric network [66]. The liquid metal elastomer films can be stretched upon application of strain while maintaining its electrical conductivity (Video S1, Supplementary Materials) because of the preserved network of the percolated liquid metal electrode. As shown in Figure 6a,b, the effective electrical resistance increased as a function of strain parallel to the direction of the conductive trace because of increased length and narrowed cross-sectional area of the electrode upon strains. The effective resistance change of the conductive trace upon applying a strain of 600% parallel to the electrode was observed to be five times higher than that of the conductive trace without strain. However, the conductive trace exhibits a slightly increased electrical resistance (1.8 times) as the strain of 600% perpendicular to the electrode is applied due to the nearly constant length of the electrode. We used this anisotropic conductivity of the ExSil 100-LM upon strains to manipulate the locomotion of a robot enabled by routing the power signals between the battery and the driving motor of a walking robot as shown in Figure 6c–e. The robot rapidly moves when powered on due to the electrical conductivity of the LM circuit without strains (Figure 6c and Video S2, Supplementary Materials); however, locomotion can be manipulated by strain along the direction of the conductive trace. When the ExSil 100-LM is strained parallel to the LM electrode trace, the robot moves slowly owing to increased effective resistance upon strain (Figure 6d). However, the locomotion gets enhanced as the ExSil 100-LM is strained perpendicular to the LM electrode (Figure 6e).

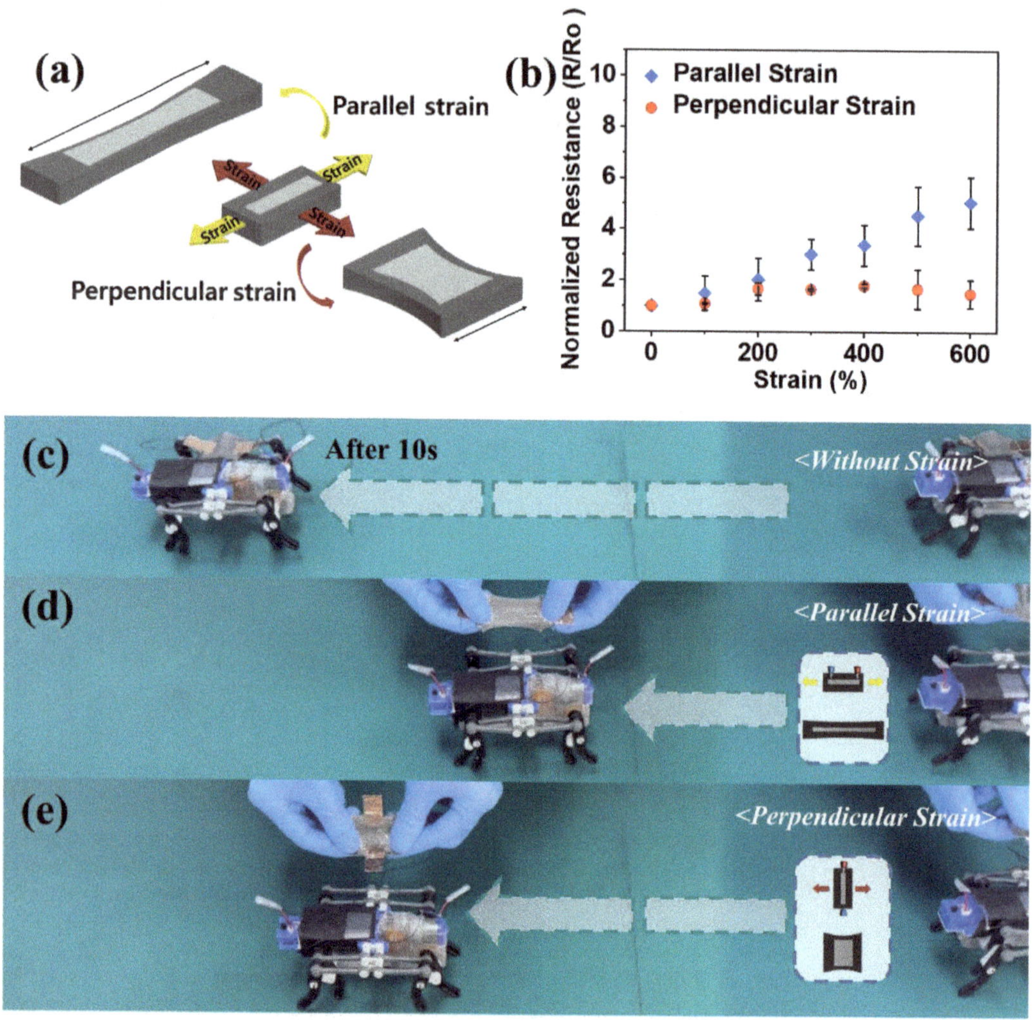

Figure 6. Liquid metal elastomer films exhibiting anisotropic conductivity. (**a**) Schematic showing the geometrical change of the liquid metal elastomer film with handwritten circuit upon strains parallel and perpendicular to the direction of the conductive trace. (**b**) Normalized resistance versus strain for the conductive trace upon parallel and perpendicular strains. (**c**–**e**) Photographs showing the locomotion of a robot enabled by the liquid metal elastomer film connected (**c**) without strain, (**d**) with parallel strain along the direction of the conductive trace, and (**e**) with perpendicular strain along the direction of the conductive trace.

The liquid metal elastomer films can be utilized as soft and wearable strain sensors to monitor various human motions, as shown in Figure 7. The elastomer used in this study has a high degree of deformation and adhesion to the surfaces due to highly entangled polymeric network and a small amount of uncured silicone oil that prevents the polymer from adhering to the curvilinear surfaces; thus, it is suitable for use as wearable strain sensors by direct attachment to the human body. As shown in Figure 7a, the liquid metal elastomer film attached to the elbow exhibited change in the effective resistance induced by mechanical bending. A higher bending angle generated higher change in the effective resistance (Figure 7b). Similarly, liquid metal elastomer film based wearable strain sensors can change effective resistance caused by movement in joints when they are attached to the wrist, knee, throat, and neck (Figure 7c–f). The liquid metal elastomer film is strained by the movement of the uvula on the throat, thus demonstrating the sensing ability of the LM conductive trace to monitor tiny body motions (Figure 7e).

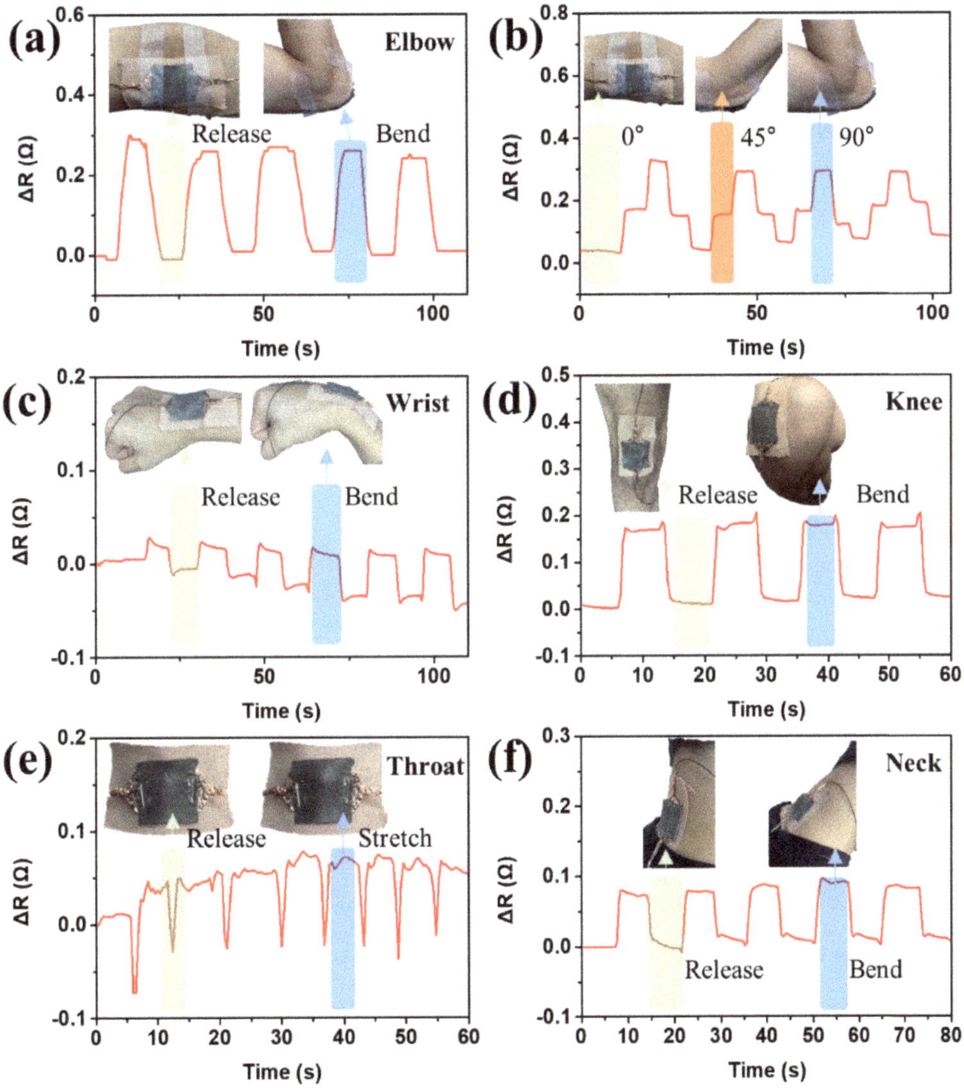

Figure 7. Liquid metal elastomer films utilized soft and wearable strain sensors monitoring various body motions by attaching them to the (**a**,**b**) elbow, (**c**) wrist, (**d**) knee, (**e**) throat, and (**f**) neck.

4. Conclusions

In summary, we demonstrated ultrastretchable and soft wearable strain sensors with anisotropic conductivity enabled by liquid metal fillers. The liquid metal elastomer film was prepared by shear mixing the elastomer with liquid metal fillers, followed by thermal curing. When the liquid metal elastomer film was mixed with liquid metal, the metal was rendered into small droplets stabilized by an oxide, resulting in a non-conductive elastomer film. However, the liquid metal elastomer film can store electrical conductivity by rupturing the oxide layer of the liquid metal droplets and sintering them by applying localized pressure using a stylus to the liquid metal elastomer film. The handwritten conductive trace in the liquid metal elastomer film can maintain the conductivity when strained; however, the conductivity is anisotropic upon strains parallel and perpendicular to the conductive trace. We utilized the anisotropic conductivity of the circuits to manipulate the locomotion of a robot by routing the power signals between the battery and the driving motor of a walking robot. The liquid metal elastomer film has a high degree of deformation and adhesion to the surface; thus, it can be utilized as a soft and wearable strain sensor to monitor various body motions. The ultrastretchable and soft liquid metal elastomer films developed in this study could be used in soft robotics, stretchable electronics, and wearable devices.

Supplementary Materials: The following supporting information can be downloaded at: https://www.mdpi.com/article/10.3390/mi14010017/s1, Video S1: Liquid metal elastomer film showing electrical conductivity while strained. Video S2: Manipulated locomotion of a robot enabled by liquid metal elastomer film with anisotropic conductivity.

Author Contributions: Conceptualization, M.C. and S.P.; Methodology, M.C., D.S., P.B., S.L., and S.P.; Writing—original draft preparation, M.C.; Writing—review and editing, S.P.; Visualization, M.C. and S.P.; Supervision, S.P.; Project administration, H.J. and S.P.; Funding acquisition, H.J. and S.P. All authors have read and agreed to the published version of the manuscript.

Funding: This work was supported by grants from the Hyundai Motor Group.

Data Availability Statement: The data presented in this study are available on request from the corresponding author.

Acknowledgments: This study was supported by grants from the Hyundai Motor Group.

Conflicts of Interest: The authors declare no conflict of interest.

References

1. Nasiri, S.; Khosravani, M.R. Progress and Challenges in Fabrication of Wearable Sensors for Health Monitoring. *Sens. Actuators A Phys.* **2020**, *312*, 112105. [CrossRef]
2. Yin, R.; Wang, D.; Zhao, S.; Lou, Z.; Shen, G. Wearable Sensors-Enabled Human-Machine Interaction Systems: From Design to Application. *Adv. Funct. Mater.* **2021**, *31*, 2008936. [CrossRef]
3. Hu, W.; Lum, G.Z.; Mastrangeli, M.; Sitti, M. Small-Scale Soft-Bodied Robot with Multimodal Locomotion. *Nature* **2018**, *554*, 81–85. [CrossRef] [PubMed]
4. Whitesides, G.M. Soft Robotics. *Angew. Chemie—Int. Ed.* **2018**, *57*, 4258–4273. [CrossRef]
5. Yang, J.C.; Mun, J.; Kwon, S.Y.; Park, S.; Bao, Z.; Park, S. Electronic Skin: Recent Progress and Future Prospects for Skin-Attachable Devices for Health Monitoring, Robotics, and Prosthetics. *Adv. Mater.* **2019**, *31*, 1904765. [CrossRef] [PubMed]
6. Wang, S.; Xu, J.; Wang, W.; Wang, G.J.N.; Rastak, R.; Molina-Lopez, F.; Chung, J.W.; Niu, S.; Feig, V.R.; Lopez, J.; et al. Skin Electronics from Scalable Fabrication of an Intrinsically Stretchable Transistor Array. *Nature* **2018**, *555*, 83–88. [CrossRef] [PubMed]
7. Li, J.; Fang, L.; Sun, B.; Li, X.; Kang, S.H. Review—Recent Progress in Flexible and Stretchable Piezoresistive Sensors and Their Applications. *J. Electrochem. Soc.* **2020**, *167*, 037561. [CrossRef]
8. Cheng, Y.; Ma, Y.; Li, L.; Zhu, M.; Yue, Y.; Liu, W.; Wang, L.; Jia, S.; Li, C.; Qi, T.; et al. Bioinspired Microspines for a High-Performance Spray Ti3C2Tx MXene-Based Piezoresistive Sensor. *ACS Nano* **2020**, *14*, 2145–2155. [CrossRef]
9. Amjadi, M.; Kyung, K.U.; Park, I.; Sitti, M. Stretchable, Skin-Mountable, and Wearable Strain Sensors and Their Potential Applications: A Review. *Adv. Funct. Mater.* **2016**, *26*, 1678–1698. [CrossRef]
10. Wang, C.; Li, X.; Hu, H.; Zhang, L.; Huang, Z.; Lin, M.; Zhang, Z.; Yin, Z.; Huang, B.; Gong, H.; et al. Monitoring of the Central Blood Pressure Waveform via a Conformal Ultrasonic Device. *Nat. Biomed. Eng.* **2018**, *2*, 687–695. [CrossRef]
11. Cheng, S.; Narang, Y.S.; Yang, C.; Suo, Z.; Howe, R.D. Stick-On Large-Strain Sensors for Soft Robots. *Adv. Mater. Interfaces* **2019**, *6*, 1900985. [CrossRef]
12. Wang, H.; Totaro, M.; Beccai, L. Toward Perceptive Soft Robots: Progress and Challenges. *Adv. Sci.* **2018**, *5*, 1800541. [CrossRef] [PubMed]
13. Xu, K.; Lu, Y.; Takei, K. Multifunctional Skin-Inspired Flexible Sensor Systems for Wearable Electronics. *Adv. Mater. Technol.* **2019**, *4*, 1800628. [CrossRef]

14. Wang, C.; Xia, K.; Wang, H.; Liang, X.; Yin, Z.; Zhang, Y. Advanced Carbon for Flexible and Wearable Electronics. *Adv. Mater.* **2019**, *31*, 1801072. [CrossRef] [PubMed]
15. Wang, P.; Hu, M.; Wang, H.; Chen, Z.; Feng, Y.; Wang, J.; Ling, W.; Huang, Y. The Evolution of Flexible Electronics: From Nature, Beyond Nature, and To Nature. *Adv. Sci.* **2020**, *7*, 2001116. [CrossRef]
16. Tolvanen, J.; Hannu, J.; Jantunen, H. Stretchable and Washable Strain Sensor Based on Cracking Structure for Human Motion Monitoring. *Sci. Rep.* **2018**, *8*, 13241. [CrossRef]
17. Yang, Y.; Huang, Q.; Payne, G.F.; Sun, R.; Wang, X. A Highly Conductive, Pliable and Foldable Cu/Cellulose Paper Electrode Enabled by Controlled Deposition of Copper Nanoparticles. *Nanoscale* **2019**, *11*, 725–732. [CrossRef]
18. Yun, G.; Tang, S.-Y.; Lu, H.; Zhang, S.; Dickey, M.D.; Li, W. Hybrid-Filler Stretchable Conductive Composites: From Fabrication to Application. *Small Sci.* **2021**, *1*, 2000080. [CrossRef]
19. Ho, M.D.; Ling, Y.; Yap, L.W.; Wang, Y.; Dong, D.; Zhao, Y.; Cheng, W. Percolating Network of Ultrathin Gold Nanowires and Silver Nanowires toward "Invisible" Wearable Sensors for Detecting Emotional Expression and Apexcardiogram. *Adv. Funct. Mater.* **2017**, *27*, 1700845. [CrossRef]
20. Choi, S.; Han, S.I.; Kim, D.; Hyeon, T.; Kim, D.H. High-Performance Stretchable Conductive Nanocomposites: Materials, Processes, and Device Applications. *Chem. Soc. Rev.* **2019**, *48*, 1566–1595. [CrossRef]
21. Oluwalowo, A.; Nguyen, N.; Zhang, S.; Park, J.G.; Liang, R. Electrical and Thermal Conductivity Improvement of Carbon Nanotube and Silver Composites. *Carbon N. Y.* **2019**, *146*, 224–231. [CrossRef]
22. Wang, Y.; Zhu, C.; Pfattner, R.; Yan, H.; Jin, L.; Chen, S.; Molina-Lopez, F.; Lissel, F.; Liu, J.; Rabiah, N.I.; et al. A Highly Stretchable, Transparent, and Conductive Polymer. *Sci. Adv.* **2017**, *3*, e160207. [CrossRef] [PubMed]
23. Teng, L.; Ye, S.; Handschuh-Wang, S.; Zhou, X.; Gan, T.; Zhou, X. Liquid Metal-Based Transient Circuits for Flexible and Recyclable Electronics. *Adv. Funct. Mater.* **2019**, *29*, 1808739. [CrossRef]
24. Zhang, B.; Dong, Q.; Korman, C.E.; Li, Z.; Zaghloul, M.E. Flexible Packaging of Solid-State Integrated Circuit Chips with Elastomeric Microfluidics. *Sci. Rep.* **2013**, *3*, 1098. [CrossRef]
25. Yamaguchi, A.; Mashima, Y.; Iyoda, T. Reversible Size Control of Liquid-Metal Nanoparticles under Ultrasonication. *Angew. Chem. Int. Ed.* **2015**, *54*, 12809–12813. [CrossRef]
26. Khan, M.R.; Eaker, C.B.; Bowden, E.F.; Dickey, M.D. Giant and Switchable Surface Activity of Liquid Metal via Surface Oxidation. *Proc. Natl. Acad. Sci. USA* **2014**, *111*, 14047–14051. [CrossRef]
27. Hong, K.; Choe, M.; Kim, S.; Lee, H.M.; Kim, B.J.; Park, S. An Ultrastretchable Electrical Switch Fiber with a Magnetic Liquid Metal Core for Remote Magnetic Actuation. *Polymers* **2021**, *13*, 2407. [CrossRef]
28. Sin, D.; Singh, V.K.; Bhuyan, P.; Wei, Y.; Lee, H.M.; Kim, B.J.; Park, S. Ultrastretchable Thermo- and Mechanochromic Fiber with Healable Metallic Conductivity. *Adv. Electron. Mater.* **2021**, *7*, 2100146. [CrossRef]
29. Yang, J.; Zhou, T.; Zhang, L.; Zhu, D.; Handschuh-Wang, S.; Liu, Z.; Kong, T.; Liu, Y.; Zhang, J.; Zhou, X. Defect-Free, High Resolution Patterning of Liquid Metals Using Reversibly Sealed, Reusable Polydimethylsiloxane Microchannels for Flexible Electronic Applications. *J. Mater. Chem. C* **2017**, *5*, 6790–6797. [CrossRef]
30. Lin, Y.; Gordon, O.; Khan, M.R.; Vasquez, N.; Genzer, J.; Dickey, M.D. Vacuum Filling of Complex Microchannels with Liquid Metal. *Lab Chip* **2017**, *17*, 3043–3050. [CrossRef]
31. Kent, T.A.; Ford, M.J.; Markvicka, E.J.; Majidi, C. Soft Actuators Using Liquid Crystal Elastomers with Encapsulated Liquid Metal Joule Heaters. *Multifunct. Mater.* **2020**, *3*, 025003. [CrossRef]
32. Guo, C.; Yu, Y.; Liu, J. Rapidly Patterning Conductive Components on Skin Substrates as Physiological Testing Devices via Liquid Metal Spraying and Pre-Designed Mask. *J. Mater. Chem. B* **2014**, *2*, 5739–5745. [CrossRef] [PubMed]
33. Bhuyan, P.; Cho, D.; Choe, M.; Lee, S.; Park, S. Liquid Metal Patterned Stretchable and Soft Capacitive Sensor with Enhanced Dielectric Property Enabled by Graphite Nanofiber Fillers. *Polymers* **2022**, *14*, 710. [CrossRef]
34. Cho, D.; Bhuyan, P.; Sin, D.; Kim, H.; Kim, E.; Park, S. Stretchable, Soft, and Variable Stiffness Elastomer Foam with Positive and Negative Piezoresistivity Enabled by Liquid Metal Inclusion. *Adv. Mater. Technol.* **2022**, *14*, 2101092. [CrossRef]
35. Mou, L.; Qi, J.; Tang, L.; Dong, R.; Xia, Y.; Gao, Y.; Jiang, X. Highly Stretchable and Biocompatible Liquid Metal-Elastomer Conductors for Self-Healing Electronics. *Small* **2020**, *16*, 2005336. [CrossRef]
36. Veerapandian, S.; Jang, W.; Seol, J.B.; Wang, H.; Kong, M.; Thiyagarajan, K.; Kwak, J.; Park, G.; Lee, G.; Suh, W.; et al. Hydrogen-Doped Viscoplastic Liquid Metal Microparticles for Stretchable Printed Metal Lines. *Nat. Mater.* **2021**, *20*, 533–540. [CrossRef] [PubMed]
37. Kim, S.; Kim, S.; Hong, K.; Dickey, M.D.; Park, S. Liquid-Metal-Coated Magnetic Particles toward Writable, Nonwettable, Stretchable Circuit Boards, and Directly Assembled Liquid Metal-Elastomer Conductors. *ACS Appl. Mater. Interfaces* **2022**, *14*, 37110–37119. [CrossRef] [PubMed]
38. Bartlett, M.D.; Fassler, A.; Kazem, N.; Markvicka, E.J.; Mandal, P.; Majidi, C. Stretchable, High-k Dielectric Elastomers through Liquid-Metal Inclusions. *Adv. Mater.* **2016**, *28*, 3726–3731. [CrossRef]
39. Pan, C.; Markvicka, E.J.; Malakooti, M.H.; Yan, J.; Hu, L.; Matyjaszewski, K.; Majidi, C. A Liquid-Metal–Elastomer Nanocomposite for Stretchable Dielectric Materials. *Adv. Mater.* **2019**, *31*, 1900663. [CrossRef]
40. Rahim, M.A.; Centurion, F.; Han, J.; Abbasi, R.; Mayyas, M.; Sun, J.; Christoe, M.J.; Esrafilzadeh, D.; Allioux, F.M.; Ghasemian, M.B.; et al. Polyphenol-Induced Adhesive Liquid Metal Inks for Substrate-Independent Direct Pen Writing. *Adv. Funct. Mater.* **2021**, *31*, 2007336. [CrossRef]

41. Li, Y.; Feng, S.; Cao, S.; Zhang, J.; Kong, D. Printable Liquid Metal Microparticle Ink for Ultrastretchable Electronics. *ACS Appl. Mater. Interfaces* **2020**, *12*, 50852–50859. [CrossRef] [PubMed]
42. Zhang, W.; Naidu, B.S.; Ou, J.Z.; O'Mullane, A.P.; Chrimes, A.F.; Carey, B.J.; Wang, Y.; Tang, S.Y.; Sivan, V.; Mitchell, A.; et al. Liquid Metal/Metal Oxide Frameworks with Incorporated Ga_2O_3 for Photocatalysis. *ACS Appl. Mater. Interfaces* **2015**, *7*, 1943–1948. [CrossRef] [PubMed]
43. Taccardi, N.; Grabau, M.; Debuschewitz, J.; Distaso, M.; Brandl, M.; Hock, R.; Maier, F.; Papp, C.; Erhard, J.; Neiss, C.; et al. Gallium-Rich Pd-Ga Phases as Supported Liquid Metal Catalysts. *Nat. Chem.* **2017**, *9*, 862–867. [CrossRef] [PubMed]
44. Tutika, R.; Kmiec, S.; Tahidul Haque, A.B.M.; Martin, S.W.; Bartlett, M.D. Liquid Metal-Elastomer Soft Composites with Independently Controllable and Highly Tunable Droplet Size and Volume Loading. *ACS Appl. Mater. Interfaces* **2019**, *11*, 17873–17883. [CrossRef] [PubMed]
45. Yang, Y.; Han, J.; Huang, J.; Sun, J.; Wang, Z.L.; Seo, S.; Sun, Q. Stretchable Energy-Harvesting Tactile Interactive Interface with Liquid-Metal-Nanoparticle-Based Electrodes. *Adv. Funct. Mater.* **2020**, *30*, 1909652. [CrossRef]
46. Liu, S.; Reed, S.N.; Higgins, M.J.; Titus, M.S.; Kramer-Bottiglio, R. Oxide Rupture-Induced Conductivity in Liquid Metal Nanoparticles by Laser and Thermal Sintering. *Nanoscale* **2019**, *11*, 17615–17629. [CrossRef]
47. Liu, S.; Yuen, M.C.; White, E.L.; Boley, J.W.; Deng, B.; Cheng, G.J.; Kramer-Bottiglio, R. Laser Sintering of Liquid Metal Nanoparticles for Scalable Manufacturing of Soft and Flexible Electronics. *ACS Appl. Mater. Interfaces* **2018**, *10*, 28232–28241. [CrossRef]
48. Krisnadi, F.; Nguyen, L.L.; Ankit; Ma, J.; Kulkarni, M.R.; Mathews, N.; Dickey, M.D. Directed Assembly of Liquid Metal–Elastomer Conductors for Stretchable and Self-Healing Electronics. *Adv. Mater.* **2020**, *32*, 2001642. [CrossRef]
49. Liu, Y.; Ji, X.; Liang, J. Rupture Stress of Liquid Metal Nanoparticles and Their Applications in Stretchable Conductors and Dielectrics. *npj Flex. Electron.* **2021**, *5*, 11. [CrossRef]
50. Boley, J.W.; White, E.L.; Kramer, R.K. Mechanically Sintered Gallium-Indium Nanoparticles. *Adv. Mater.* **2015**, *27*, 2355–2360. [CrossRef]
51. Markvicka, E.J.; Bartlett, M.D.; Huang, X.; Majidi, C. An Autonomously Electrically Self-Healing Liquid Metal-Elastomer Composite for Robust Soft-Matter Robotics and Electronics. *Nat. Mater.* **2018**, *17*, 618–624. [CrossRef] [PubMed]
52. Tutika, R.; Haque, A.B.M.T.; Bartlett, M.D. Self-Healing Liquid Metal Composite for Reconfigurable and Recyclable Soft Electronics. *Commun. Mater.* **2021**, *2*, 64. [CrossRef]
53. Fassler, A.; Majidi, C. Liquid-Phase Metal Inclusions for a Conductive Polymer Composite. *Adv. Mater.* **2015**, *27*, 1928–1932. [CrossRef] [PubMed]
54. Mohammed, M.G.; Kramer, R. All-Printed Flexible and Stretchable Electronics. *Adv. Mater.* **2017**, *29*, 1604965. [CrossRef]
55. Ford, M.J.; Ambulo, C.P.; Kent, T.A.; Markvicka, E.J.; Pan, C.; Malen, J.; Ware, T.H.; Majidi, C. A Multifunctional Shape-Morphing Elastomer with Liquid Metal Inclusions. *Proc. Natl. Acad. Sci. USA* **2019**, *116*, 21438–21444. [CrossRef]
56. Lin, Y.; Cooper, C.; Wang, M.; Adams, J.J.; Genzer, J.; Dickey, M.D. Handwritten, Soft Circuit Boards and Antennas Using Liquid Metal Nanoparticles. *Small* **2015**, *11*, 6397–6403. [CrossRef]
57. Ford, M.J.; Patel, D.K.; Pan, C.; Bergbreiter, S.; Majidi, C. Controlled Assembly of Liquid Metal Inclusions as a General Approach for Multifunctional Composites. *Adv. Mater.* **2020**, *32*, 2002929. [CrossRef]
58. Xin, Y.; Zhang, S.; Lou, Y.; Xu, J.; Zhang, J. Determinative Energy Dissipation in Liquid Metal Polymer Composites for Advanced Electronic Applications. *Adv. Mater. Technol.* **2020**, *5*, 2000018. [CrossRef]
59. Park, S.; Mondal, K.; Treadway, R.M.; Kumar, V.; Ma, S.; Holbery, J.D.; Dickey, M.D. Silicones for Stretchable and Durable Soft Devices: Beyond Sylgard-184. *ACS Appl. Mater. Interfaces* **2018**, *10*, 11261–11268. [CrossRef]
60. Keene, B.J. Review of Data for the Surface Tension of Pure Metals. *Int. Mater. Rev.* **1993**, *38*, 157–192. [CrossRef]
61. Handschuh-Wang, S.; Stadler, F.J.; Zhou, X. Critical Review on the Physical Properties of Gallium-Based Liquid Metals and Selected Pathways for Their Alteration. *J. Phys. Chem. C* **2021**, *125*, 20113–20142. [CrossRef]
62. Handschuh-Wang, S.; Chen, Y.; Zhu, L.; Zhou, X. Analysis and Transformations of Room-Temperature Liquid Metal Interfaces—A Closer Look through Interfacial Tension. *ChemPhysChem* **2018**, *19*, 1584–1592. [CrossRef] [PubMed]
63. Dobosz, A.; Plevachuk, Y.; Sklyarchuk, V.; Sokoliuk, B.; Gancarz, T. Thermophysical Properties of the Liquid Ga-Sn-Zn Eutectic Alloy. *Fluid Phase Equilib.* **2018**, *465*, 1–9. [CrossRef]
64. Style, R.W.; Tutika, R.; Kim, J.Y.; Bartlett, M.D. Solid-Liquid Composites for Soft Multifunctional Materials. *Adv. Funct. Mater.* **2021**, *31*, 2005804. [CrossRef]
65. Kazem, N.; Bartlett, M.D.; Majidi, C. Extreme Toughening of Soft Materials with Liquid Metal. *Adv. Mater.* **2018**, *30*, 1706594. [CrossRef]
66. Markvicka, E.J.; Tutika, R.; Bartlett, M.D.; Majidi, C. Soft Electronic Skin for Multi-Site Damage Detection and Localization. *Adv. Funct. Mater.* **2019**, *29*, 1900160. [CrossRef]

MDPI
St. Alban-Anlage 66
4052 Basel
Switzerland
Tel. +41 61 683 77 34
Fax +41 61 302 89 18
www.mdpi.com

Micromachines Editorial Office
E-mail: micromachines@mdpi.com
www.mdpi.com/journal/micromachines

www.ingramcontent.com/pod-product-compliance
Lightning Source LLC
Chambersburg PA
CBHW061837210926
51850CB00006B/910
* 9 7 8 3 0 3 6 5 8 4 6 7 6 *